INVESTIGATIONS GUIDE

Soils, Rocks, and Landforms

FOSS Next Generation

Full Option Science System
Developed at the Lawrence Hall of Science, University of California, Berkeley
Published and Distributed by Delta Education

FOSS Lawrence Hall of Science Team
Larry Malone and Linda De Lucchi, FOSS Project Codirectors and Lead Developers
Kathy Long, FOSS Assessment Director; David Lippman, Program Manager; Carol Sevilla, Publications Design Coordinator; Susan Stanley, Illustrator; John Quick, Photographer
FOSS Curriculum Developers: Brian Campbell, Teri Lawson, Alan Gould, Susan Kaschner Jagoda, Ann Moriarty, Jessica Penchos, Kimi Hosoume, Virginia Reid, Joanna Snyder, Erica Beck Spencer, Joanna Totino, Diana Velez, Natalie Yakushiji
Susan Ketchner, Technology Project Manager
FOSS Technology Team: Dan Bluestein, Christopher Cianciarulo, Matthew Jacoby, Kate Jordan, Frank Kusiak, Nicole Medina, Jonathan Segal, Dave Stapley, Shan Tsai

Delta Education Team
Bonnie A. Piotrowski, Editorial Director, Elementary Science
Project Team: Jennifer Apt, Sandra Burke, Joann Hoy, Kristen Mahoney, Jennifer McKenna, Angela Miccinello

Thank you to all FOSS Grades K-5 Trial Teachers
Heather Ballard, Wilson Elementary, Coppell, TX; Mirith Ballestas De Barroso, Treasure Forest Elementary, Houston, TX; Terra L. Barton, Harry McKillop Elementary, Melissa, TX; Rhonda Bernard, Frances E. Norton Elementary, Allen, TX; Theresa Bissonnette, East Millbrook Magnet Middle School, Raleigh, NC; Peter Blackstone, Hall Elementary School, Portland, ME; Tiffani Brisco, Seven Hills Elementary, Newark, TX; Darrow Brown, Lake Myra Elementary School, Wendell, NC; Heather Callaghan, Olive Chapel Elementary, Apex, NC; Katie Cannon, Las Colinas Elementary, Irving, TX; Elaine M. Cansler, Brassfield Road Elementary School, Raleigh, NC; Kristy Cash, Wilson Elementary, Coppell, TX; Monica Coles, Swift Creek Elementary School, Raleigh, NC; Shirley Conner, Ocean Avenue Elementary School, Portland, ME; Sally Connolly, Cape Elizabeth Middle School, Cape Elizabeth, ME; Melissa Cook-Airhart, Harry McKillop Elementary, Melissa, TX; Melissa Costa, Olive Chapel Elementary, Apex, NC; Hillary P. Croissant, Harry McKillop Elementary, Melissa, TX; Rene Custeau, Hall Elementary School, Portland, ME; Nancy Davis, Martha and Josh Morriss Mathematics and Engineering Elementary School, Texarkana, TX; Nancy Deveneau, Wilson Elementary, Coppell, TX; Karen Diaz, Las Colinas Elementary, Irving, TX; Marlana Dumas, Las Colinas Elementary, Irving, TX; Mary Evans, R.E. Good Elementary School, Carrollton, TX; Jacquelyn Farley, Moss Haven Elementary, Dallas, TX; Corinna Ferrier, Oak Forest Elementary, Humble, TX; Allison Fike, Wilson Elementary, Coppell, TX; Barbara Fugitt, Martha and Josh Morriss Mathematics and Engineering Elementary School, Texarkana, TX; Colleen Garvey, Farmington Woods Elementary, Cary, NC; Judy Geller, Bentley Elementary School, Oakland, CA; Erin Gibson, Las Colinas Elementary, Irving, TX; Kelli Gobel, Melissa Ridge Intermediate School, Melissa, TX; Dollie Green, Melissa Ridge Intermediate School, Melissa, TX; Brenda Lee Harrigan, Bentley Elementary School, Oakland, CA; Cori Harris, Samuel Beck Elementary, Trophy Club, TX; Kim Hayes, Martha and Josh Morriss Mathematics and Engineering Elementary School, Texarkana, TX; Staci Lynn Hester, Lacy Elementary School, Raleigh, NC; Amanda Hill, Las Colinas Elementary, Irving, TX; Margaret Hillman, Ocean Avenue Elementary School, Portland, ME; Cindy Holder, Oak Forest Elementary, Humble, TX; Sarah Huber, Hodge Road Elementary, Knightdale, NC; Susan Jacobs, Granger Elementary, Keller, TX; Carol Kellum, Wallace Elementary, Dallas, TX; Jennifer A. Kelly, Hall Elementary School, Portland, ME; Brittani Kern, Fox Road Elementary, Raleigh, NC; Jodi Lay, Lufkin Road Middle School, Apex, NC; Melissa Lourenco, Lake Myra Elementary School, Wendell, NC; Ana Martinez, RISD Academy, Dallas, TX; Shaheen Mavani, Las Colinas Elementary, Irving, TX; Mary Linley McClendon, Math Science Technology Magnet School, Richardson, TX; Adam McKay, Davis Drive Elementary, Cary, NC; Leslie Meadows, Lake Myra Elementary School, Wendell, NC; Anne Mechler, J. Erik Jonsson Community School, Dallas, TX; Anne Miller, J. Erik Jonsson Community School, Dallas, TX; Shirley Diann Miller, The Rice School, Houston, TX; Keri Minier, Las Colinas Elementary, Irving, TX; Stephanie Renee Nance, T.H. Rogers Elementary, Houston, TX; Cynthia Nilsen, Peaks Island School, Peaks Island, ME; Elizabeth Noble, Las Colinas Elementary, Irving, TX; Courtney Noonan, Shadow Oaks Elementary School, Houston, TX; Sarah Peden, Aversboro Elementary School, Garner, NC; Carrie Prince, School at St. George Place, Houston, TX; Marlaina Pritchard, Melissa Ridge Intermediate School, Melissa, TX; Alice Pujol, J. Erik Jonsson Community School, Dallas, TX; Claire Ramsbotham, Cape Elizabeth Middle School, Cape Elizabeth, ME; Paul Rendon, Bentley Elementary, Oakland, CA; Janette Ridley, W.H. Wilson Elementary School, Coppell, TX; Kristina (Crickett) Roberts, W.H. Wilson Elementary School, Coppell, TX; Heather Rogers, Wendell Creative Arts & Science Magnet Elementary School, Wendell, NC; Alissa Royal, Melissa Ridge Intermediate School, Melissa, TX; Megan Runion, Olive Chapel Elementary, Apex, NC; Christy Scheef, J. Erik Jonsson Community School, Dallas, TX; Samrawit Shawl, T.H. Rogers School, Houston, TX; Nicole Spivey, Lake Myra Elementary School, Wendell, NC; Ashley Stephenson, J. Erik Jonsson Community School, Dallas, TX; Jolanta Stern, Browning Elementary School, Houston, TX; Gale Stimson, Bentley Elementary, Oakland, CA; Ted Stoeckley, Hall Middle School, Larkspur, CA; Cathryn Sutton, Wilson Elementary, Coppell, TX; Camille Swander, Ocean Avenue Elementary School, Portland, ME; Brandi Swann, Westlawn Elementary School, Texarkana, TX; Robin Taylor, Arapaho Classical Magnet, Richardson, TX; Michael C. Thomas, Forest Lane Academy, Dallas, TX; Jomarga Thompkins, Lockhart Elementary, Houston, TX; Mary Timar, Madera Elementary, Lake Forest, CA; Helena Tongkeamha, White Rock Elementary, Dallas, TX; Linda Trampe, J. Erik Jonsson Community School, Dallas, TX; Charity VanHorn, Fred A. Olds Elementary, Raleigh, NC; Kathleen VanKeuren, Lufkin Road Middle School, Apex, NC; Valerie Vassar, Hall Elementary School, Portland, ME; Megan Veron, Westwood Elementary School, Houston, TX; Mary Margaret Waters, Frances E. Norton Elementary, Allen, TX; Stephanie Robledo Watson, Ridgecrest Elementary School, Houston, TX; Lisa Webb, Madisonville Intermediate, Madisonville, TX; Matt Whaley, Cape Elizabeth Middle School, Cape Elizabeth, ME; Nancy White, Canyon Creek Elementary, Austin, TX; Barbara Yurick, Oak Forest Elementary, Humble, TX; Linda Zittel, Mira Vista Elementary, Richmond, CA

Photo Credits: © Evgeniya Moroz/Shutterstock (cover); © Daniel Lohmer/Shutterstock; © Laurie Meyer; © Zach Smith

Published and Distributed by Delta Education, a member of the School Specialty Family
The FOSS program was developed in part with the support of the National Science Foundation grant nos. MDR-8751727 and MDR-9150097. However, any opinions, findings, conclusions, statements, and recommendations expressed herein are those of the authors and do not necessarily reflect the views of NSF. FOSSmap was developed in collaboration between the BEAR Center at UC Berkeley and FOSS at the Lawrence Hall of Science.

Copyright © 2019 by The Regents of the University of California

Standards cited herein from NGSS Lead States. 2013. *Next Generation Science Standards: For States, By States.* Washington, DC: The National Academies Press. Next Generation Science Standards is a registered trademark of Achieve. Neither Achieve nor the lead states and partners that developed the Next Generation Science Standards was involved in the production of, and does not endorse, this product.

All rights reserved. Any part of this work may be reproduced or transmitted in any form or by any means, electronic or mechanical, including photocopying and recording, or by an information storage or retrieval system without prior written permission. For permission please write to: FOSS Project, Lawrence Hall of Science, University of California, Berkeley, CA 94720 or foss@berkeley.edu.

Soils, Rocks, and Landforms
Investigations Guide, 1487576
978-1-62571-345-2
Printing 9 – 5/2018
Webcrafters, Madison, WI

INVESTIGATIONS GUIDE

Soils, Rocks, and Landforms

TABLE OF CONTENTS

Overview	1
Framework and NGSS	31
Materials	57
Technology	69
Investigation 1: Soils and Weathering	83
Part 1: Soil Composition	94
Part 2: Physical Weathering	110
Part 3: Chemical Weathering	120
Part 4: Schoolyard Soils	134
Investigation 2: Landforms	147
Part 1: Erosion and Deposition	158
Part 2: Stream-Table Investigations	170
Part 3: Schoolyard Erosion and Deposition	183
Part 4: Fossil Evidence	190
Investigation 3: Mapping Earth's Surface	207
Part 1: Making a Topographic Map	218
Part 2: Drawing a Profile	229
Part 3: Mount St. Helens Case Study	241
Part 4: Rapid Changes	250
Investigation 4: Natural Resources	265
Part 1: Introduction to Natural Resources	274
Part 2: Making Concrete	284
Part 3: Earth Materials in Use	292
Assessment	307

Welcome to FOSS® Next Generation™

Getting Started with FOSS Next Generation for Grades 3–5

Whether you're new to hands-on science or a FOSS veteran, you'll be up and running in no time and ready to lead your students on a fantastic voyage through the wonders of the natural and designed world.

Watch our short video series or browse the next few pages to get started!

Getting Started with FOSS: Meet Your Module video

Scan here or visit deltaeducation.com/goFOSS

Three-Dimensional Active Science

It's time to experience the three dimensions of the NGSS—**disciplinary core ideas**, **crosscutting concepts**, and **science and engineering practices**. Engage in rich investigations that immerse your students in real-world applications of important scientific phenomena, supported by just-in-time teaching tips and strategies.

Getting Started with Your Equipment Kit

Meet Your FOSS Module!

Your FOSS module includes one or more large boxes, called drawers, and two smaller boxes for the Teacher Toolkit, student books, and other equipment. Each drawer has a label on the front listing its contents. Your packing list is always in Drawer 1.

Permanent Equipment

Your equipment kit includes enough permanent equipment for up to 8 groups (32 students). This equipment is classroom-tested and expected to last 7–10 years.

Consumable Equipment

Your kit also includes consumable materials for three class uses. Convenient refill kits provide materials for three additional uses and are available through Delta Education.

Easy Set-up and Clean-up!

FOSS Next Generation equipment drawers are packed by investigation to facilitate prep and to make packing up for the next use a snap!

Drawer sections include:

- Unique materials needed for one investigation
- Common equipment used in multiple investigations
- Consumable materials—when it's empty you know it's time to refill!

Order Refills Online

deltaeducation.com/refillcenter

Live Organisms

Some investigations require live organisms. Schools are encouraged to purchase these organisms from a local biological supply company to minimize both transit time and the impact of adverse weather on the health of the organisms.

If living material cards are purchased from Delta Education, they will be shipped separately in a green and white envelope. Keep these cards in a safe place until it's time to redeem them for the investigation.

Call Delta Education at 800-258-1302 at least three weeks before you need your organisms.

Premium Student eBook Access

If your school purchased a premium class license for the *FOSS Science Resources* student eBook, your access codes will be shipped separately in a blue and white striped envelope. Use this access code on FOSSweb to unlock student eBook access.

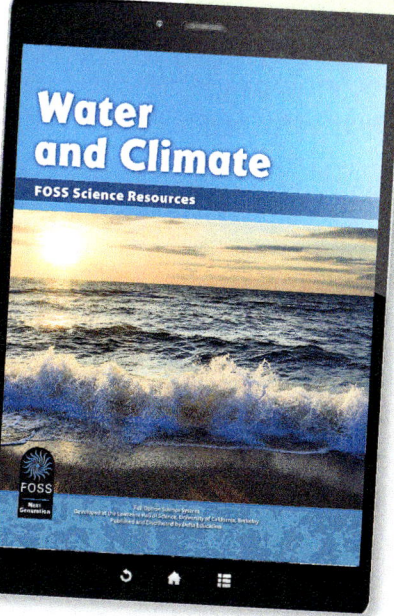

Getting Started with Your Teacher Toolkit

The Teacher Toolkit is the most important part of the FOSS program. There are three parts of the Teacher Toolkit—the **Investigations Guide**, **Teacher Resources**, and the student **Science Resources** book. It's here that all the wisdom and experience from years of research and classroom development comes together to support teachers with lesson facilitation and in-depth strategies for taking investigations to the next level.

1. Investigations Guide

The **Investigations Guide** is your roadmap to prepare for and lead the FOSS investigations. Chapters are tabbed for easy access to important module information.

The module **Overview** chapter gives you a high-level look at the 10–12 weeks of instruction in each module including a summary matrix, schedule for the module, and product support contacts.

Framework and the NGSS chapter provides a complete overview of NGSS connections, learning progressions, and background to support the conceptual framework for the module.

The **Materials** chapter is a must-read resource that helps you get your student equipment ready for first-time use and shares helpful tips for getting your classroom ready for FOSS.

The **Technology** chapter provides an overview for each digital resource in the module and gets you up and running on FOSSweb.com, complete with technical support.

Each **Investigation** chapter includes an At-a-Glance overview, science background content with NGSS connections, and in-depth guidance for preparing and facilitating instruction.

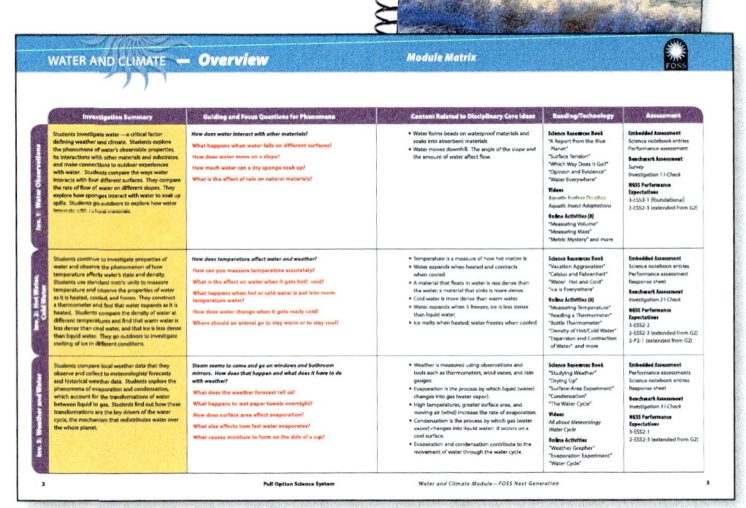

Module matrix

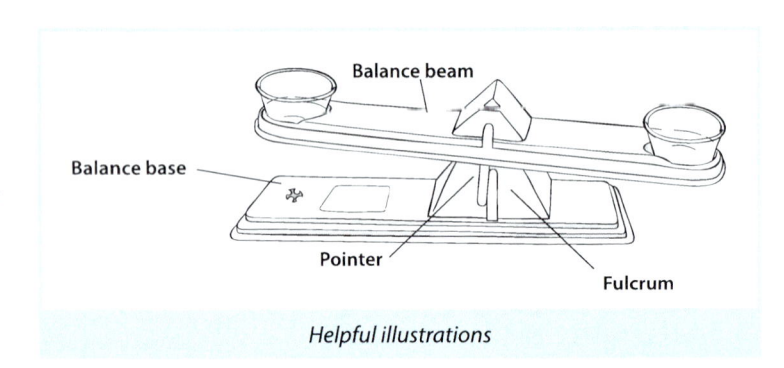

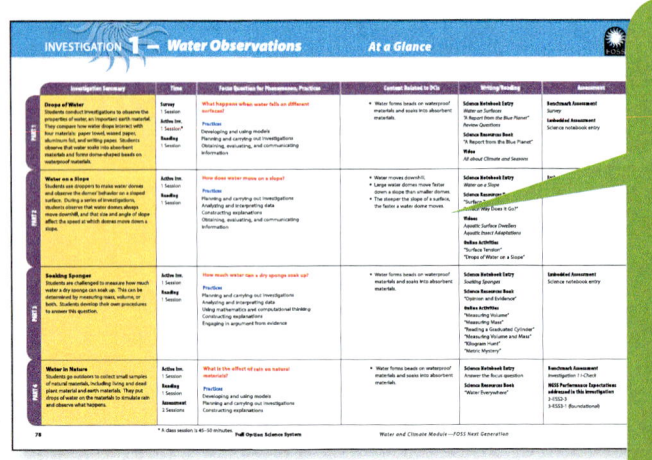

Helpful illustrations

The At-a-Glance chart includes:
- Summaries and pacing for investigation scheduling
- Focus questions for investigative phenomena
- Connections to disciplinary core ideas
- Reading, writing, and technology integration opportunities
- Embedded and benchmark assessments

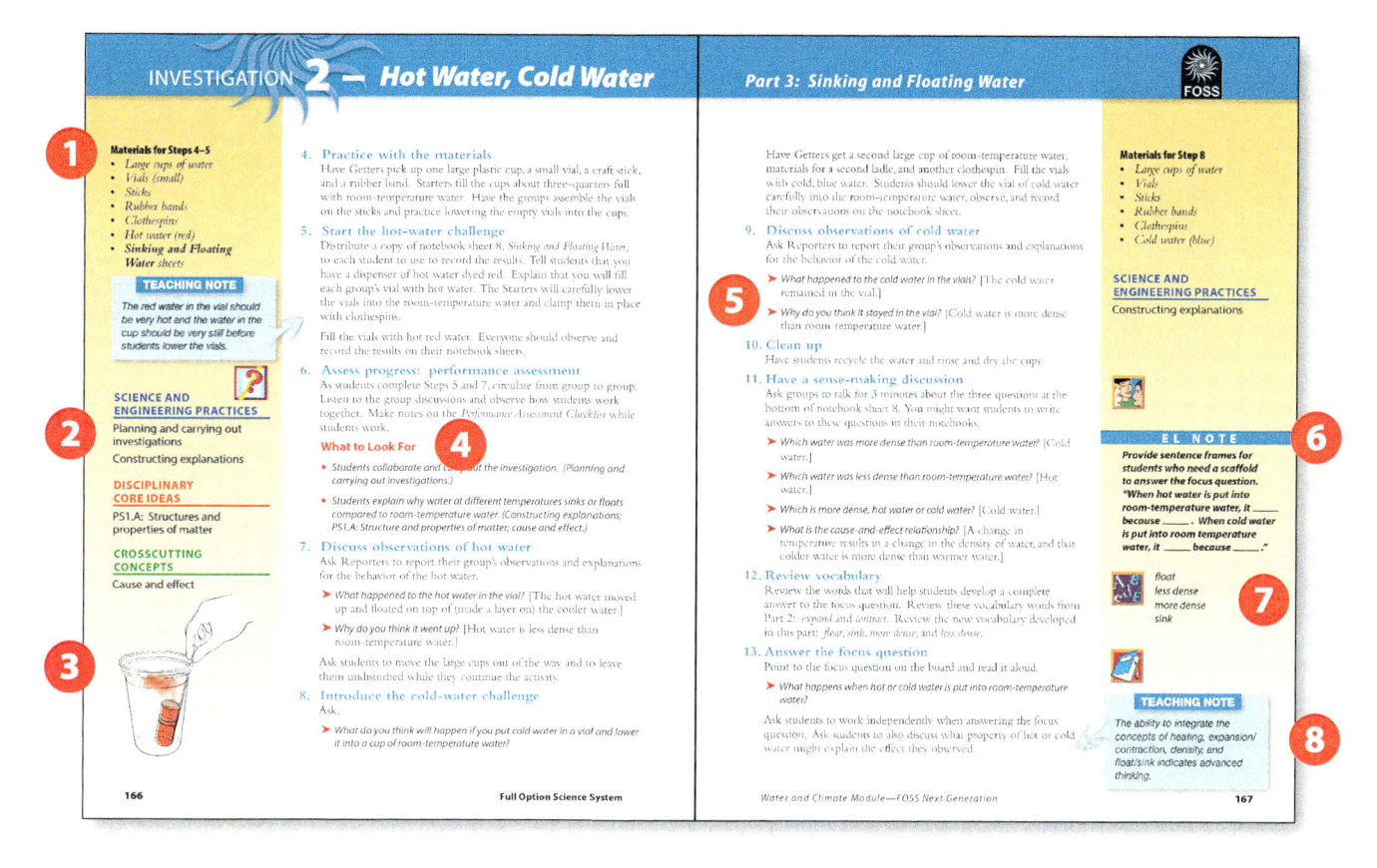

FOSS investigations provide the right support, when you need it with point-of-use guidance.

1. Materials used in the current steps
2. Key three-dimensional highlights
3. Helpful drawings and diagrams
4. Embedded assessment "What to Look For"
5. Discussion questions and model responses
6. Strategies to support English learners
7. Vocabulary review
8. Teaching notes to facilitate instruction

The **Assessment** chapter gives you an in-depth look at the research-based components of the FOSS Assessment System, guidance on assessing for the NGSS, and generalized next-step strategies to use in your classroom. Find duplication masters, assessment charts, coding guides, and specific next-step strategies on FOSSweb.com.

Getting Started with Your Teacher Toolkit

2. Teacher Resources

Your *Investigations Guide* tells you how to facilitate each investigation of a module. The **Teacher Resources** provides guidance on how to do it at your grade level across three modules throughout the year with effective practices and strategies derived from research and extensive field-testing.

A grade-level **Planning Guide** provides an overview to your three modules and an introduction to three-dimensional teaching and learning.

The **Science Notebooks** chapter provides age-appropriate methods to support students in developing productive science notebooks. Access powerful research-based next-step strategies to maximize the effectiveness of the notebook as a formative assessment tool.

Science-Centered Language Development is a collection of standards-aligned strategies to support and enhance literacy development in the context of science—reading, writing, speaking, listening, and vocabulary development.

In **Taking FOSS Outdoors**, find guidance for managing the space, time, and materials needed to provide authentic, real-world learning experiences in students' schoolyards and local communities.

Teacher Resources also includes:

- Grade-level connections to Common Core ELA and Math standards
- Module-specific notebook, teacher, and assessment blackline masters.

Check FOSSweb for the latest updates to chapters in *Teacher Resources*.

3. FOSS Science Resources Student Book

The Teacher Toolkit includes one copy of the student book. Reading is an integral part of science learning. Reading informational text critically and effectively is an important component of today's ELA standards. Once students have engaged with phenomena firsthand, they go more in-depth with articles in *FOSS Science Resources*.

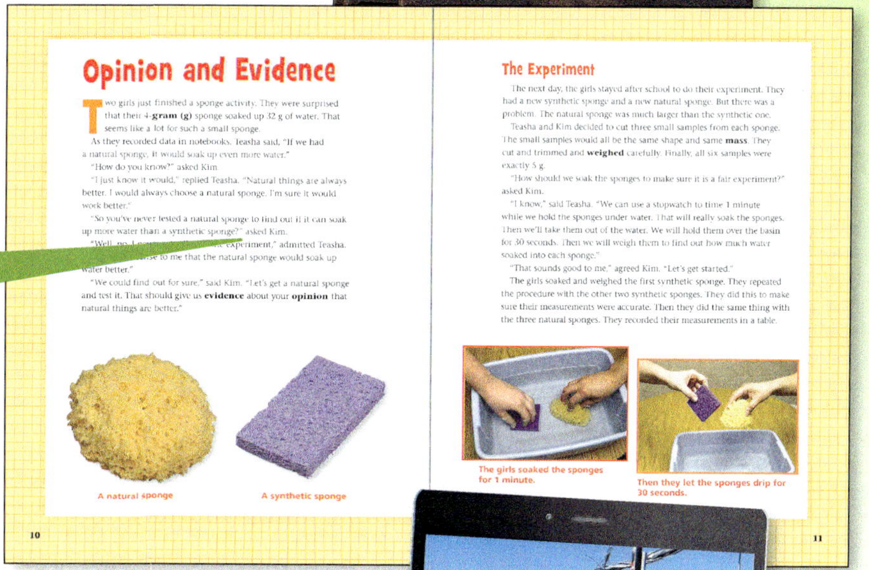

Articles from FOSS *Science Resources* complement and enhance the active investigations, giving students opportunities to:

- Ask and answer questions
- Use evidence to support their ideas
- Use text to acquire information
- Draw information from multiple sources
- Interpret illustrations to build understanding

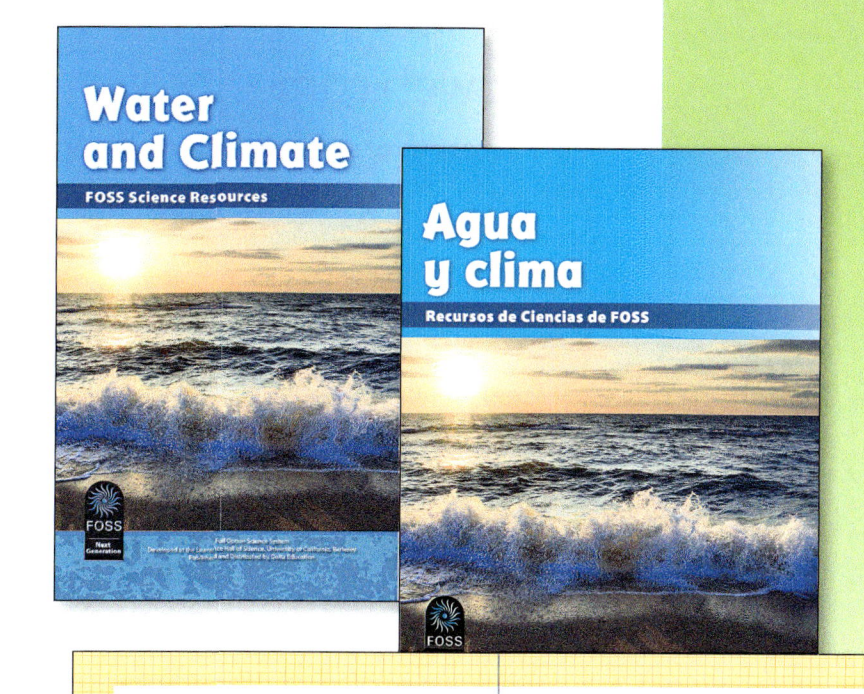

Interactive eBooks

FOSS Science Resources is available as a convenient, platform-neutral interactive student eBook with integrated audio, highlighted text, and links to videos and online activities. Student access to eBooks is available as an additional purchase.

Getting Started with Technology

FOSSweb.com

Easy access to program support resources

FOSSweb.com is your home for accessing the complete portfolio of digital resources in the FOSS program. Easily manage each of your modules, create class pages, and keep helpful references at your fingertips.

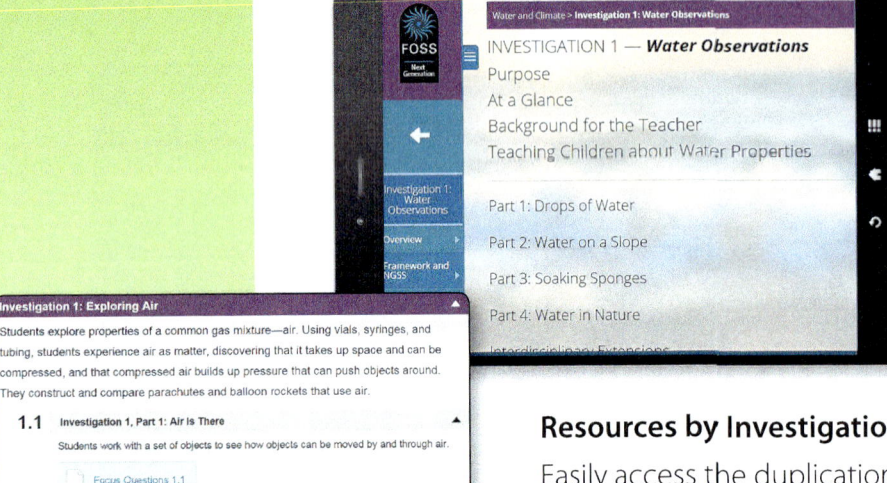

eInvestigations Guide

This easy-to-use interactive version of the *Investigations Guide* is mobile-friendly and offers simplified navigation, collapsible sections, and the ability to add customized notes.

Resources by Investigation

Easily access the duplication masters, online activities, and streaming videos needed for the current investigation part.

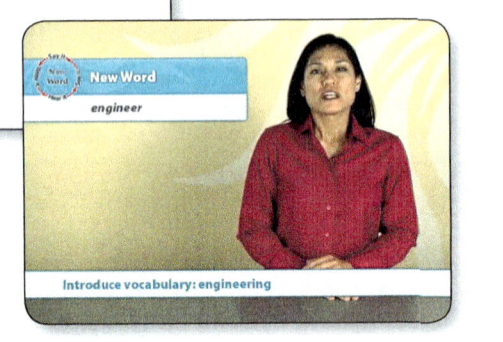

Teacher Preparation Videos

Videos provide helpful equipment setup instructions, safety information, and a summary of what students will do and learn throughout a part.

Interactive Whiteboard Lessons

Developed for SMART™ or Promethean boards, these resources help you facilitate each part of every investigation and give the class a visual reference.

Online Activities for Differentiating Instruction

FOSSweb digital resources provide engaging, interactive virtual investigations and tutorials that offer additional content and skill support for students. These experiences also help students who were absent catch up with class.

Streaming Videos

Videos are available on FOSSweb to support many investigations and often take students "on location" around the world or showcase experiments that would be too messy, expensive, or dangerous for the classroom.

FOSSmap Online Assessment

Students in grades 3–5 can take benchmark assessments online with automatic coding of most responses to save you valuable time and provide instant feedback. A variety of student- and class-level reports help you identify where extra attention is needed and support communication with families and administration. FOSSmap is accessible through the FOSSweb portal.

Three-Dimensional Active Learning

The FOSS program has always placed student learning of science *practices* on equal footing with science *concepts and core ideas* and the NGSS and *Framework for K–12 Science Education* have provided a new language with which to articulate this. In each **FOSS Next Generation** investigation, students are engaged in the three dimensions of the NGSS to develop increasingly complex knowledge and understanding.

Science and engineering practices are the cognitive tools scientists and engineers use to answer questions and design solutions. FOSS students use these tools to gather evidence and to explain real-world phenomena.

Grade-level appropriate **disciplinary core ideas** are the concepts and established ideas of science. FOSS students develop these building blocks throughout investigations to make sense of phenomena.

Crosscutting concepts help students to connect the varied concepts and disciplines of science. FOSS students apply these concepts to different situations in order to make connections and develop comprehensive understanding.

FOSS Forward Thinking

The FOSS Vision

When the Full Option Science System (FOSS) began, the founders envisioned a science curriculum that was enjoyable, logical, and intuitive for teachers, and stimulating, provocative, and informative for students. Achieving this vision was informed by research in cognitive science, learning theory, and critical study of effective practice. The modular design of the FOSS product allowed users to select topics that aligned with district or state learning objectives, or simply resonated with their perception of comprehensive and reasonable science instruction. The original design of the FOSS Program was comprehensive in terms of coverage. FOSS was designed to provide real and meaningful student experience with important scientific ideas and to nurture developmentally appropriate knowledge of the objects, organisms, systems, and principles governing, the natural world.

The FOSS Next Generation Program

But the developers never envisioned FOSS to be a static curriculum, and now the Full Option Science System has evolved into a fully realized 21st century science program with authentic connection to the *Next Generation Science Standards (NGSS)*. The FOSS science curriculum is a comprehensive science program, featuring instructional guidance, student equipment, student reading materials, digital resources, and an embedded assessment system. The FOSS philosophy has always taken very seriously the teaching of good, comprehensive, accurate, science content using the methods of inquiry to advance that science knowledge. But the *Framework for K–12 Science Education*, on which the NGSS are based has allowed us to articulate our mission in a more coherent manner, using the vocabulary established by the authors of the *Framework*. The FOSS instructional design now strives to

a. communicate the disciplinary core ideas (content) of science, while

b. guiding and encouraging students to engage in or exercise the science and engineering practices (inquiry methods) to develop knowledge of the disciplinary core ideas, and

c. help students apprehend the crosscutting concepts (themes that unite core ideas, overarching concepts) that connect the learning experiences within a discipline and bridge meaningfully across disciplines as students gain more and more knowledge of the natural world.

> The Full Option Science System has evolved into a fully realized 21st century science program with authentic connection to the Next Generation Science Standards (NGSS).

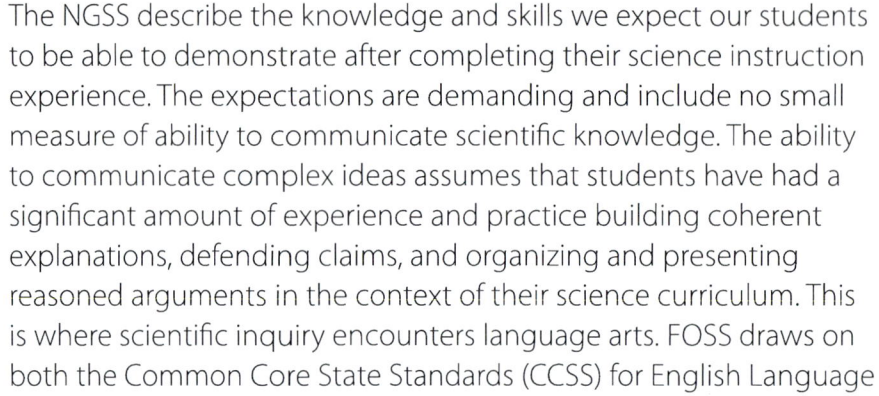

The NGSS describe the knowledge and skills we expect our students to be able to demonstrate after completing their science instruction experience. The expectations are demanding and include no small measure of ability to communicate scientific knowledge. The ability to communicate complex ideas assumes that students have had a significant amount of experience and practice building coherent explanations, defending claims, and organizing and presenting reasoned arguments in the context of their science curriculum. This is where scientific inquiry encounters language arts. FOSS draws on both the Common Core State Standards (CCSS) for English Language Arts and research data regarding the productive use of student science notebooks. FOSS developers realize that the most effective science program must seamlessly integrate science instruction goals and language arts skills. Science is one of the most engaging and productive arenas for introducing and exercising language arts skills: vocabulary, nonfiction (informational) reading, cause-and-effect relationships, on and on.

FOSS is strongly grounded in the realities of the classroom and the interests and experiences of the learners. The content in FOSS is teachable and learnable over multiple grade levels as students increase in their abilities to reason about and integrate complex ideas within and between disciplines.

FOSS is crafted with a structured, yet flexible, teaching philosophy that embraces the much-heralded 21st century skills; collaborative teamwork, critical thinking, and problem solving. The FOSS curriculum design promotes a classroom culture that allows both teachers and students to assume prominent roles in the management of the learning experience.

FOSS is built on the assumptions that understanding of core scientific knowledge and how science functions is essential for citizenship, that all teachers can teach science, and that all students can learn science. Formative assessment in FOSS creates a community of reflective practice. Teachers and students make up the community and establish norms of mutual support, trust, respect, and collaboration. The goal of the community is that everyone will demonstrate progress and will learn and grow.

SOILS, ROCKS, AND LANDFORMS — *Overview*

Contents

Introduction	1
Module Matrix	2
FOSS Components	6
FOSS Instructional Design	10
Differentiated Instruction for Access and Equity	18
FOSS Investigation Organization	21
Establishing a Classroom Culture	24
Safety in the Classroom and Outdoors	28
Scheduling the Module	29
FOSS Contacts	30

The NGSS Performance Expectations bundled in this module include:

Earth and Space Sciences
4-ESS1-1
4-ESS2-1
4-ESS2-2
4-ESS3-1
4-ESS3-2

Engineering, Technology, and Applications of Science
3–5-ETS1-1
3–5-ETS1-2

▶ **NOTE**
The three modules for grade 4 in FOSS Next Generation are

Energy
Soils, Rocks, and Landforms
Environments

INTRODUCTION

Geology is the study of our planet's earth materials and natural resources. Because they are so ubiquitous and abundant, they are often taken for granted. The **Soils, Rocks, and Landforms Module** provides students with firsthand experiences with soils and rocks and modeling experiences using tools such as topographic maps and stream tables to engage with the anchor phenomenon of the surface of Earth's landscape—the shape and the composition of landforms. The driving questions for the module are What are Earth's land surface made of? and Why are landforms not the same everywhere?

This module has four investigations that focus on the phenomena that weathering by water, ice, wind, living organisms, and gravity breaks rocks into smaller pieces, erosion (water, ice, and wind) transports earth materials to new locations, and deposition is the result of that transport process that builds new land. Students conduct controlled experiments by incrementally changing specific environmental conditions to determine the impact of changing the variables of slope and amount of water in stream tables. Students interpret data from diagrams and visual representations to build explanations from evidence and make predictions of future events. They develop model mountains and represent the landforms from different perspectives to look for change. Students gain experiences that will contribute to the understanding of crosscutting concepts of patterns; cause and effect; scale, proportion, and quantity; systems and system models; structure and function; and stability and change.

FOSS Full Option Science System

SOILS, ROCKS, AND LANDFORMS — *Overview*

Module Summary	Guiding and Focus Questions for Phenomena
Inv. 1: Soils and Weathering Students engage firsthand with the phenomenon of soils. They investigate properties of soil by comparing four different soils. They learn that soils are composed of essentially the same types of materials (inorganic earth materials and humus), but the amounts of the materials vary. They begin to explore how rocks break into smaller pieces through physical and chemical weathering. Students go outdoors to explore and compare properties of local soils.	*How do soils form?* What is soil? What causes big rocks to break down into smaller rocks? How are rocks affected by acid rain? What's in our schoolyard soils?
Inv. 2: Landforms Students engage with the phenomena of erosion and deposition of weathered earth material by flowing water. They use stream-table models to observe that water moves earth materials from one location to another. They investigate the variables of slope and water quantity and plan and conduct their own stream-table investigations. Students look for evidence of erosion and deposition outdoors. Students pursue explanations for the phenomenon of fossils found in layers of sedimentary rock. They think about what happens to sediments over long periods of time as sediments layer on top of each other. They learn about the different processes that can result in fossils and how fossils provide evidence of life and landscapes from the ancient past.	*How do erosion and deposition impact landforms?* *What do the location of fossils in rock layers tell us about past life on Earth?* How do weathered rock pieces move from one place to another? How does slope affect erosion and deposition? How do floods affect erosion and deposition? Where are erosion and deposition happening in our schoolyard? How do fossils get in rocks and what can they tell us about the past?

Module Matrix

Content Related to Disciplinary Core Ideas	Reading/Technology	Assessment
• Soils can be described by their properties. • Soils are composed of different kinds and amounts of earth materials and humus. • Weathering is the breakdown of rocks and minerals at or near Earth's surface. • The physical-weathering processes of abrasion and freezing break rocks and minerals into smaller pieces. • Chemical weathering occurs when exposure to water and air changes rocks and minerals into something new.	**Science Resources Book** "What Is Soil?" "Weathering" **Videos** Weathering and Erosion Soils **Online Activities** "Tutorial: Weathering" "Virtual Investigation: Water Retention of Soils"	**Embedded Assessment** Science notebook entry Response sheet Performance assessment **Benchmark Assessment** Survey Investigation 1 I-Check **NGSS Performance Expectation** 4-ESS2-1
• Weathered rock material can be reshaped into new landforms by the slow processes of erosion and deposition. • Erosion is the transport (movement) of weathered rock material (sediments) by moving water or wind. • Deposition is the settling of sediments when the speed of moving water or wind declines. • The rate and volume of erosion relate directly to the amount of energy in moving water or wind. • The energy of moving water depends on the mass of water in motion and its velocity. The greater the mass and velocity, the greater the energy. • Fossils provide evidence of organisms that lived long ago as well as clues to changes in the landscape and past environments.	**Science Resources Book** "Erosion and Deposition" "Landforms Photo Album" "Fossils Tell a Story" "Pieces of a Dinosaur Puzzle" **Videos** Weathering and Erosion Fossils **Online Activities** "Geology Lab: Stream Tables" "Tutorial—Stream Tables: Slope and Flood" "Virtual Investigation: Stream Tables" "Tutorial—Soil Formation" "Tutorial—Fossils"	**Embedded Assessment** Science notebook entries Performance assessment Response sheet **Benchmark Assessment** Investigation 2 I-Check **NGSS Performance Expectations** 4-ESS1-1 4-ESS2-1 4-ESS2-2

SOILS, ROCKS, AND LANDFORMS — *Overview*

Module Summary	Guiding and Focus Questions for Phenomena
Inv. 3: Mapping Earth's Surface — Students engage with the phenomena of Earth's mountains. They are introduced to the study of topography by building a model of the mountain landform. Students use the foam model of Mount Shasta to create a topographic map, and use this map to produce another representation of the landforms— a profile of the mountain. Students learn about volcanoes; they use the topographer's tools to analyze the impact of the Mount St. Helens eruption. Students are introduced to phenomena that cause rapid changes to Earth's surface: landslides, earthquakes, floods, and volcanoes, and generate ideas that engineers and scientists might use to reduce the impact of these Earth changes.	*How do maps help us observe Earth's surface features?* *What might reduce the impact of catastrophic Earth surface events?* How can we represent the different elevations of landforms? How can we draw a profile of a mountain from a topographic map? How can scientists and engineers help reduce the impacts that events like volcanic eruptions might have on people? What events can change Earth's surface quickly?
Inv. 4: Natural Resources — Students engage with the phenomena of natural resources and how they are used. Students start by reviewing what they have learned in Investigations 1–3. Then they focus on earth materials as renewable and nonrenewable natural resources. They learn the importance of earth materials as resources. The class makes a stepping stone out of concrete and goes on a schoolyard walk to find objects and structures and consider what natural resources were used to construct them.	*What makes rock a natural resource and how is this resource used by people?* What are natural resources and what is important to know about them? How are natural resources used to make concrete? How do people use natural resources to make or build things?

Module Matrix

Content Related to Disciplinary Core Ideas	Reading/Technology	Assessment
- A topographic map uses contour lines to show the shape and elevation of the land. - The change in elevation between two adjacent contour lines is always uniform. The closer the contour lines, the steeper the slope and vice versa. - A profile is a side view or cross-section representation of a landform, and can be derived from the information on a topographic map. - The surface of Earth is constantly changing; sometimes those changes take a long time to occur and sometimes they happen rapidly. - Catastrophic events have the potential to change Earth's surface quickly. - Scientists and engineers can do things to reduce the impacts of natural Earth processes on humans.	**Science Resources Book** "Topographic Maps" "The Story of Mount Shasta" "It Happened So Fast!" **Videos** *Volcanoes* *Mount St. Helens Impact* *All about Earthquakes* **Online Activity** "Topographer"	**Embedded Assessment** Science notebook entries Response sheet Performance assessment **Benchmark Assessment** *Investigation 3 I-Check* **NGSS Performance Expectations** 4-ESS2-2 4-ESS3-2 3–5-ETS1-2
- Natural resources are natural materials taken from the environment and used by humans. - Rocks and minerals are natural resources important for shelter and transportation. - Concrete is an important building material made from earth materials (limestone to make cement, sand and gravel for aggregates, and water for mixing). - Some natural resources are renewable (sunlight, air and wind, water, soil, plants, and animals) and some are nonrenewable (minerals and fossil fuels). - Alternative sources of energy include solar, wind, and geothermal energy. - Scientists and engineers work together to improve the use of natural resources to make them more durable and useful.	**Science Resources Book** "Monumental Rocks" "Geoscientists at Work" "Making Concrete" "Earth Materials in Art" "Where Do Rocks Come From?" (optional) **Video** *Natural Resources* **Online Activities** "Resource ID" "Virtual Investigation: Natural Resources"	**Embedded Assessment** Response sheet Science notebook entry Performance assessment **Benchmark Assessment** *Posttest* **NGSS Performance Expectations** 4-ESS3-1 3–5-ETS1-1

Soils, Rocks, and Landforms Module—FOSS Next Generation

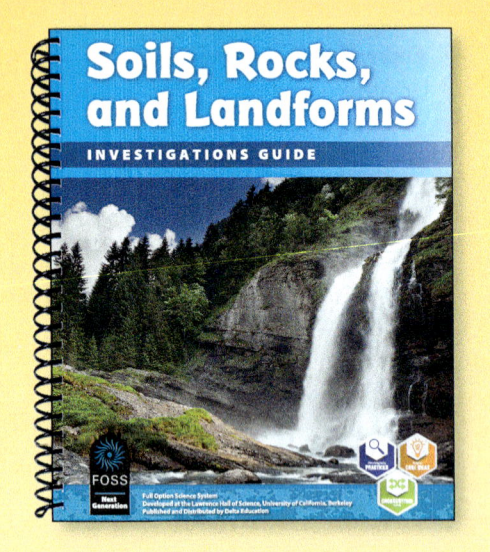

SOILS, ROCKS, AND LANDFORMS — *Overview*

FOSS COMPONENTS

Teacher Toolkit for Each Module

The FOSS Next Generation Program has three modules for grade 4—Energy, Environments, and Soils, Rocks, and Landforms.

Each module comes with a *Teacher Toolkit* for that module. The *Teacher Toolkit* is the most important part of the FOSS Program. It is here that all the wisdom and experience contributed by hundreds of educators has been assembled. Everything we know about the content of the module, how to teach the subject, and the resources that will assist the effort are presented here. Each toolkit has three parts.

Investigations Guide. This spiral-bound document contains these chapters.

- Overview
- Framework and NGSS
- Materials
- Technology
- Investigations (four in this module)
- Assessment

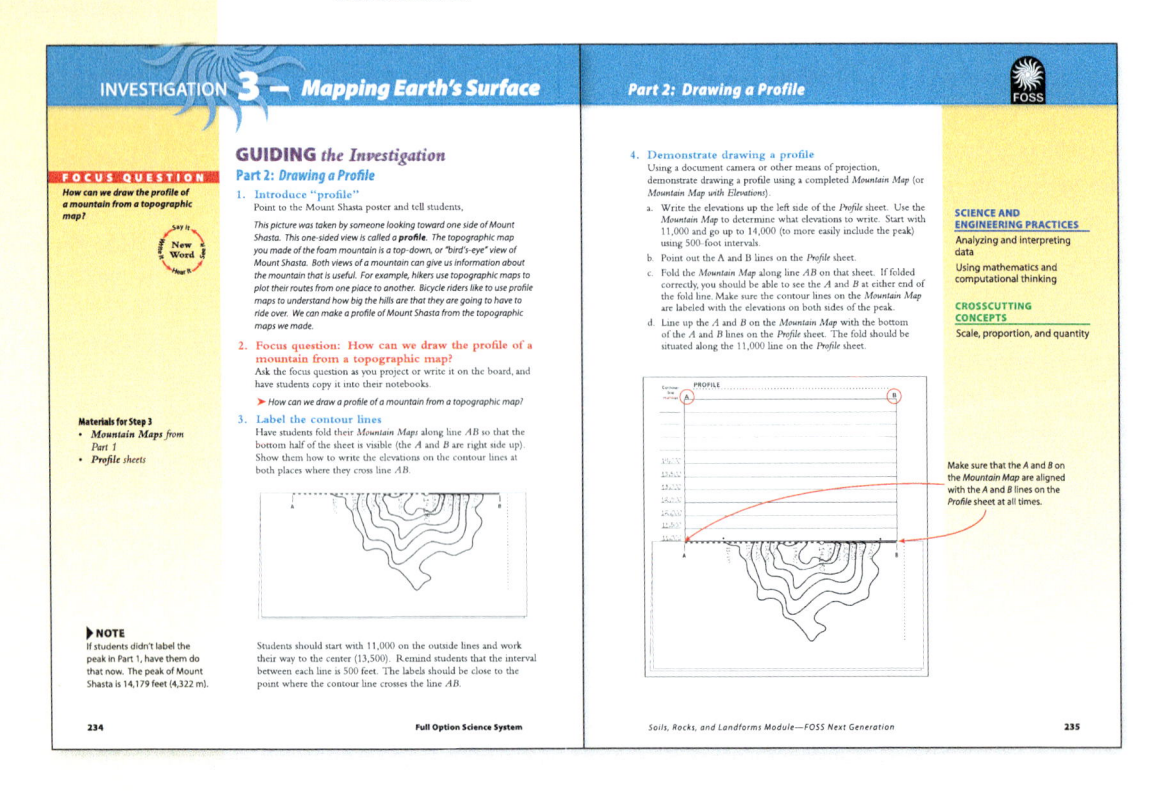

FOSS Components

FOSS Science Resources book. One copy of the student book of readings is included in the *Teacher Toolkit*.

Teacher Resources. These chapters can be downloaded from FOSSweb and are also in the bound *Teacher Resources* book.

- FOSS Program Goals
- Planning Guide—Grade 4
- Science and Engineering Practices—Grade 4
- Crosscutting Concepts—Grade 4
- Sense-Making Discussions for Three-Dimensional Learning—Grade 4
- Access and Equity
- Science Notebooks in Grades 3–5
- Science-Centered Language Development
- FOSS and Common Core ELA—Grade 4
- FOSS and Common Core Math—Grade 4
- Taking FOSS Outdoors
- Science Notebook Masters
- Teacher Masters
- Assessment Masters

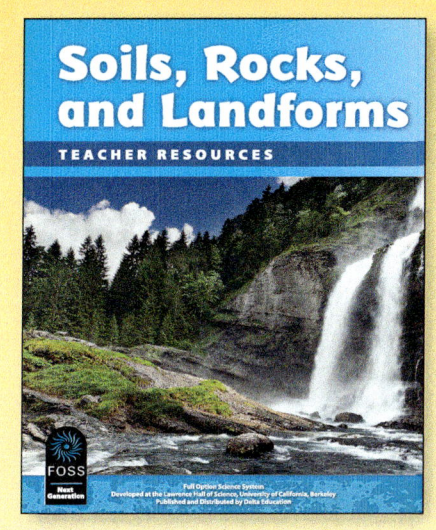

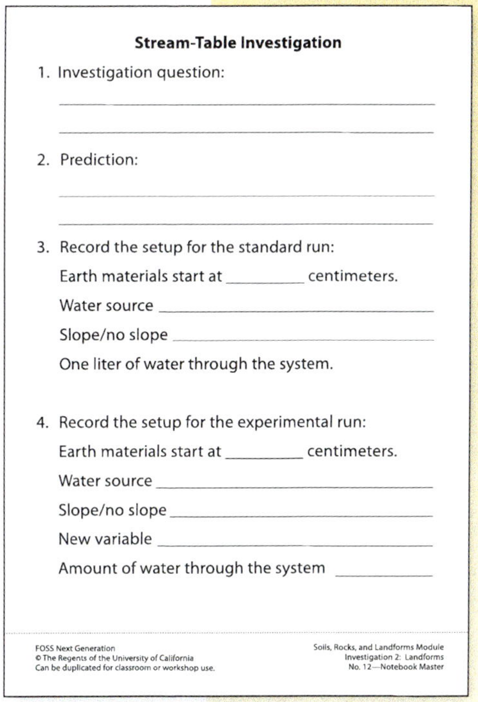

Equipment for Each Module or Grade Level

The FOSS Program provides the materials needed for the investigations, in sturdy, front-opening drawer-and-sleeve cabinets. Inside, you will find high-quality materials packaged for a class of 32 students. Consumable materials are supplied for three uses before you need to resupply. Teachers may be asked to supply small quantities of common classroom materials.

Delta Education can assist you with materials management strategies for schools, districts, and regional consortia.

Soils, Rocks, and Landforms Module—FOSS Next Generation

SOILS, ROCKS, AND LANDFORMS — Overview

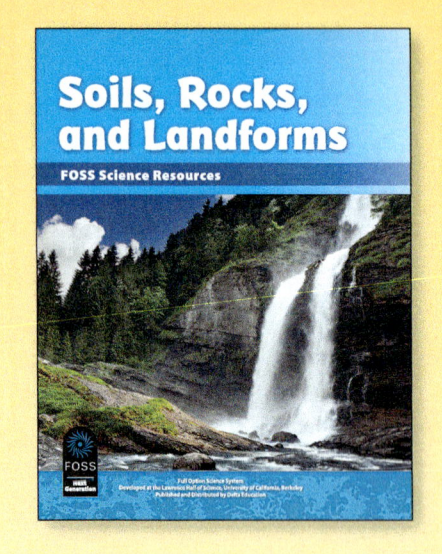

FOSS Science Resources Books

FOSS Science Resources: Soils, Rocks, and Landforms is a book of original readings developed to accompany this module. The readings are referred to as articles in *Investigations Guide*. Students read the articles in the book as they progress through the module. The articles cover specific concepts, usually after the concepts have been introduced in the active investigation.

The articles in *FOSS Science Resources* and the discussion questions provided in *Investigations Guide* help students make connections to the science concepts introduced and explored during the active investigations. Concept development is most effective when students are allowed to experience organisms, objects, and phenomena firsthand before engaging the concepts in text. The text and illustrations help make connections between what students experience concretely and the ideas that explain their observations.

Chemical Weathering

Chemical weathering happens when minerals in rocks are changed by chemicals in water and air. The starting minerals change into new substances.

Many rocks contain iron. When oxygen in air comes in contact with iron, the iron in the rock can rust. Rust is iron oxide. Iron oxide is softer than other iron minerals. This causes the rock to break apart faster.

Carbon dioxide gas in the air **dissolves** in water droplets. This makes **acid**. The acid droplets can fall as rain. The acid causes the **calcite** in **limestone** and **marble** to make holes. This is a chemical change. Monuments, buildings, and gravestones made of marble or limestone change and weaken when exposed to acid rain.

Salt can cause chemical weathering. Salt water can **react** with minerals in rocks to make new minerals. When the new substances are softer than the original mineral, holes can form. The weak rock breaks and falls apart more easily.

Chemical weathering of a rock containing iron

Chemical weathering of marble by acid rain

Chemical weathering of sandstone by salt water

Erosion and Deposition

A trip to the beach is fun. One of the best parts is playing in the sand. And there is so much sand. Where did it all come from? Was it made right there, or did it come from some other place? Much of the sand on the beach came from mountains. **Erosion** moved the sand from the mountains to the beach. Erosion is the taking away of weathered rock. After rocks have weathered into small pieces, they can be carried away by gravity, water, or wind. Most of the sand shown here was carried to the beach by water flowing in rivers and streams.

As long as water keeps flowing, the bits of sand keep moving downstream. When the river enters the ocean, the water slows down. The sand settles to the bottom of the ocean. The settling of **sediments** is called **deposition**. Deposits of sand form beaches all over the world.

A beach

Full Option Science System

FOSS Components

Technology

The FOSS website opens new horizons for educators, students, and families, in the classroom or at home. Each module has digital resources for students and families—interactive simulations, virtual investigations, and online activities. For teachers, FOSSweb provides online teacher *Investigations Guides*; grade-level planning guides (with connections to ELA and math); materials management strategies; science teaching and professional development tools; contact information for the FOSS Program developers; and technical support. In addition, FOSSweb provides digital access to PDF versions of the *Teacher Resources* component of the *Teacher Toolkit*, digital-only instructional resources that supplement the print and kit materials, and access to FOSSmap, the online assessment and reporting system for grades 3–8.

With an educator account, you can customize your homepage, set up easy access to the digital components of the modules you teach, and create class pages for your students with access to tutorials and online assessments.

▶ **NOTE**
To access all the teacher resources and to set up customized pages for using FOSS, log in to FOSSweb through an educator account. See the Technology chapter in this guide for more specifics.

Ongoing Professional Learning

The Lawrence Hall of Science and Delta Education strive to develop long-term partnerships with districts and teachers through thoughtful planning, effective implementation, and ongoing teacher support. FOSS has a strong network of consultants who have rich and experienced backgrounds in diverse educational settings using FOSS.

▶ **NOTE**
Look for professional development opportunities and online teaching resources on www.FOSSweb.com.

Soils, Rocks, and Landforms Module—FOSS Next Generation

SOILS, ROCKS, AND LANDFORMS — *Overview*

FOSS INSTRUCTIONAL DESIGN

FOSS is designed around active investigation that provides engagement with science concepts and science and engineering practices. Surrounding and supporting those firsthand investigations are a wide range of experiences that help build student understanding of core science concepts and deepen scientific habits of mind.

The Elements of the FOSS Instructional Design

Active Investigation

- Using Formative Assessment
- Integrating Science Notebooks
- Taking FOSS Outdoors
- Engaging in Science–Centered Language Development
- Accessing Technology
- Reading *FOSS Science Resources* Books

10 Full Option Science System

FOSS Instructional Design

Each FOSS investigation follows a similar design to provide multiple exposures to science concepts. The design includes these pedagogies.

- Active investigation in collaborative groups: firsthand experiences with phenomena in the natural and designed worlds
- Recording in science notebooks to answer a focus question dealing with the scientific phenomenon under investigation
- Reading informational text in *FOSS Science Resources* books
- Online activities to acquire data or information or to elaborate and extend the investigation
- Outdoor experiences to collect data from the local environment or to apply knowledge
- Assessment to monitor progress and inform student learning

In practice, these components are seamlessly integrated into a curriculum designed to maximize every student's opportunity to learn.

A **learning cycle** employs an instructional model based on a constructivist perspective that calls on students to be actively involved in their own learning. The model systematically describes both teacher and learner behaviors in a coherent approach to science instruction.

A popular model describes a sequence of five phases of intellectual involvement known as the 5Es: engage, explore, explain, elaborate, and evaluate. The body of foundational knowledge that informs contemporary learning-cycle thinking has been incorporated seamlessly and invisibly into the FOSS curriculum design.

Engagement with real-world **phenomena** is at the heart of FOSS. In every part of every investigation, the investigative phenomenon is referenced implicitly in the focus question that guides instruction and frames the intellectual work. The focus question is a prominent part of each lesson and is called out for the teacher and student. The investigation Background for the Teacher section is organized by focus question—the teacher has the opportunity to read and reflect on the phenomenon in each part in preparing for the lesson. Students record the focus question in their science notebooks, and after exploring the phenomenon thoroughly, explain their thinking in words and drawings.

In science, a phenomenon is a natural occurrence, circumstance, or structure that is perceptible by the senses—an observable reality. Scientific phenomena are not necessarily phenomenal (although they may be)—most of the time they are pretty mundane and well within the everyday experience. What FOSS does to enact an effective engagement with the NGSS is thoughtful selection of scientific phenomena for students to investigate.

▶ **NOTE**
The anchor phenomena establish the storyline for the module. The investigative phenomena guide each investigation part. Related examples of everyday phenomena are incorporated into the readings, videos, discussions, formative assessments, outdoor experiences, and extensions.

Soils, Rocks, and Landforms Module—FOSS Next Generation

SOILS, ROCKS, AND LANDFORMS — *Overview*

Active Investigation

Active investigation is a master pedagogy. Embedded within active learning are a number of pedagogical elements and practices that keep active investigation vigorous and productive. The enterprise of active investigation includes

- context: sharing prior knowledge, questioning, and planning;
- activity: doing and observing;
- data management: recording, organizing, and processing;
- analysis: discussing and writing explanations.

Context: sharing, questioning, and planning. Active investigation requires focus. The context of an inquiry can be established with a focus question about a phenomenon or challenge from you or, in some cases, from students. (How do weathered rock pieces move from one place to another?) At other times, students are asked to plan a method for investigation. This might start with a teacher demonstration or presentation. Then you challenge students to plan an investigation, such as to find out how the number of winds of wire around a core affect the strength of an electromagnet. In either case, the field available for thought and interaction is limited. This clarification of context and purpose results in a more productive investigation.

Activity: doing and observing. In the practice of science, scientists put things together and take things apart, observe systems and interactions, and conduct experiments. This is the core of science—active, firsthand experience with objects, organisms, materials, and systems in the natural world. In FOSS, students engage in the same processes. Students often conduct investigations in collaborative groups of four, with each student taking a role to contribute to the effort.

The active investigations in FOSS are cohesive, and build on each other to lead students to a comprehensive understanding of concepts. Through investigations and readings, students gather meaningful data.

Data management: recording, organizing, and processing. Data accrue from observation, both direct (through the senses) and indirect (mediated by instrumentation). Data are the raw material from which scientific knowledge and meaning are synthesized. During and after work with materials, students record data in their science notebooks. Data recording is the first of several kinds of student writing.

Students then organize data so they will be easier to think about. Tables allow efficient comparison. Organizing data in a sequence (time) or series (size) can reveal patterns. Students process some data into graphs, providing visual display of numerical data. They also organize data and process them in the science notebook.

FOSS Instructional Design

Analysis: discussing and writing explanations. The most important part of an active investigation is extracting its meaning. This constructive process involves logic, discourse, and prior knowledge. Students share their explanations for phenomena, using evidence generated during the investigation to support their ideas. They conclude the active investigation by writing a summary of their learning as well as questions raised during the activity in their science notebooks.

Science Notebooks

Research and best practice have led FOSS to place more emphasis on the student science notebook. Keeping a notebook helps students organize their observations and data, process their data, and maintain a record of their learning for future reference. The process of writing about their science experiences and communicating their thinking is a powerful learning device for students. The science-notebook entries stand as credible and useful expressions of learning. The artifacts in the notebooks form one of the core exhibitions of the assessment system.

You will find the duplication masters for grades 1–5 presented in notebook format. They are reduced in size (two copies to a standard sheet) for placement (glue or tape) into a bound composition book. Full-sized masters for grades 3–5 that can be filled in electronically and are suitable for display are available on FOSSweb. Student work is entered partly in spaces provided on the notebook sheets and partly on adjacent blank sheets in the composition book. Look to the chapter in *Teacher Resources* called Science Notebooks in Grades 3–5 for more details on how to use notebooks with FOSS.

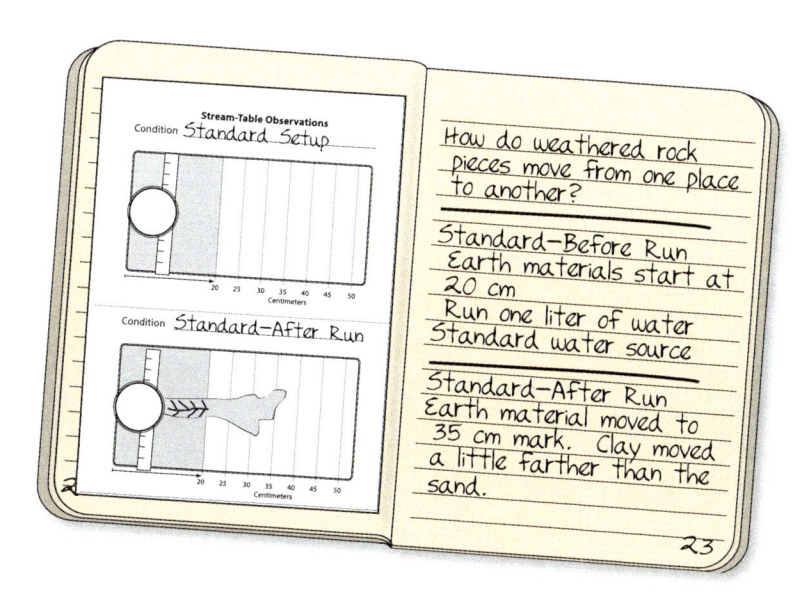

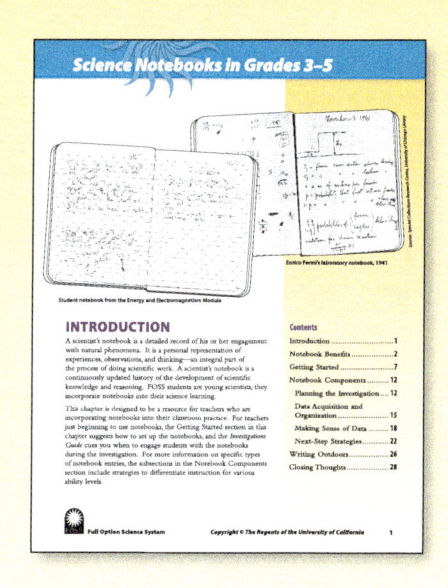

Soils, Rocks, and Landforms Module—FOSS Next Generation

13

SOILS, ROCKS, AND LANDFORMS — Overview

Reading in *FOSS Science Resources*

The *FOSS Science Resources* books, available in print and interactive eBooks, are primarily devoted to expository articles and biographical sketches. FOSS suggests that the reading be completed during language-arts time to connect to the Common Core State Standards for ELA. When language-arts skills and methods are embedded in content material that relates to the authentic experience students have had during the FOSS active learning sessions, students are interested, and they get more meaning from the text material.

Recommended strategies to engage students in reading, writing, speaking, and listening around the articles in the *FOSS Science Resources* books are included in the flow of Guiding the Investigation. In addition, a library of resources is described in the Science-Centered Language Development chapter in *Teacher Resources*.

The FOSS and the Common Core ELA—Grade 4 chapter in *Teacher Resources* shows how FOSS provides opportunities to develop and exercise the Common Core ELA practices through science. A detailed table identifies these opportunities in the three FOSS modules for the fourth grade.

Engaging in Online Activities through FOSSweb

The simulations and online activities on FOSSweb are designed to support students' learning at specific times during instruction. Digital resources include streaming videos that can be viewed by the class or small groups. Resources may also include virtual investigations and tutorials that students can use to review the active investigations and to support students who need more time with the concepts or who have been absent and missed the active investigations.

The Technology chapter provides details about the online activities for students and the tools and resources for teachers to support and enrich instruction. There are many ways for students to engage with the digital resources—in class as individuals, in small groups, or as a whole class, and at home with family and friends.

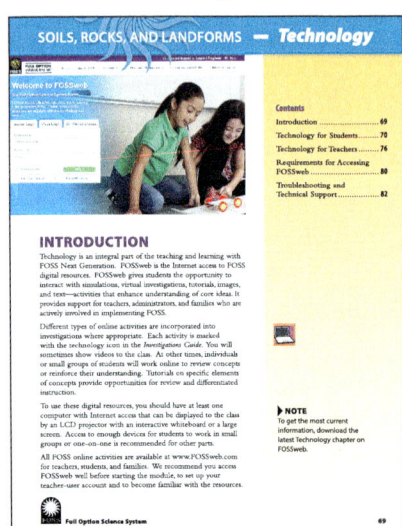

FOSS Instructional Design

Assessing Progress

The FOSS assessment system includes both formative and summative assessments. Formative assessment monitors learning during the process of instruction. It measures progress, provides information about learning, and is predominantly diagnostic. Summative assessment looks at the learning after instruction is completed, and it measures achievement.

Formative assessment in FOSS, called **embedded assessment**, is an integral part of instruction, and occurs on a daily basis. You observe action during class in a performance assessment or review notebooks after class. Performance assessments look at students' engagement in science and engineering practices or their recognition of crosscutting concepts, and are indicated with the second assessment icon. Embedded assessment provides continuous monitoring of students' learning and helps you make decisions about whether to review, extend, or move on to the next idea to be covered.

The embedded assessments are based on authentic work produced by students during the course of participating in the FOSS activities. Students do their science, and you observe the actions and look at their notebook entries. Bullet points in the Guiding the Investigation tell you specifically what students should know and be able to communicate.

Benchmark assessments include the *Survey*, I-Checks, *Posttest*, and interim assessments. The *Survey* is given before instruction begins. It provides information about students' prior knowledge. **I-Checks** are actually hybrid tools: they can provide summative information about students' achievement, but more importantly, they can be used formatively as well to provide diagnostic information. Reviewing specific items on an I-Check with the class provides additional opportunities for students to clarify their thinking. The *Posttest* is a summative assessment given after instruction is complete.

Interim assessments give students practice with items specifically designed to measure three-dimensional learning described by NGSS performance expectations. Interim assessment tasks generally begin with a scenario, and ask students to apply practices and crosscutting concepts as well as disciplinary core ideas to respond to the item. Interim assessment tasks can be administered during a module, at the end of the module, or as an end-of-year grade-level assessment.

All benchmark items are carefully designed to be valid, reliable, and accessible to all students. They focus on assessment *for* learning, and when accompanied by thoughtful self-assessment activities and feedback, contribute to the development of a growth mindset.

▶ **TECHNOLOGY COMPONENTS OF THE FOSS ASSESSMENT SYSTEM**
FOSSmap for teachers and online assessment for students are the technology components of the FOSS assessment system. Students in grades 3–5 can take assessments online. FOSSmap provides the tools for you to review those assessments online so you can determine next steps for the class or differentiated instruction for individual students based on assessment performance. See the Assessment chapter for more information on these technology components.

Soils, Rocks, and Landforms Module—FOSS Next Generation

SOILS, ROCKS, AND LANDFORMS — *Overview*

Taking FOSS Outdoors

FOSS throws open the classroom door and proclaims the entire school campus to be the science classroom. The true value of science knowledge is its usefulness in the real world and not just in the classroom. Taking regular excursions into the immediate outdoor environment has many benefits. First of all, it provides opportunities for students to apply things they learned in the classroom to novel situations. When students are able to transfer knowledge of scientific principles to natural systems, they experience a sense of accomplishment.

In addition to transfer and application, students can learn things outdoors that they are not able to learn indoors. The most important object of inquiry outdoors is the outdoors itself. To today's youth, the outdoors is something to pass through as quickly as possible to get to the next human-managed place. For many, engagement with the outdoors and natural systems must be intentional, at least at first. With repeated visits to familiar outdoor learning environments, students may first develop comfort in the outdoors, and then a desire to embrace and understand natural systems.

The last part of most investigations is an outdoor experience. Venturing out will require courage the first time or two you mount an outdoor expedition. It will confuse students as they struggle to find the right behavior that is a compromise between classroom rigor and diligence and the freedom of recreation. With persistence, you will reap rewards. You will be pleased to see students' comportment develop into proper field-study habits, and you might be amazed by the transformation of students with behavior issues in the classroom who become your insightful observers and leaders in the schoolyard environment.

Teaching outdoors is the same as teaching indoors—except for the space. You need to manage the same four core elements of classroom teaching: time, space, materials, and students. Because of the different space, new management procedures are required. Students can get farther away. Materials have to be transported. The space has to be defined and honored. Time has to be budgeted for getting to, moving around in, and returning from the outdoor study site. All these and more issues and solutions are discussed in the Taking FOSS Outdoors chapter in *Teacher Resources*.

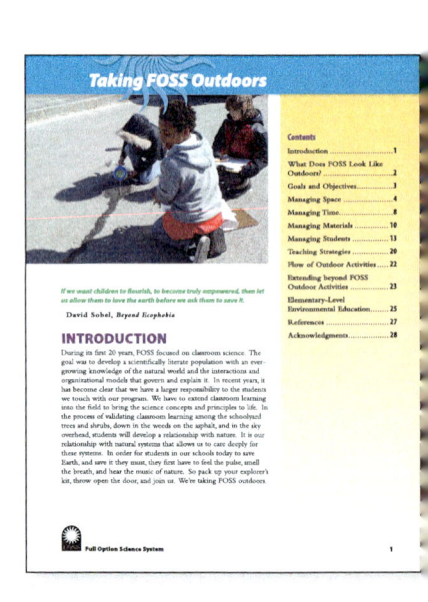

▶ **NOTE**
The kit includes a set of four *Conservation* posters so you can discuss the importance of natural resources with students.

FOSS Instructional Design

Science-Centered Language Development and Common Core State Standards for ELA

The FOSS active investigations, science notebooks, *FOSS Science Resources* articles, and formative assessments provide rich contexts in which students develop and exercise thinking and communication. These elements are essential for effective instruction in both science and language arts—students experience the natural world in real and authentic ways and use language to inquire, process information, and communicate their thinking about scientific phenomena. FOSS refers to this development of language process and skills within the context of science as science-centered language development.

In the Science-Centered Language Development chapter in *Teacher Resources*, we explore the intersection of science and language and the implications for effective science teaching and language development. Language plays two crucial roles in science learning: (1) it facilitates the communication of conceptual and procedural knowledge, questions, and propositions, and (2) it mediates thinking—a process necessary for understanding. For students, language development is intimately involved in their learning about the natural world. Science provides a real and engaging context for developing literacy and language-arts skills identified in contemporary standards for English language arts.

The most effective integration depends on the type of investigation, the experience of students, the language skills and needs of students, and the language objectives that you deem important at the time. The Science-Centered Language Development chapter is a library of resources and strategies for you to use. The chapter describes how literacy strategies are integrated purposefully into the FOSS investigations, gives suggestions for additional literacy strategies that both enhance students' learning in science and develop or exercise English-language literacy skills, and develops science vocabulary with scaffolding strategies for supporting all learners. We identify effective practices in language-arts instruction that support science learning and examine how learning science content and engaging in science and engineering practices support language development.

Specific methods to make connections to the Common Core State Standards for English Language Arts are included in the flow of Guiding the Investigation. These recommended methods are linked to the CCSS ELA through ELA Connection notes. In addition, the FOSS and the Common Core ELA chapter in *Teacher Resources* summarizes all of the connections to each standard at the given grade level.

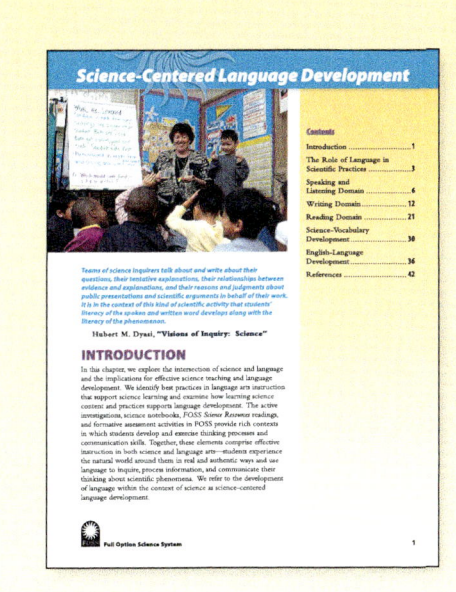

Soils, Rocks, and Landforms Module—FOSS Next Generation

SOILS, ROCKS, AND LANDFORMS — *Overview*

DIFFERENTIATED INSTRUCTION FOR ACCESS AND EQUITY

Learning from Experience

The roots of FOSS extend back to the mid-1970s and the Science Activities for the Visually Impaired and Science Enrichment for Learners with Physical Handicaps projects (SAVI/SELPH Program). As this special-education science program expanded into fully integrated (mainstreamed) settings in the 1980s, hands-on science proved to be a powerful medium for bringing all students together. The subject matter is universally interesting, and the joy and satisfaction of discovery are shared by everyone. Active science by itself provides part of the solution to full inclusion and provides many opportunities at the same time for differentiated instruction.

Many years later, FOSS began a collaboration with educators and researchers at the Center for Applied Special Technology (CAST), where principles of Universal Design for Learning (UDL) had been developed and applied. FOSS continues to learn from our colleagues about ways to use new media and technologies to improve instruction. Here are the UDL guiding principles.

Principle 1. Provide multiple means of representation. Give learners various ways to acquire information and demonstrate knowledge.

Principle 2. Provide multiple means of action and expression. Offer students alternatives for communicating what they know.

Principle 3. Provide multiple means of engagement. Help learners get interested, be challenged, and stay motivated.

"Active science by itself provides part of the solution to full inclusion and provides many opportunities at the same time for differentiated instruction."

Differentiated Instruction for Access and Equity

FOSS for All Students

The FOSS Program has been designed to maximize the science learning opportunities for all students, including those who have traditionally not had access to or have not benefited from equitable science experiences—students with special needs, ethnically diverse learners, English learners, students living in poverty, girls, and advanced and gifted learners. FOSS is rooted in a 30-year tradition of multisensory science education and informed by recent research on UDL and culturally and linguistically responsive teaching and learning. Procedures found effective with students with special needs and students who are learning English are incorporated into the materials and strategies used with all students during the initial instruction phase. In addition, the **Access and Equity** chapter in *Teacher Resources* (or go to FOSSweb to download this chapter) provides strategies and suggestions for enhancing the science and engineering experiences for each of the specific groups noted above.

Throughout the FOSS investigations, students experience multiple ways of interacting with phenomena and expressing their understanding through a variety of modalities. Each student has multiple opportunities to demonstrate his or her strengths and needs, thoughts, and aspirations.

The challenge is then to provide appropriate follow-up experiences or enhancements appropriate for each student. For some students, this might mean more time with the active investigations or online activities. For other students, it might mean more experience and/or scaffolds for developing models, building explanations, or engaging in argument from evidence.

For some students, it might mean making vocabulary and language structures more explicit through new concrete experiences or through reading to students. It may help them identify and understand relationships and connections through graphic organizers.

For other students, it might be designing individual projects or small-group investigations. It might be more opportunities for experiencing science outside the classroom in more natural, outdoor environments or defining problems and designing solutions in their communities.

Soils, Rocks, and Landforms Module—FOSS Next Generation

SOILS, ROCKS, AND LANDFORMS — *Overview*

Assessment and Extensions

The next-step strategies suggested during the self-assessment sessions following I-Checks provide opportunities for differentiated instruction. These strategies can also be used to provide targeted and strategic instruction for students who need additional specific support. For more on next-step strategies, see the Assessment chapter.

There are additional approaches and strategies for ensuring access and equity by providing appropriate differentiated instruction. The FOSS Program provides tools and strategies so that you know what students are thinking throughout the module. Based on that knowledge and what you know about your students' cultural and linguistic background, as well as their individual strengths and needs, read through the extension activities for additional ways to enhance the learning experience for your students. Interdisciplinary extensions in the arts, social studies, math, and language arts, as well as more advanced projects are listed at the end of each investigation. Use these ideas to meet the individual needs and interests of your students. In addition, online activities including tutorials and virtual investigations are effective tools to provide differentiated levels of instruction.

English Learners

The FOSS Program provides a rich laboratory for language development for English learners. A variety of techniques to make science concepts clear and concrete, including modeling, visuals, and active investigations in small groups. Instruction is guided and scaffolded through carefully designed lesson plans, and students are supported throughout.

Science vocabulary and language structures are introduced in authentic contexts while students engage in hands-on learning and collaborative discussion. Strategies for helping all students read, write, speak, and listen are described in the Science-Centered Language Development chapter. A specific section on English learners provides suggestions for both integrating English language development (ELD) approaches during the investigation and for developing designated (targeted and strategic) ELD-focused lessons that support science learning.

FOSS INVESTIGATION ORGANIZATION

Modules are subdivided into **investigations** (four in this module). Investigations are further subdivided into three to five **parts**. Each investigation has a general guiding question for the phenomenon students investigate, and each part of each investigation is driven by a **focus question**. The focus question, usually presented as the part begins, engages the student with the phenomenon and signals the challenge to be met, mystery to be solved, or principle to be uncovered. The focus question guides students' actions and thinking and makes the learning goal of each part explicit for teachers. Each part concludes with students recording an answer to the focus question in their notebooks.

The investigation is summarized for the teacher in the At-a-Glance chart at the beginning of each investigation.

Investigation-specific **scientific background** information for the teacher is presented in each investigation chapter organized by the focus questions.

The **Teaching Children about** section makes direct connections to the NGSS foundation boxes for the grade level—Disciplinary Core Ideas, Science and Engineering Practices, and Crosscutting Concepts. This information is later presented in color-coded sidebar notes to identify specific places in the flow of the investigation where connections to the three dimensions of science learning appear. The Teaching Children about section ends with information about teaching and learning and a conceptual-flow graphic of the content.

The **Materials** and **Getting Ready** sections provide scheduling information and detail exactly how to prepare the materials and resources for conducting the investigation.

Teaching notes and **ELA Connections** appear in blue boxes in the sidebars. These notes comprise a second voice in the curriculum—an educative element. The first (traditional) voice is the message you deliver to students. The second (educative) voice, shared as a teaching note, is designed to help you understand the science content and pedagogical rationale at work behind the instructional scene. ELA Connection boxes provide connections to the Common Core State Standards for English Language Arts.

FOCUS QUESTION

How do weathered rock pieces move from one place to another?

SCIENCE AND ENGINEERING PRACTICES
Developing and using models

DISCIPLINARY CORE IDEAS
ESS2.A: Earth materials and systems

CROSSCUTTING CONCEPTS
Cause and effect

TEACHING NOTE

This focus question can be answered with a simple yes or no, but the question has power when students support their answers with evidence. Their answers should take the form "Yes, because ____ ."

SOILS, ROCKS, AND LANDFORMS — *Overview*

The **Getting Ready** and **Guiding the Investigation** sections have several features that are flagged in the sidebars. These include several icons to remind you when a particular pedagogical method is suggested, as well as concise bits of information in several categories.

The **safety** icon alerts you to potential safety issues related to chemicals, allergic reactions, and the use of safety goggles.

The small-group **discussion** icon asks you to pause while students discuss data or construct explanations in their groups.

The **new-word** icon alerts you to a new vocabulary word or phrase that should be introduced thoughtfully.

The **vocabulary** icon indicates where students should review recently introduced vocabulary.

The **recording** icon points out where students should make a science-notebook entry.

The **reading** icon signals when the class should read a specific article in the *FOSS Science Resources* book.

The **technology** icon signals when the class should use a digital resource on FOSSweb.

FOSS Investigation Organization

The **assessment** icons appear when there is an opportunity to assess student progress by using embedded or benchmark assessments. Some are performance assessments—observations of science and engineering practices, indicated by a second icon which includes a beaker and ruler.

The **outdoor** icon signals when to move the science learning experience into the schoolyard.

The **engineering** icon indicates opportunities for an experience incorporating engineering practices.

The **math** icon indicates an opportunity to engage in numerical data analysis and mathematics practice.

The **crosscutting concepts** icon indicates an opportunity to expand on the concept by going to *Teacher Resources*, Crosscutting Concepts chapter. This chapter provides details on how to engage students with that concept in the context of the investigation.

The **EL note** provides a specific strategy to use to assist English learners in developing science concepts.

EL NOTE

To help with pacing, you will see icons for **breakpoints**. Some breakpoints are essential, and others are optional.

POSSIBLE BREAKPOINT

Soils, Rocks, and Landforms Module—FOSS Next Generation

SOILS, ROCKS, AND LANDFORMS — *Overview*

ESTABLISHING A CLASSROOM CULTURE

Working in Collaborative Groups

Collaboration is important in science. Scientists usually collaborate on research enterprises. Groups of researchers often contribute to the collection of data, the analysis of findings, and the preparation of the results for publication.

Collaboration is expected in the science classroom, too. Some tasks call for everyone to have the same experience, either taking turns or doing the same things simultaneously. At other times, group members may have different experiences that they later bring together.

Research has shown that students learn better and are more successful when they collaborate. Working together promotes student interest, participation, learning, language development, and self-confidence. FOSS investigations use collaborative groups extensively. Here are a few general guidelines for collaborative learning that have proven successful over the years.

For most activities in upper-elementary grades, collaborative groups of four in which students take turns assuming specific responsibilities work best. Groups can be identified completely randomly (first four names drawn from a hat constitute group 1), or you can assemble groups to ensure diversity and inclusion. Thoughtfully constituted groups tend to work better.

Groups can be maintained for extended periods of time, or they can be reconfigured frequently. Six to nine weeks seems about optimum, so students might stay together throughout an entire module.

Functional roles within groups can be determined by the members themselves, or they can be assigned in one of several ways. Each member in a collaborative group can be assigned a number or a color. Then you need only announce which color or number will perform a certain task for the group at a certain time. Compass points can also be used: the person seated on the east side of the table will be the Reporter for this investigation.

The functional roles used in the investigations follow. If you already use other names for functional roles in your class, use them in place of those in the investigations.

The **Getter** is responsible for materials. One person from each group gets equipment from the materials station, and later returns the equipment.

Full Option Science System

Establishing a Classroom Culture

One person is the **Starter** for each task. This person supervises setting up the equipment and makes sure that everyone gets a turn and that everyone has an opportunity to contribute ideas to the investigation.

The **Recorder** makes sure that everyone has recorded information in his or her science notebook and, as appropriate, records the group data and ideas.

The **Reporter** reports group data to the class or transcribes it to the board or class chart, and shares the main points of the group discussion during class discussion.

Getting started with collaborative groups requires patience, but the rewards are great. Once collaborative groups are in place, you will be able to engage students in more meaningful conversations about science content. You are free to "cruise" the groups, to observe and listen to students as they work, and to interact with individuals and small groups as needed.

Norms for Sense-Making Discussions

Setting up norms for discussion and holding yourself and your students accountable is the first step toward creating a culture of productive talk in the classroom that supports engagement in the science and engineering practices. Students need to feel free to express their ideas, and to provide and receive criticism from others as they work toward understanding of the disciplinary core ideas of science and methods of engineering.

Establish norms at the beginning of the school year. It is recommended that this be done together as a class activity. However, presenting a poster of norms to students and asking them to discuss why each one is important can also be effective. Before each sense-making discussion, review the norms. Review what it will look like, sound like, and feel like when everyone is following the agreements. You might have students work on one or two at a time as they are developing their oral discourse skills. After discussion, save a few minutes for reflection on how well the group or the class adhered to the norms and what they can do better next time. More strategies for supporting academic discourse can be found in the Sense-Making Discussions for Three-Dimensional Learning and Science-Centered Language Development chapters in *Teacher Resources* (also available as downloadable PDFs on FOSSweb).

This poster is an example of student responsibilities that the class discussed and adopted as their norms.

Soils, Rocks, and Landforms Module—FOSS Next Generation

SOILS, ROCKS, AND LANDFORMS — Overview

Managing Materials

The Materials section lists the items in the equipment kit and any teacher-supplied materials. It also describes things to do to prepare a new kit and how to check and prepare the kit for your classroom. Classroom volunteers and helpful students can also assist in setting up and preparing the materials.

Individual photos of each piece of FOSS equipment are available for printing from FOSSweb, and can help students and you identify each item. They can be used to support emerging readers and English language learners to acquire and use new vocabulary words necessary to engage in the investigations. (Photo equipment cards are available in English and Spanish formats.)

The FOSS Program designers suggest using a central materials distribution system. You organize all the materials for an investigation at a single location called the materials station. As the investigation progresses, one member of each group gets materials as they are needed, and another returns the materials when the investigation is complete. You place the equipment and resources at the station, and students do the rest. Students can also be involved in cleaning and organizing the materials at the end of a session. Be sure to work with students to Reduce, Reuse, and Recycle the materials used in science.

When Students Are Absent

When a student is absent for a session, give him or her a chance to spend some time with the materials at a center. Another student might act as a peer tutor. Allow the student to bring home a *FOSS Science Resources* book to read with a family member. There are suggested interactive reading strategies for each article as well as embedded review items that the student can respond to verbally or in writing.

There is a set of two or three virtual investigations for each FOSS module for grades 3–5. Students who have been absent from certain investigations can access these simulations online through FOSSweb. The virtual investigations require students to record data and answer concluding questions in their science notebooks. Sometimes the notebook sheet that was used in the classroom investigation is also used for the virtual investigation.

Establishing a Classroom Culture

Collaborative Teaching and Learning

Collaborative learning requires a collective as well as individual growth mindset. A growth mindset is when people believe that their most basic abilities can be developed through dedication and hard work (see the research of Carol Dweck and her book *Mindset: The Psychology of Success*). As students work together to make sense of phenomena and develop their inquiry and discourse skills, it's important to recognize and value their efforts to try new approaches and their willingness to make their thinking visible. Remind students that everyone in the classroom, including you the teacher, will be learning new ideas and ways to think about the world. Where there is productive struggle, there is learning. Here are a few ways to help students develop a growth mind-set for science and engineering.

- **Praise effort, not right answers**. When students are successful at a task, provide positive feedback about their level of engagement and effort in the practices, e.g., the efforts they put into careful observations, how well they organized and interpreted their data, the relevancy of their questions, how well they connected or applied new concepts, and their use of precise vocabulary, etc. Also, try to provide feedback that encourages students to continue to improve their learning and exploring, e.g., is there another way to approach this question? Have you thought about _____? What evidence is there to support _____?

- **Foster and validate divergent thinking**. During sense-making discussions, continually emphasize how important it is to share emerging ideas and to be open to the ideas of others in order to build understanding. Model for students how you refine and revise your thinking based on new information. Make it clear to students that the point is not for them to show they have the right answer, but rather to help each other arrive at new understandings. Point out positive examples of students expressing and revising their ideas.

Establishing a classroom culture that supports three-dimensional teaching and learning centers on collaboration. Collaborative groupings, materials management, and norms are structures you can put into place to foster collaboration. These structures along with the expectations that students will be negotiating meaning together as a community of learners, creates a learning environment where students are compelled to work, think, and communicate like scientists and engineers to help one another learn.

Soils, Rocks, and Landforms Module—FOSS Next Generation

SOILS, ROCKS, AND LANDFORMS — *Overview*

SAFETY IN THE CLASSROOM AND OUTDOORS

Following the procedures described in each investigation will make for a very safe experience in the classroom. You should also review your district safety guidelines and make sure that everything you do is consistent with those guidelines. Two posters are included in the kit: *Science Safety* for classroom use and *Outdoor Safety* for outdoor activities.

Look for the safety icon in the Getting Ready and Guiding the Investigation sections that will alert you to safety considerations throughout the module.

Safety Data Sheets (SDS) for materials used in the FOSS Program can be found on FOSSweb. If you have questions regarding any SDS, call Delta Education at 1-800-258-1302 (Monday–Friday, 8 a.m.–5 p.m. ET).

Science Safety in the Classroom

General classroom safety rules to share with students are listed here.

1. Listen carefully to your teacher's instructions. Follow all directions. Ask questions if you don't know what to do.
2. Tell your teacher if you have any allergies.
3. Never put any materials in your mouth. Do not taste anything unless your teacher tells you to do so.
4. Never smell any unknown material. If your teacher tells you to smell something, wave your hand over the material to bring the smell toward your nose.
5. Do not touch your face, mouth, ears, eyes, or nose while working with chemicals, plants, or animals.
6. Always protect your eyes. Wear safety goggles when necessary. Tell your teacher if you wear contact lenses.
7. Always wash your hands with soap and warm water after handling chemicals, plants, or animals.
8. Never mix any chemicals unless your teacher tells you to do so.
9. Report all spills, accidents, and injuries to your teacher.
10. Treat animals with respect, caution, and consideration.
11. Clean up your work space after each investigation.
12. Act responsibly during all science activities.

Full Option Science System

Scheduling the Module

SCHEDULING THE MODULE

Below is a suggested teaching schedule for the module. The investigations are numbered and should be taught in order, as the concepts build upon each other from investigation to investigation. We suggest that a minimum of nine weeks be devoted to this module.

Active-investigation (A) sessions include hands-on work with materials active thinking about experiences, small-group discussion, writing in science notebooks, and learning new vocabulary in context.

Reading (R) sessions involve reading *FOSS Science Resources* articles. Reading can be completed during language-arts time to make connections to Common Core State Standards for ELA (CCSS ELA).

During **Wrap-Up/Warm-Up (W)** sessions, students share notebook entries and engage in connections to CCSS ELA. These sessions can also be completed during language-arts time.

I-Checks are short summative assessments at the end of each investigation. Students have a short notebook review session the day before and a self-assessment of selected items the following day. (See the Assessment chapter for the next-step strategies for self-assessment.)

> **NOTE**
> The Getting Ready section for each part of an investigation helps you prepare. It provides information on scheduling the activities and introduces the tools and techniques used in the activity. Be prepared—read the Getting Ready section thoroughly and review the teacher preparation video on FOSSweb.

Week	Day 1	Day 2	Day 3	Day 4	Day 5
	Survey				
1	START Inv. 1 Part 1			START Inv. 1 Part 2	
	A	A	A/R/W	A	A/W
2	START Inv. 1 Part 3				START Inv. 1 Part 4
	A	A	A	R/W	A
3				START Inv. 2 Part 1	
	A/Review	I-Check	Self-assess	A	R/W
4	START Inv. 2 Part 2				START Inv. 2 Part 3
	A	A	A	A/W	A/W
5	START Inv. 2 Part 4				
	A	A	R/Review	I-Check 2	Self-assess
6	START Inv. 3 Part 1		START Inv. 3 Part 2		
	A	R/W	A	R	R/W
7	START Inv. 3 Part 3	START Inv. 3 Part 4			
	A	A	R	Review	I-Check 3
8		START Inv. 4 Part 1			START Inv. 4 Part 2
	Self-assess	A	A/R	R/W	A
9		START Inv. 4 Part 3			
	R/W	A/R	R	Review	Posttest

Soils, Rocks, and Landforms Module—FOSS Next Generation

SOILS, ROCKS, AND LANDFORMS — Overview

FOSS CONTACTS

General FOSS Program information

www.FOSSweb.com

www.DeltaEducation.com/FOSS

Developers at the Lawrence Hall of Science

foss@berkeley.edu

Customer Service at Delta Education

www.DeltaEducation.com/contact.aspx

Phone: 1-800-258-1302, 8:00 a.m.–5:00 p.m. ET

FOSSmap (online component of FOSS assessment system)

http://FOSSmap.com/

FOSSweb account questions/access codes/help logging in

techsupport.science@schoolspecialty.com

Phone: 1-800-258-1302, 8:00 a.m.–5:00 p.m. ET

School Specialty online support

loginhelp@schoolspecialty.com

Phone: 1-800-513-2465, 8:30 a.m.–6:00 p.m. ET

FOSSweb tech support

support@fossweb.com

Professional development

www.FOSSweb.com/Professional-Development

Safety issues

www.DeltaEducation.com/SDS

Phone: 1-800-258-1302, 8:00 a.m.–5:00 p.m. ET

For chemical emergencies, contact Chemtrec 24 hours per day.

Phone: 1-800-424-9300

Sales and replacement parts

www.DeltaEducation.com/FOSS/buy

Phone: 1-800-338-5270, 8:00 a.m.–5:00 p.m. ET

SOILS, ROCKS, AND LANDFORMS – Framework and NGSS

INTRODUCTION TO PERFORMANCE EXPECTATIONS

"The NGSS are standards or goals, that reflect what a student should know and be able to do; they do not dictate the manner or methods by which the standards are taught. . . . Curriculum and assessment must be developed in a way that builds students' knowledge and ability toward the PEs [performance expectations]" (*Next Generation Science Standards*, 2013, page xiv).

This chapter shows how the NGSS Performance Expectations were bundled in the **Soils, Rocks, and Landforms Module** to provide a coherent set of instructional materials for teaching and learning.

This chapter also provides details about how this FOSS module fits into the matrix of the FOSS Program (page 41). Each FOSS module K–5 and middle school course 6–8 has a functional role in the FOSS conceptual frameworks that were developed based on a decade of research on science education and the influence of *A Framework for K–12 Science Education* (2012) and *Next Generation Science Standards* (NGSS, 2013).

The FOSS curriculum provides a coherent vision of science teaching and learning in the three ways described by the NRC *Framework*. First, FOSS is designed around learning as a developmental progression, providing experiences that allow students to continually build on their initial notions and develop more complex science and engineering knowledge. Students develop functional understanding over time by building on foundational elements (intermediate knowledge). That progression is detailed in the conceptual frameworks.

Second, FOSS limits the number of core ideas, choosing depth of knowledge over broad shallow coverage. Those core ideas are addressed at multiple grade levels in ever greater complexity. FOSS investigations at each grade level focus on elements of core ideas that are teachable and learnable at that grade level.

Third, FOSS investigations integrate engagement with scientific ideas (content) and the practices of science and engineering by providing firsthand experiences.

Teach the module with the confidence that the developers have carefully considered the latest research and have integrated into each investigation the three dimensions of the *Framework* and NGSS, and have designed powerful connections to the Common Core State Standards for English Language Arts.

Contents

Introduction to Performance Expectations 31

FOSS Conceptual Framework 40

Background for the Conceptual Framework in Soils, Rocks, and Landforms ... 42

Connections to NGSS by Investigation 52

Recommended FOSS Next Generation K–8 Scope and Sequence 56

The NGSS Performance Expectations bundled in this module include

Earth and Space Sciences
4-ESS1-1
4-ESS2-1
4-ESS2-2
4-ESS3-1
4-ESS3-2

Engineering, Technology, and Applications of Science
3–5 ETS1-1
3–5 ETS1-2

FOSS Full Option Science System

SOILS, ROCKS, AND LANDFORMS — Framework and NGSS

DISCIPLINARY CORE IDEAS

A Framework for K–12 Science Education has three core ideas in Earth and space sciences.

ESS1: Earth's place in the universe

ESS2: Earth's systems

ESS3: Earth and human activity

The questions and descriptions of the core ideas in the text on these pages are taken from the NRC *Framework* for the grades 3–5 grade band to keep the core ideas in a rich and useful context.

The performance expectations related to each core idea are taken from the NGSS for grade 4.

Disciplinary Core Ideas Addressed

The **Soils, Rocks, and Landforms Module** connects with the NRC *Framework* 3–5 grade band and the NGSS performance expectations for grade 4. The module focuses on core ideas for Earth sciences.

Earth and Space Sciences

Framework core idea ESS1: Earth's place in the universe—What is the universe, and what is Earth's place in it?

- **ESS1.C: The history of planet Earth**
  How do people reconstruct and date events in Earth's planetary history? [Earth has changed over time. Understanding how landforms develop, are weathered, and erode can help to infer the history of the current landscape.]

The following NGSS Performance Expectation for ESS1 is derived from the Framework disciplinary core idea above.

- 4-ESS1-1. Identify evidence from patterns in rock formations and fossils in rock layers to support an explanation for changes in a landscape over time. [Clarification Statement: Examples of evidence from patterns could include rock layers with marine shell fossils above rock layers with plant fossils and no shells, indicating a change from land to water over time; and, a canyon with different rock layers in the walls and a river in the bottom, indicating that over time a river cut through the rock.] [Assessment Boundary: Assessment does not include specific knowledge of the mechanism of rock formation or memorization of specific rock formations and layers. Assessment is limited to relative time.]

Framework core idea ESS2: Earth's systems—How and why is Earth constantly changing?

- **ESS2.A: Earth materials and systems**
  How do the major Earth systems interact? [Rainfall, water, ice, wind, living organisms, and gravity break rocks, soils, and sediments into smaller particles and move them around.]

- **ESS2.B: Plate tectonics and large-scale system interactions**
  Why do the continents move, and what causes earthquakes and volcanoes? [The locations of mountain ranges, deep ocean trenches, ocean floor structures, earthquakes, and volcanoes occur in patterns. Earthquakes happen near or deep below Earth's surface, volcanoes are found on the continents and ocean floor, major mountain

Introduction to Performance Expectations

chains form inside continents or near edges. Maps can help locate the different land and water features where people live and in other areas of Earth.]

- **ESS2:C The role of water in Earth's surface processes**
  How do the properties and movements of water shape Earth's surface and affect its systems? [The downhill movement of water shapes the appearance of the land.]

- **ESS2.E: Biogeology**
  How do living organisms alter Earth's processes and structures? [Many types of rocks and minerals are formed from the remains of organisms or are altered by their activities.]

The following NGSS Performance Expectations for ESS2 is derived from the Framework disciplinary core ideas above.

- 4-ESS2-1. Make observations and/or measurements to provide evidence of the effects of weathering or the rate of erosion by water, ice, wind, or vegetation. [Clarification Statement: Examples of variables to test could include angle of slope in the downhill movement of water, amount of vegetation, speed of wind, relative rate of deposition, cycles of freezing and thawing of water, cycles of heating and cooling, and volume of water flow.] [Assessment Boundary: Assessment is limited to a single form of weathering or erosion.]

- 4-ESS2-2. Analyze and interpret data from maps to describe patterns of Earth's features. [Clarification Statement: Maps can include topographic maps of Earth's land and ocean floor, as well as maps of the locations of mountains, continental boundaries, volcanoes, and earthquakes.]

Framework core idea ESS3: Earth and human activity—How do Earth's surface processes and human activities affect each other?

- **ESS3.A: Natural resources**
  How do humans depend on Earth's resources? [All materials, energy, and fuels that humans use are derived from natural sources, and their use affects the environment in multiple ways. Some resources are renewable over time, and others are not.]

- **ESS3.B: Natural hazards**
  How do natural hazards affect individuals and societies? [A variety of hazards result from natural processes (e.g., earthquakes, tsunamis, volcanic eruptions, severe weather, floods, coastal erosion). Humans cannot eliminate natural hazards but can take steps to reduce their impacts.]

▶ **REFERENCES**

National Research Council. *A Framework for K–12 Science Education: Practices, Crosscutting Concepts, and Core Ideas.* Washington, DC: The National Academies Press, 2012.

NGSS Lead States. *Next Generation Science Standards: For States, By States.* Washington, DC: The National Academies Press, 2013.

▶ **NOTE**

Core idea ESS3.A is addressed in this module by looking at materials, specifically earth materials, as natural resources.

SOILS, ROCKS, AND LANDFORMS — Framework and NGSS

> **NOTE**
> NGSS performance expectation 4-ESS3-1 as it relates to energy and fuels is addressed in the grade 4 module: **Energy**. In the **Soils, Rocks, and Landforms Module**, students explore how materials are natural resources and they describe resources as renewable and nonrenewable. The experiences in this module are considered foundational to the performance expectation.

The following NGSS Performance Expectations for ESS3 are derived from the Framework disciplinary core ideas above.

- 4-ESS3-1. Obtain and combine information to describe that energy and fuels are derived from natural resources and their uses affect the environment. [Clarification Statement: Examples of renewable energy resources could include wind energy, water behind dams, and sunlight; non-renewable energy resources are fossil fuels and fissile materials. Examples of environmental effects could include loss of habitat due to dams, loss of habitat due to surface mining, and air pollution from burning of fossil fuels.]

- 4-ESS3-2. Generate and compare multiple solutions to reduce the impacts of natural Earth processes on humans. [Clarification Statement: Examples of solutions could include designing an earthquake resistant building and improving monitoring of volcanic activity.] [Assessment Boundary: Assessment is limited to earthquakes, floods, tsunamis, and volcanic eruptions.]

Introduction to Performance Expectations

Engineering, Technology, and Applications of Science
Framework core idea ETS1: Engineering design—How do engineers solve problems?

- **ETS1.A: Defining and delimiting an engineering problem**
  What is a design for? What are the criteria and constraints of a successful solution? [Possible solutions to a problem are limited by available materials and resources (constraints). The success of a designed solution is determined by considering the desired features of a solution (criteria). Different proposals for solutions can be compared on the basis of how well each one meets the specified criteria for success or how well each takes the constraints into account.]

- **ETS1.B: Developing possible solutions**
  What is the process for developing potential design solutions? [Research on a problem should be carried out before beginning to design a solution. An often productive way to generate ideas is for people to work together to brainstorm, test, and refine possible solutions. Testing a solution involves investigating how well it performs under a range of likely conditions. Tests are often designed to identify failure points or difficulties, which suggest the elements of a design that need to be improved. At whatever stage, communication with peers about proposed solutions is an important part of the design process, and shared ideas can lead to improved designs.]

The following NGSS Grades 3–5 Performance Expectations for ETS1 are derived from the Framework disciplinary core ideas above.

- 3-5-ETS1-1: Define a simple design problem reflecting a need or a want that includes specified criteria for success and constraints on materials, time, or cost.

- 3-5-ETS1-2: Generate and compare multiple possible solutions to a problem based on how well each is likely to meet the criteria and constraints of the problem.

DISCIPLINARY CORE IDEAS

A Framework for K–12 Science Education has two core ideas in engineering, technology, and applications of science.

ETS1: Engineering design

ETS2: Links among engineering, technology, science, and society

NOTE: Only one of these core ideas, ETS1, is represented in the NGSS performance expectations for grade 4.

The questions and descriptions of the core ideas in the text on these pages are taken from the *NRC Framework* 3–5 grade band to keep the core ideas in a rich and useful context.

The performance expectations related to each core idea are taken from the NGSS for grade 3–5.

The **Soils, Rocks, and Landforms Module** contributes to ETS1-1 and ETS1-2; the grade 4 module **Energy** focuses on all three ETS performance expectations.

SOILS, ROCKS, AND LANDFORMS — Framework and NGS

SCIENCE AND ENGINEERING PRACTICES

A Framework for K–12 Science Education (National Research Council, 2012) describes eight science and engineering practices as essential elements of a K–12 science and engineering curriculum. All eight practices are incorporated into the learning experiences in the **Soils, Rocks, and Landforms Module**.

The learning progression for this dimension of the framework is addressed in *Next Generation Science Standards* (National Academies Press, 2013), volume 2, appendix F. Elements of the learning progression for practices recommended for grade 4 as described in the performance expectations appear in bullets below each practice.

Science and Engineering Practices Addressed

1. **Asking questions and defining problems**
   - Ask questions that can be investigated based on patterns such as cause-and-effect relationships.

2. **Developing and using models**
   - Identify limitations of models.
   - Develop or use models to describe and predict phenomena.
   - Use a model to test cause-and-effect relationships or interactions concerning the functioning of a natural system.

3. **Planning and carrying out investigations**
   - Plan and conduct an investigation collaboratively to produce data to serve as the basis for evidence, using fair tests in which variables are controlled and the number of trials considered.
   - Make observations and/or measurements to produce data to serve as the basis for evidence for an explanation of a phenomenon or test a design solution.
   - Make predictions about what would happen if a variable changes.

4. **Analyzing and interpreting data**
   - Analyze and interpret data to make sense of phenomena using logical reasoning.
   - Compare and contrast data collected by different groups in order to discuss similarities and differences in their findings.

5. **Using mathematics and computational thinking**
   - Organize simple data sets to reveal patterns that suggest relationships.

6. **Constructing explanations and designing solutions**
   - Construct an explanation of observed relationships (e.g., the distribution of plants in the backyard).
   - Use evidence (e.g., measurements, observations, patterns) to construct or support an explanation or design a solution to a problem.
   - Identify the evidence that supports particular points in an explanation.
   - Apply scientific ideas to solve design problems.

Introduction to Performance Expectations

7. **Engaging in argument from evidence**
   - Compare and refine arguments based on an evaluation of the evidence presented.
   - Respectfully provide and receive critiques from peers about a proposed procedure, explanation, or model by citing relevant evidence and posing specific questions.
   - Construct an argument with evidence, data, and/or a model.
   - Use data to evaluate claims about cause and effect.

8. **Obtaining, evaluating, and communicating information**
   - Read and comprehend grade-appropriate complex texts and/or other reliable media to summarize and obtain scientific and technical ideas and describe how they are supported by evidence.
   - Obtain and combine information from books and/or other reliable media to explain phenomena or solutions to a design problem.
   - Communicate scientific and/or technical information orally and/or in written formats, including various forms of media, such as tables, diagrams, and charts.

SOILS, ROCKS, AND LANDFORMS — Framework and NGS

CROSSCUTTING CONCEPTS

A Framework for K–12 Science Education describes seven crosscutting concepts as essential elements of a K–12 science and engineering curriculum. The crosscutting concepts listed here are those recommended for grade 4 in the NGSS and are incorporated into the learning opportunities in the **Soils, Rocks, and Landforms Module**.

The learning progression for this dimension of the framework is addressed in volume 2, appendix G, of the NGSS. Elements of the learning progression for crosscutting concepts recommended for grade 4, as described in the performance expectations, appear after bullets below each concept.

Crosscutting Concepts Addressed

Patterns
- Similarities and differences in patterns can be used to sort and classify natural phenomena. Patterns of change can be used to make predictions.

Cause and effect
- Cause-and-effect relationships are routinely identified and used to explain change.

Scale, proportion, and quantity
- Observable phenomena exist from very short to very long time periods.

Systems and system models
- A system can be described in terms of its components and their interactions.

Structure and function
- Different materials have different substructures which can sometimes be observed.
- Substructures have shapes and parts that serve functions.

Stability and change
- Change is measured in terms of differences over time and may occur at different rates.
- Some systems appear stable, but over long periods of time will eventually change.

Introduction to Performance Expectations

Connections: Understandings about the Nature of Science

Scientific investigations use a variety of methods.

- Scientific methods are determined by questions. Scientific investigations use a variety of methods, tools, and techniques.

Scientific knowledge is based on empirical evidence.

- Science findings are based on recognizing patterns.
- Scientists use tools and technologies to make accurate measurements and observations.

Scientific knowledge assumes an order and consistency in natural systems.

- Science assumes consistent patterns in natural systems.

Science is a human endeavor.

- Men and women from all cultures and backgrounds choose careers as scientists and engineers. Most scientists and engineers work in teams. Science affects everyday life. Creativity and imagination are important to science.

Connections to Engineering, Technology, and Applications of Science

- **Interdependence of science, engineering, and technology.** Science and technology support each other. Knowledge of relevant scientific concepts and research findings is important in engineering. Tools and instruments are used to answer scientific questions, while scientific discoveries lead to the development of new technologies.

CONNECTIONS

See volume 2, appendix H and appendix J, in the NGSS for more on these connections.

For details on learning connections to Common Core State Standards English Language Arts and Math, see the chapters FOSS and Common Core ELA—Grade 4 and FOSS and Common Core Math—Grade 4 in *Teacher Resources*.

SOILS, ROCKS, AND LANDFORMS – Framework and NGS

FOSS CONCEPTUAL FRAMEWORK

In the last half decade, teaching and learning research has focused on learning progressions. The idea behind a learning progression is that **core ideas** in science are complex and wide-reaching, requiring years to develop fully—ideas such as the structure of matter or the relationship between the structure and function of organisms. From the age of awareness throughout life, matter and organisms are important to us. There are things students can and should understand about these core ideas in primary school years, and progressively more complex and sophisticated things they should know as they gain experience and develop cognitive abilities. When we as educators can determine those logical progressions, we can develop meaningful and effective curriculum for students.

FOSS has elaborated learning progressions for core ideas in science for kindergarten through grade 8. Developing a learning progression involves identifying successively more sophisticated ways of thinking about a core idea over multiple years.

If mastery of a core idea in a science discipline is the ultimate educational destination, then well-designed learning progressions provide a map of the routes that can be taken to reach that destination. . . . Because learning progressions extend over multiple years, they can prompt educators to consider how topics are presented at each grade level so that they build on prior understanding and can support increasingly sophisticated learning. (National Research Council, *A Framework for K–12 Science Education*, 2012, p. 26)

> **TEACHING NOTE**
>
> FOSS has conceptual structure at the module and strand levels. The concepts are carefully selected and organized in a sequence that makes sense to students when presented as intended.

The FOSS modules are organized into three domains: physical science, earth science, and life science. Each domain is divided into two strands, as shown in the table "FOSS Next Generation—K–8 Sequence." Each strand represents a core idea in science and has a conceptual framework.

- Physical Science: matter; energy and change
- Earth and Space Science: dynamic atmosphere; rocks and landforms
- Life Science: structure and function; complex systems

The sequence in each strand relates to the core ideas described in the NRC *Framework*. Modules at the bottom of the table form the foundation in the primary grades. The core ideas develop in complexity as you proceed up the columns.

FOSS Conceptual Framework

Information about the FOSS learning progression appears in the **conceptual framework** (page 43), which shows the structure of scientific knowledge taught and assessed in this module, and the **content sequence** (pages 50–51), a graphic and narrative description that puts this single module into a K–8 strand progression.

FOSS is a research-based curriculum designed around the core ideas described in the NRC *Framework*. The FOSS module sequence provides opportunities for students to develop understanding over time by building on foundational elements or intermediate knowledge leading to the understanding of core ideas. Students develop this understanding by engaging in appropriate science and engineering practices and exposure to crosscutting concepts. The FOSS conceptual frameworks therefore are *more detailed* and *finer-grained* than the set of goals described by the NGSS performance expectations (PEs). The following statement reinforces the difference between the standards as a blueprint for assessment and a curriculum, such as FOSS.

Some reviewers of both public drafts [of NGSS] *requested that the standards specify the intermediate knowledge necessary for scaffolding toward eventual student outcomes. However, the NGSS are a set of goals. They are PEs for the end of instruction—not a curriculum. Many different methods and examples could be used to help support student understanding of the DCIs and science and engineering practices, and the writers did not want to prescribe any curriculum or constrain any instruction. It is therefore outside the scope of the standards to specify intermediate knowledge and instructional steps.* (Next Generation Science Standards, 2013, volume 2, p. 342)

FOSS Next Generation—K–8 Sequence

	PHYSICAL SCIENCE		EARTH SCIENCE		LIFE SCIENCE	
	MATTER	ENERGY AND CHANGE	ATMOSPHERE AND EARTH	ROCKS AND LANDFORMS	STRUCTURE/ FUNCTION	COMPLEX SYSTEMS
6–8	Waves; Gravity and Kinetic Energy / Chemical Interactions / Electromagnetic Force		Planetary Science / Earth History / Weather and Water		Heredity and Adaptation / Populations and Ecosystems / Diversity of Life; Human Systems Interactions	
5	Mixtures and Solutions		Earth and Sun		Living Systems	
4		Energy		Soils, Rocks, and Landforms	Environments	
3	Motion and Matter		Water and Climate		Structures of Life	
2	Solids and Liquids			Pebbles, Sand, and Silt	Insects and Plants	
1		Sound and Light	Air and Weather		Plants and Animals	
K	Materials and Motion		Trees and Weather		Animals Two by Two	

SOILS, ROCKS, AND LANDFORMS — Framework and NGS

BACKGROUND FOR THE CONCEPTUAL FRAMEWORK
in Soils, Rocks, and Landforms

Origin of Earth

The rocks that form mountains; the water flowing in rivers, filling the ocean and forming clouds; the oxygen, carbon dioxide, and other gases that compose the atmosphere—these are all examples of what scientists call earth materials. All life on Earth depends on these materials, for they, with the addition of the Sun's energy, provide the food, air, and water that every living plant and animal needs to survive.

The study of the atmosphere, Earth's crust and interior, and its rivers and ocean is the task of the earth scientist. This module focuses on the properties of the solid materials that form Earth—the minerals, the rocks, and the landforms.

Geology is simply the study of Earth. Physical geology focuses on the materials that Earth is made of and the processes, such as erosion and mountain building, that occur at and below Earth's surface. Historical geology focuses on understanding Earth's origin, its physical development, and its future.

The latest theories suggest that Earth began as a protoplanet (a planet in the making), condensing out of the debris surrounding the protosun. Once formed, Earth fortuitously settled into orbit at a favorable distance from the thermonuclear reactions of the Sun, and had a gravitational field that allowed an atmosphere to form.

During the early history of Earth, the decay of radioactive elements and the heat released by colliding particles within Earth melted its interior. The heavier elements, mostly iron and nickel, sank to the center of Earth, while the lighter elements that compose most of Earth's crust migrated upward. Layers of different materials formed inside Earth.

Today, geologists describe four basic layers of Earth's interior: the inner core, a mostly solid iron-rich zone about 1,216 km thick; the outer core, a molten metallic layer 2,270 km thick; the mantle, a semi-solid, extremely viscous layer of molten-rock materials (minerals) of varying thickness up to 2,885 km thick; and the crust, a thin outer skin of rock ranging from 5 to 40 km thick. It is with the crust that we are mostly concerned.

FOSS Conceptual Framework

CONCEPTUAL FRAMEWORK
Earth Science, Rocks and Landforms:
Soils, Rocks, and Landforms

Structure of Earth

Concept A The geosphere (lithosphere) has properties that can be observed and quantified.

- Soil is made partly from weathered rock and partly from organic materials. Soils vary from place to place and differ in their ability to support plants.
- Landforms and bodies of water can be represented in models and maps.

Concept B The hydrosphere has properties that can be observed and quantified.

- Water exists in three states on Earth: solid, liquid, and gas.

Concept C Humans depend on Earth's land, ocean, atmosphere, and biosphere for many different resources.

- People use earth materials to construct things. Properties of different earth materials make them suitable for specific uses.
- Humans can use scientific knowledge and engineering design to reduce the impact of Earth's natural hazards.

Earth Interactions

Concept A All Earth processes are the result of energy flowing and matter cycling within and among the planet's systems.

- Weathered rock materials can be reshaped into new landforms by the slow processes of erosion and deposition. Water plays a large role in these processes.
- Earth changes over time. Some catastrophic Earth events happen very quickly; others occur very slowly over a time period much longer than one can observe.

Concept B Understanding how landforms develop, are weathered, and erode can help infer Earth's history.

- Sediments deposited by water usually form flat, horizontal layers. Sediments turn into solid rock over time. The presence and location of certain fossil types indicate the order in which rock layers were formed.
- Fossils provide evidence of organisms that lived long ago and the nature of their environments.

Soils, Rocks, and Landforms Module—FOSS Next Generation

SOILS, ROCKS, AND LANDFORMS — Framework and NGS

Soil

What is soil? To the farmer, soil is the layer of earth material in which plants anchor their roots and from which they get the nutrients and water they need to grow. To a geologist, soil is the layer of earth materials at Earth's surface that has been produced by weathering of rocks and sediments. To an engineer, soil is any ground that can be dug up by earth-moving equipment and requires no blasting. And to students, soil is dirt. In FOSS, *soil* is defined as a mixture of different-sized earth materials, such as gravel, sand, and silt, and organic material called humus. Humus is the dark, musty-smelling stuff derived from the decayed and decomposed remains of plant and animal life. The proportions of these materials that make up soil differ from one location to another. All life depends on a dozen or so elements that must ultimately be derived from Earth's crust. Soil has been called the bridge between earth material and life; only after minerals have been broken down and incorporated into the soil can plants process the nutrients and make them available to people and other animals.

Erosion Shapes the Land

Anyone who has stood at the edge of Grand Canyon in Arizona has witnessed the results of awesome forces that interacted and changed the face of Earth. Few landforms on Earth's surface rival Grand Canyon in size, beauty, and grandeur. Its dimensions—over 1.6 km deep in places, over 14 km wide, and almost 350 km long—are mind-boggling. At the bottom of the canyon is a café-au-lait ribbon of water that looks insignificant from the vantage point of the canyon rim. The ribbon is the Colorado River, and it can take most of the credit for creating this awe-inspiring landform. Grand Canyon can serve as our example to explore erosion.

This magnificent canyon began its life as a simple gully crossing a landform called the Colorado Plateau. Until maybe 7 million years ago the land was fairly flat, and water in the streams moved sluggishly. Then the Colorado Plateau was uplifted. As it rose, it tilted, increasing the slope of the land over which the ancient Colorado River flowed. This change in the slope multiplied the river's ability to erode the rock and to carry sediments. Sediments are earth materials that have been eroded, transported, and eventually deposited. Just as a saw cuts through a piece of wood, the Colorado River cut through thousands of meters of rock, exposing what are today the walls of Grand Canyon.

FOSS Conceptual Framework

Erosion takes several forms in Grand Canyon. Running water plays the most important role. The Colorado River carries a heavy load of earth materials. Some minerals, such as salts and carbonates, dissolve and travel in solution down the river. Tiny particles, such as silt, clay, and sand, travel in suspension, and heavier particles, such as gravel, cobbles, and boulders, roll and tumble down the riverbed. The materials carried by the river flow, slide, and roll along the sides and bottom of the stream channel, further eroding the riverbed as they pound, scrape, and scour, knocking off and carrying away additional rock particles.

Water can erode rock in another way. When water infiltrates cracks in rocks and the temperature falls below 0°C, the water freezes, expands, and breaks already weakened rock along the cracks. Huge blocks of rock can be wedged off the cliffs, accelerating the rate of erosion.

Gravity's effect on large masses of rock also contributes to erosion. Gravity is the force behind rockfalls—large masses of broken rocks falling to the bottom of a cliff. This rock debris (talus) may continue to move imperceptibly downslope under the force of gravity.

Wind also erodes the canyon. Cliffs are literally sandblasted by strong winds carrying particles of sand. The particles wear away material that in turn is carried away by the wind.

The shape of Grand Canyon is also affected by differential erosion. Some rocks are more resistant to erosion than others. The results are often striking and may appear as caps of erosion-resistant rock perched on pinnacles of weaker rock, forming what people call hoodoos, goblins, and other names that pique the imagination. Western-movie buffs will recognize these landforms as traditional background scenery.

Although they haven't played a part in shaping Grand Canyon, glaciers have produced dramatic changes to Earth's surface in many areas. Huge continental glaciers gouged out the basins of the Great Lakes in the Midwest and the Finger Lakes in New York, and rounded the peaks and hills of most of the northern areas of the United States and Canada. Ancient valley glaciers carved Yosemite Valley in California, and valley glaciers are still at work in Alaska and on the slopes of mountains such as Mount Rainier (Washington) and Mount Shasta (California). Glaciers form from an accumulation of ice and snow that doesn't melt from year to year. The weight of the ice eventually makes the glacier flow and push its way over the land. The cold and the weight of the ice freeze pieces of rock to the bottom surface of the glacier and tear them away from the bedrock. This turns the glacier into a gigantic sandpaper block, scraping and polishing the rock surfaces over which it moves.

Soils, Rocks, and Landforms Module—FOSS Next Generation

SOILS, ROCKS, AND LANDFORMS — Framework and NGS

The following list describes just a few of the great variety of landforms caused by erosion and deposition.

Alluvial fan: a fan-shaped deposit of earth materials formed where a stream flows from a steep slope onto flatter land.

Canyon: a V-shaped gorge with steep sides eroded by a stream.

Delta: a fan-shaped deposit of earth materials at the mouth of a stream.

Dune: a mound, ridge, or hill of wind-blown sand.

Floodplain: land that gets covered with water during a flood.

Hill: an isolated elevation in the land, usually no more than 30 m from base to peak.

Levee: an embankment along a stream that protects land from flooding. Levees can be natural or constructed.

Meander: a curve or loop in a river or a stream.

Mesa: an isolated, broad, flat-topped hill having at least one steep cliff.

Mouth: where a stream enters another body of water.

Valley: a low area between hills and mountains, where a stream often flows.

Waterfall: a steep to vertical descent of a stream channel.

Deposition Creates Other Landforms

The material eroded and transported by water, wind, and ice is the raw material of new landforms. When a stream slows down or dries up and can no longer carry its load, when the wind stops blowing, or when a glacier melts, the materials it carried are deposited.

Deposition is the process by which eroded earth materials settle out of the air, water, or ice. The amount of material carried by water depends primarily on the energy in the stream. Anything that increases the velocity of the water—an increase in slope or the amount of water, or a reduction in the friction of the water in contact with the riverbed due to a greater volume of water—increases the stream's capacity for carrying a load. Anything that decreases the velocity or volume of water—a decrease in slope with no increase in water volume, an obstruction such as a dam, a stream joining a larger body of water—will cause a stream to drop its burden as sediments. The larger, heavier sediments are deposited first, then the finer sand and silt. The salts and other minerals in solution are usually not deposited until the body of water in which they are dissolved evaporates.

Using Models to Study Earth Processes

Most earth-shaping events happen on such a large scale and in such inaccessible places that it is impossible for students to experience them directly. But models—small, simplified representations of the larger objects and events—are used by scientists to better comprehend the larger realities. In this module, students set up stream tables to investigate some of the conditions that affect erosion and deposition in stream systems, and build model mountains to learn mapmaking.

Models and Maps

Maps are a kind of portable model. The first maps were drawn on pieces of bone over 15,000 years ago. The oldest map known today was discovered at Mezhirich, Ukraine, and seems to depict the area around the site where it was found.

Most maps provide a bird's-eye view of an area of Earth's surface with notations and symbols to assist the viewer. Modern maps are drawn for a number of purposes: highway maps help us find our way, resource maps show a nation's mineral supplies, maps show population sizes around the world, and so forth. Teams of specialists work together to create these maps. An aerial photographer takes the pictures of Earth's surface that will act as a base for the map. The surveyor works

FOSS Conceptual Framework

in the field making detailed and precise measurements of landforms and human structures. The cartographer uses the data provided by the surveyor and the photos taken by the photographer to create the final product, a map with symbols and a key.

Topographic Maps

The map we are most interested in for the study of landforms is the topographic map. It represents the shape, location, and elevation of the land. Its unique characteristic is the use of contour lines. Contour lines connect all the points on Earth's surface that have the same elevation or height above sea level. They represent the vertical dimension of the land on a flat piece of paper. The change in vertical distance from one contour line to the next is always the same; this uniform step in elevation is the contour interval. Closely spaced lines represent steep slopes and widely spaced lines represent nearly level land. With practice and imagination, one can begin to transform the shape and the spacing of the contour lines into a three-dimensional vision of the landforms they represent.

Landforms

Geographers, the scientists who study the features of Earth's surface, describe and classify landforms in terms of several features and properties. These features include what landforms are made of, the process that formed them, and their relative age. One classification scheme describes three basic landform types. Mountains are the high, steeply sloped, uplifted areas of Earth's surface. Some examples are the Rocky Mountains, the Andes, and the Himalayas. Plateaus are high, nearly level, uplifted areas composed of horizontal layers of rock. The Colorado Plateau through which the Colorado River flows and the Columbia Plateau in the northwest United States are examples of plateaus. Plains are low areas of Earth that either have been eroded nearly level or are formed of flat-lying sediments. The Great Plains in the midsection of the United States and the coastal plains along the continental boundaries are examples of this landform.

SOILS, ROCKS, AND LANDFORMS — *Framework and NGS*

Rapid Versus Slow Change

Most processes that cause Earth's surface to change happen at quite a bit less than a snail's pace, with the effects accumulating over millions of years. Weathering, erosion, deposition, and movement of glaciers usually happen at such slow rates that only detailed records of observations of rocks, sediments, and landforms and good inferences provide support that change has occurred. But sometimes changes happen rapidly. Periodically intense rains result in massive flows of water, or floods. During floods, a stream's speed increases dramatically. Giant boulders can roll down watercourses, and the force of the water undermines and carries away trees. Rivers can overflow their banks and change their course. Huge amounts of earth material can erode from one place and deposit in another during floods.

During these powerful natural events, human structures take a beating. Houses, washed from their foundations, end up in the rivers. Railroads and highways that follow a river's floodplain wash away periodically. Reservoirs can become clogged with the sediments washed in by a river. Bridges, dams, and other structures fight against the natural processes that work to change the shape of the land.

Earthquakes, landslides, and volcanoes also produce relatively fast changes to the landscape. An understanding of the causes and effects of these various earth movements is important to anyone living in areas impacted by these processes. The effects are both beautiful landforms and scenery, and disasters to structures and lifestyle at the same time.

Earth Materials and Human Uses

Humans from the earliest of times have used earth materials to fashion a wide variety of tools: mortars for grinding; axes, carving tools, digging sticks, hammers, and sickles for domestic and agricultural use; and daggers and arrowheads for weapons. Rocks and minerals have been used in many forms of art, from sculptures to decorative paints and jewelry. And ore minerals are the source of most useful metals. Copper was used to make beads and pins even before 5000 BCE.

Today earth materials continue to be a part of our lives at all levels. Children use the earth to make mud pies and sand castles. Adults put earth materials together to build roads, bridges, and buildings.

FOSS Conceptual Framework

Earth Processes and Elementary Students

Throughout this module students learn the basic ideas that contribute to some of the big ideas and inferences geologists have conceived to explain the structure of Earth's interior and crust and the internal and external processes that build and change them. The theory of plate tectonics has unified many observations and verified relationships geologists have made and discovered about Earth's surface.

The study of earth science offers a number of challenges. The planet Earth is an extremely large laboratory with many of its secrets not yet discovered or understood. Identifying all the variables that influence the creation of a rock or the eruption of a volcano is a formidable task. The processes that create earth materials take millions and millions of years, making them impossible to replicate in the laboratory or classroom. Yet students can undertake investigations in the classroom to prepare themselves to better understand the structure of their world.

Because Earth-shaping events happen on such a large scale, it is difficult for students to experience them directly. So students also set up investigations to simulate processes of chemical and physical weathering. They view images and read about places where these processes occur in nature. Students set up stream tables to investigate conditions that affect erosion and deposition in stream systems. From these experiences, it is expected that students will see relationships between their models and landforms in the world around them.

SOILS, ROCKS, AND LANDFORMS — Framework and NGS

Rocks and Landforms Content Sequence

This table shows the four FOSS modules and courses that address the content sequence "rocks and landforms" for grades K–8. Running through the sequence are the two progressions—structure of Earth and Earth interactions. The supporting elements in each module (somewhat abbreviated) are listed. The elements for the **Soils, Rocks, and Landforms Module** are expanded to show how they fit into the sequence.

Module or course	ROCKS AND LANDFORMS	
	Structure of Earth	**Earth interactions**
Earth History (middle school)	• The geological time scale, interpreted from rock strata and fossils, provides a way to organize Earth's history. Lower layers are older than higher layers—superposition. • Earth's crust is fractured into plates that move over, under, and past one another. • Volcanoes and earthquakes occur along plate boundaries. • The rock cycle is a way to describe the process by which new rock is created.	• Landforms are shaped by slow, persistent processes driven by weathering, erosion, deposition, and plate tectonics. • Water's movement changes Earth's surface. • Energy is derived from the Sun and Earth's hot interior. • All Earth processes are the result of energy flowing and matter cycling within and among Earth's systems. • Evolution is shaped by geological conditions.
Soils, Rocks, and Landforms (grade 4)		
Water and Climate (grade 3)	• Water is found almost everywhere on Earth (e.g., vapor, clouds, rain, snow, ice). Most of Earth's water is in the ocean. • Water expands when heated, contracts when cooled, and expands when it freezes. • Cold water is more dense than warmer water; liquid water is more dense than ice. • Scientists observe, measure, and record patterns of weather to make predictions. • Soils retain more water than rock particles alone.	• Water flows downhill. • Ice melts when heated; water freezes when cooled. • The water cycle is driven by the Sun and involves evaporation, condensation, precipitation, and runoff. • Density determines whether objects float or sink in water. • Climate is the range of an area's typical weather. • A variety of natural hazards result from weather-related phenomena.
Pebbles, Sand, and Silt (grade 2)	• Rocks are earth materials composed of minerals; rocks have properties. • Rock sizes include clay, silt, sand, gravel, pebbles, cobbles, and boulders. • The properties of different earth materials make them suitable for specific uses. • Water can be a solid, liquid, or gas. • Natural sources of water include streams, rivers, ponds, lakes, marshes, and the ocean. • Landforms and bodies of water can be represented in models and maps.	• Smaller rocks result from weathering. • Water carries soils and rocks from one place to another. • Soil is made partly from weathered rock and partly from organic material. • Soils vary from place to place. Soils differ in their ability to support plants.

FOSS Conceptual Framework

Structure of Earth	Earth interactions
• Soils are composed of different kinds and amounts of earth materials and humus; they can be described by their properties. • Water exists in three states. • Earth materials are natural resources. Some resources are renewable, others are not. • Humans can use scientific knowledge and engineering design to reduce the impact of Earth's hazards. • Landforms and bodies of water can be represented in models and maps.	• Physical and chemical weathering breaks rock into smaller pieces (sediments). • Downhill movement of water as it flows to the ocean shapes land. • Erosion is the movement of sediments; deposition is the process by which sediments come to rest in another place. • Sediments usually form flat, horizontal layers. Sediments turn into solid rock over time. The presence and location of certain fossil types indicate the order in which rock layers were formed. • Landslides, earthquakes, and volcanoes can produce significant changes in landforms in a short period of time. • Some changes to Earth's surface happen quickly, others more slowly. • Some events happen in cycles; others have a beginning and an end.

▶ **NOTE**
See the Assessment chapter at the end of this *Investigations Guide* for more details on how the FOSS embedded and benchmark assessment opportunities align to the conceptual frameworks and the learning progressions. In addition, the Assessment chapter describes specific connections between the FOSS assessments and the NGSS performance expectations.

The NGSS Performance Expectations addressed in this module include:

Earth and Space Sciences
4-ESS1-1
4-ESS2-1
4-ESS2-2
4-ESS3-1
4-ESS3-2

Engineering, Technology, and Applications of Science
3–5 ETS1-1
3–5 ETS1-2

See pages 32–35 in this chapter for more details on the Grade 4 NGSS Performance Expectations.

SOILS, ROCKS, AND LANDFORMS — Framework and NGSS

CONNECTIONS TO NGSS BY INVESTIGATION

	Science and Engineering Practices	Connections to Common Core State Standards—ELA
Inv. 1: Soils and Weathering	Asking questions Developing and using models Planning and carrying out investigations Analyzing and interpreting data Constructing explanations Engaging in argument from evidence Obtaining, evaluating, and communicating information	RI 1: Refer to details/examples when explaining what the text says and when drawing inferences from text. RI 2: Determine the main idea of a text and explain how it is supported by key details; summarize the text. RI 3: Explain procedures or concepts in a scientific text. RI 4: Determine the meaning of general academic domain-specific words or phrases. RI 5: Describe overall structure of information in a text. RI 6: Compare and contrast a firsthand and secondhand account of the same topic. RI 7: Interpret information presented visually; explain how information contributes to an understanding of the text. RI 8: Explain how an author uses reasons and evidence to support particular points in a text. W 5: Strengthen writing by revising. W 8: Gather relevant information from experiences and print, and categorize the information. SL 1: Engage in collaborative discussions. SL 4: Report on a text in an organized manner, using appropriate facts and relevant details. L 4: Determine or clarify the meaning words. L 5: Demonstrate understanding of word relationships.
Inv. 2: Landforms	Asking questions Developing and using models Planning and carrying out investigations Analyzing and interpreting data Constructing explanations Engaging in argument from evidence Obtaining, evaluating, and communicating information	RI 1: Refer to details/examples when explaining what the text says and when drawing inferences from text. RI 3: Explain procedures or concepts in a scientific text. RI 6: Compare and contrast a firsthand and secondhand account of the same topic. RI 7: Interpret information presented visually, and explain how the information contributes to an understanding of the text. RI 10: Read and comprehend science text. W 5: Strengthen writing by revising. SL 1: Engage in collaborative discussions. SL 2: Paraphrase portions of a text presented in diverse media. SL 5: Add visual displays to presentations.

Connections to NGSS by Investigation

Disciplinary Core Ideas		Crosscutting Concepts
ESS2.A: Earth materials and systems • Rainfall helps to shape the land and affects the types of living things found in a region. Water, ice, wind, living organisms, and gravity break rocks, soils, and sediments into smaller particles and move them around. **(4-ESS2-1)**	**ESS2.E: Biogeology** • Living things affect the physical characteristics of their regions. **(4-ESS2-1)**	Patterns Cause and effect Systems and system models
ESS1.C: The history of planet Earth • Local, regional, and global patterns of rock formations reveal changes over time due to Earth's forces such as earthquakes. The presence and location of certain fossil types indicate the order in which rock layers were formed. **(4-ESS1-1)** **ESS2.A: Earth materials and systems** • Rainfall helps to shape the land and affects the types of living things found in a region. Water, ice, wind, living organisms, and gravity break rocks, soils, and sediments into smaller particles and move them around. **(4-ESS2-1)**	**ESS2.B: Plate tectonics and large-scale system interactions** • The locations of mountain ranges, deep ocean trenches, ocean floor structures, earthquakes, and volcanoes occur in patterns. Most earthquakes and volcanoes occur in bands that are often along the boundaries between continents and oceans. Major mountain chains form inside continents or near their edges. Maps can help locate the different land and water features of Earth. **(4-ESS2-2)**	Patterns Cause and effect Scale, proportion, and quantity Systems and system models Stability and change

Soils, Rocks, and Landforms Module—FOSS Next Generation

SOILS, ROCKS, AND LANDFORMS — Framework and NGS

Science and Engineering Practices	Connections to Common Core State Standards—ELA
Inv. 3: Mapping Earth's Surface Developing and using models Planning and carrying out investigations Analyzing and interpreting data Using mathematics and computational thinking Constructing explanations and designing solutions Engaging in argument from evidence Obtaining, evaluating, and communicating information	RI 1: Refer to details/examples when explaining what the text says and when drawing inferences from text. RI 2: Determine the main idea of a text and explain how it is supported by key details; summarize the text. RI 3: Explain procedures or concepts in a scientific text. RI 4: Determine the meaning of general academic domain-specific words or phrases. RI 5: Describe the overall structure of information in a text. RI 7: Interpret information presented visually; explain how information contributes to an understanding of the text. RI 9: Integrate information from two texts on a topic. W 8: Gather relevant information from experiences and print, and categorize the information. SL 4: Report on a text in an organized manner, using appropriate facts and relevant details.
Inv. 4: Natural Resources Planning and carrying out investigations Constructing explanations and designing solutions Engaging in argument from evidence Obtaining, evaluating, and communicating information	RI 1: Refer to details/examples when explaining what the text says and when drawing inferences from text. RI 2: Determine the main idea of a text and explain how it is supported by key details; summarize the text. RI 3: Explain procedures, ideas, or concepts in a scientific text. RI 4: Determine the meaning of general academic domain-specific words or phrases. RI 5: Describe the overall structure of information in a text. RI 7: Interpret information presented visually; explain how information contributes to an understanding of the text. W 7: Conduct short research projects that build knowledge through investigation of different aspects of a topic. W 8: Gather relevant information from experiences and print, and categorize the information. SL 1: Engage in collaborative discussions. SL 2: Paraphrase information presented orally. SL 4: Report on a text in an organized manner, using appropriate facts and relevant details. L 4: Determine or clarify the meaning of words.

Connections to NGSS by Investigation

Disciplinary Core Ideas		Crosscutting Concepts
ETS1.B: Developing possible solutions • At whatever stage, communicating with peers about proposed solutions is an important part of the design process, and shared ideas can lead to improved designs. **(3-5-ETS1-2)** **ESS1.C: The history of planet Earth** • Local, regional, and global patterns of rock formations reveal changes over time due to Earth's forces such as earthquakes. The presence and location of certain fossil types indicate the order in which rock layers were formed. **(4-ESS1-1)**	**ESS2.B: Plate tectonics and large-scale system interactions** • The locations of mountain ranges, deep ocean trenches, ocean floor structures, earthquakes, and volcanoes occur in patterns. Most earthquakes and volcanoes occur in bands that are often along the boundaries between continents and oceans. Major mountain chains form inside continents or near their edges. Maps can help locate the different land and water features of Earth. **(4-ESS2-2)** **ESS3.B: Natural hazards** • A variety of hazards result from natural processes. Humans cannot eliminate the hazards but can take steps to reduce their impact. **(4-ESS3-2)**	Patterns Cause and effect Scale, proportion, and quantity Stability and change
ESS3.A: Natural resources • Energy and fuels that humans use are derived from natural sources, and their use affects the environment in multiple ways. Some resources are renewable over time, and others are not. **(4-ESS3-1)**	**ETS1.A: Defining and delimiting engineering problems** • Possible solutions to a problem are limited by available materials and resources (constraints). The success of a designed solution is determined by considering the desired features of a solution (criteria). Different proposals for solutions can be compared on the basis of how well each one meets the specified criteria for success or how well each takes the constraints into account. **(3-5-ETS1-1)**	Scale, proportion, and quantity Structure and function

Soils, Rocks, and Landforms Module—FOSS Next Generation

SOILS, ROCKS, AND LANDFORMS — **Framework and NGS**

RECOMMENDED FOSS NEXT GENERATION K–8 SCOPE AND SEQUENCE

Grade	Integrated Middle Grades				
6–8	Heredity and Adaptation*	Electromagnetic Force*	Gravity and Kinetic Energy*	Waves*	Planetary Science
	Chemical Interactions		Earth History		Populations and Ecosystems
	Weather and Water		Diversity of Life		Human Systems Interactions*

*Half-length courses   ● Physical Science content   ● Earth Science content   ● Life Science content   ● Engineering content

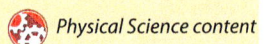

Grade	Physical Science	Earth Science	Life Science
5	Mixtures and Solutions	Earth and Sun	Living Systems
4	Energy	Soils, Rocks, and Landforms	Environments
3	Motion and Matter	Water and Climate	Structures of Life
2	Solids and Liquids	Pebbles, Sand, and Silt	Insects and Plants
1	Sound and Light	Air and Weather	Plants and Animals
K	Materials and Motion	Trees and Weather	Animals Two by Two

Full Option Science System

SOILS, ROCKS, AND LANDFORMS — *Materials*

Contents

Introduction	57
Kit Inventory List	58
Materials Supplied by the Teacher	60
Preparing a New Kit	62
Preparing the Kit for Your Classroom	63
Care, Reuse, and Recycling	67

INTRODUCTION

The Soils, Rocks, and Landforms kit contains

- *Teacher Toolkit: Soils, Rocks, and Landforms*
  1 *Investigations Guide: Soils, Rocks, and Landforms*
  1 *Teacher Resources: Soils, Rocks, and Landforms*
  1 *FOSS Science Resources: Soils, Rocks, and Landforms*
- *FOSS Science Resources: Soils, Rocks, and Landforms* (class set of student books)
- Permanent equipment for one class of 32 students
- Consumable equipment for three classes of 32 students

FOSS modules use central materials distribution. You organize all the materials for an investigation on a single table called the materials station. As the investigation progresses, one member of each group gets materials as they are needed, and another returns the materials when the investigation is completed. You place items at the station—students do the rest.

Individual photos of each piece of FOSS equipment are available online for printing. For updates to information on materials used in this module and access to the Safety Data Sheets (SDS), go to www.FOSSweb.com. Links to replacement-part lists and customer service are also available on FOSSweb.

▶ **NOTE**
To see how all of the materials in the module are set up and used, view the teacher preparation video on FOSSweb.

▶ **NOTE**
Delta Education Customer Service can be reached at 1-800-258-1302.

Full Option Science System

SOILS, ROCKS, AND LANDFORMS — *Materials*

KIT INVENTORY List

Drawer 1 of 3

Equipment Condition

* The student books, if included in your purchase, are shipped separately.

▶ **NOTE**
The teacher toolkit is shipped separately. However, there is space in drawer 1 to store your toolkit.

✪ These items might occasionally need replacement.

Print Materials

Qty	Item
1	*Teacher Toolkit: Soils, Rocks, and Landforms* (1 *Investigations Guide*, 1 *Teacher Resources*, and 1 *FOSS Science Resources: Soils, Rocks, and Landforms*)
1	Poster, *Mount Shasta*
1	Poster, *Mount St. Helens*
1	Poster set, *Conservation*, 4/set
2	Posters, *FOSS Science Safety* and *FOSS Outdoor Safety*
2	Rapid Changes card sets, 16 cards/set

Items for Investigation 1

Qty	Item
1	Beaker, 100 mL
64	Evaporation dishes ✪
8	Jars, plastic, with screw lids, 1/2 L
2	Mineral, calcite, large
100	Paper plates ✪
20	Rock, banded sandstone
10	Rock, basalt, samples, small (fits in 12-dram vial)
40	Rock, conglomerate, pieces ✪
40	Rock, granite, gray, pieces ✪
30	Rock, limestone, samples, small (fits in 12-dram vial) ✪
10	Rock, marble, samples, small (fits in 12-dram vial) ✪
10	Spoons, metal
8	FOSS® trays, clear plastic
1	Vial holder, blue plastic
25	Zip bags, 1 L

Drawer 2 of 3

Items for Investigation 2

Qty	Item
8	Containers, with lids, 1 L
1	Duct tape, roll ✪
8	Meter tapes
10	Rock, sandstone, samples, small (fits in 12-dram vial) ✪
8	Rulers, 30 cm
50	Shells, assorted

Full Option Science System

Qty	Item	Equipment Condition
8	Stream tables (tray, plastic)	
8	Water sources, flood, 1/2 L, with red dot	
8	Water sources, standard, 1/2 L, with blue dot	
16	Wood angles	

Items for Investigation 3

Qty	Item	
8	Foam-mountain sets with dowel	
8	Maps, *Mount Shasta*, topographic maps	
8	Transparencies, *Mount Shasta* (hand-drawn from topo map)	

Items for Investigation 4

Qty	Item	
1	Trowel	

Drawer 3 of 3

Shared Items

Qty	Item	
8	Basins	
9	Containers, 1/2 L	
50	Cups, plastic, 250 mL	
1	Pitcher	
8	Syringes, 50 mL	
40	Vials with caps, 12 dr.	

Consumable Items

Qty	Item	
1	Clay, ceramic, gray, 2.3 kg/bag (5 lb.)	
1	Clay, powdered, gray, bags, 0.45 kg/bag (1 lb.)	
100	Cotton swabs	
100	Cups, paper, 90 mL (3 oz.)	
1	Food coloring, set	
1	Gravel, bag, 2.3 kg/bag (5 lb.)	
2	Pebbles, bags, small, .5 kg/bag (1 lb.)	
1	Potting soil, 2 L, 2 kg/bag (4 lb.)	
2	Sand, mountain stream (coarse, containers, 1 kg/container (2 lb.))	
500	Self-stick notes	

Plastic box for earth material—permanent equipment

Qty	Item	
3	Clay, powdered, white, bags, 0.45 kg (1 lb.)/bag (stream table) ✪	
9	Sand, fine, bags, 1.35 kg (3 lb.)/bag (for stream table) ✪	

▶ **NOTE**
This module includes access to FOSSweb, which includes the streaming videos and online activities used throughout the module.

✪ These items might occasionally need replacement.

Soils, Rocks, and Landforms Module—FOSS Next Generation

SOILS, ROCKS, AND LANDFORMS — Materials

> **NOTE**
> Throughout the *Investigations Guide*, we refer to materials not provided in the kit as "teacher-supplied." These materials are generally common or consumable items that schools and/or classrooms already have, such as rulers, paper towels, and computers. If your school/classroom does not have these items, they can be provided by teachers, schools, districts, or materials centers (if applicable). You can also borrow the items from other departments or classrooms, or request these items as community donations.

MATERIALS *Supplied by the Teacher*

Each part of each investigation has a Materials section that describes the materials required for that part. It lists materials needed for each student or group of students and for the class.

Be aware that you must supply some items. These are indicated with an asterisk (★) in the materials list for each part of the investigation. Here is a summary list of those items by investigation.

For all investigations

- Chart paper and marking pen
- 1 Computer with Internet connection
- Drawing utensils (pencils, crayons, colored pencils, marking pens)
- Glue sticks
- Projection system
- Science notebooks (composition books)
- Self-stick notes (for review sessions)

For outdoor investigations

- 1 Bag for carrying materials
- 32 Clipboards or cardboard pieces with binder clips
- Containers for water (plastic gallon jugs or recycled 2 L soft-drink bottles with caps)
- Pencils
- String (optional)

Investigation 1: Soils and Weathering

- 1 Glass bottle with tight-fitting cap
- Newspaper
- 1 Permanent black marking pen
- Safety goggles
- 1 Sturdy plastic bag
- Transparent tape
- Vinegar, white, about 1.2 L
- Water
- 8 Zip bags, large

Full Option Science System

Investigation 2: Landforms
- Cameras (optional)
- 1 Dustpan and broom
- Materials for stream table (gram pieces, cardboard, craft sticks)
- Newspaper
- Paper towels
- Plaster of Paris, 1 cup for every 8 students
- 4 Empty recycled soft-drink bottles with caps, 2 L
- Transparent tape
- 1 Watering can, large outdoor (optional)

Investigation 3: Mapping Earth's Surface
- Aluminum foil (optional)
- Index cards or half-sentence strips
- Transparent tape

Investigation 4: Natural Resources
- 1 Bucket
- 1 Camera (optional)
- 1 Cardboard box, ~30 cm (1') square
- 1 Bag of ready-mix concrete, 5 kg (10 lb.)
- 1 Garbage bag or piece of plastic
- 1 Pair of rubber or gardening gloves
- Local gravel, 1 cup
- 8–16 Local pebbles
- Local sand, 1 cup

Soils, Rocks, and Landforms Module—FOSS Next Generation

SOILS, ROCKS, AND LANDFORMS — Materials

PREPARING *a New Kit*

If you are preparing a new kit for classroom use, you can do several things initially that will save time during routine preparation for instruction.

1. **Separate eight pieces of gray granite and conglomerate**
   Separate the pieces of gray granite and pieces of conglomerate into two groups: rocks that students will tumble (16 rocks of each kind), and samples that will not be tumbled but used as a control (4 rocks of each kind). This allows multiple classes to tumble the rock and still have some unmarred rocks for comparison.

2. **Obtain safety goggles**
   Check your district regulations regarding the use of safety goggles. Students will be using small amounts of vinegar in Investigations 1. Safety goggles can be purchased from Delta Education.

PREPARING *the Kit for Your Classroom*

Some preparation is required each time you use the kit. Doing these things before beginning the module will make daily setup quicker and easier.

1. Check consumable materials
A number of items in the kit are listed as consumable. Some of these items will be used up during the investigations (food coloring, gravel, gray clay, self-stick notes, potting soil), and others will wear out and need replacement (duct tape, paper plates, evaporation dishes, stream table earth material ingredients, and some of the rocks). Before throwing items out, consider ways to recycle them and get your students involved in this process.

2. Prepare or check four soil types (Investigation 1)
In Investigation 1, Part 1, students investigate four soil samples (Soil 1, Soil 2, Soil 3, Soil 4). Each group will need 1 vial about 2/3 full of each soil type.

Use the materials in the kit (gray powdered clay, gravel, pebbles, and potting soil for humus) to measure and mix these soil samples following the directions in Step 5 of Getting Ready. The recipe will make enough for one class.

There is enough materials in the kit to make the soil samples for three classes. Use only what you need for your class.

3. Check vials and evaporating dishes
The vials and evaporating dishes are reusable and should be rinsed and dried before going back into the kit. Check them for cracks and replace when necessary.

4. Use the FOSS tray
The clear plastic FOSS tray was designed for this module. Each group of four students uses one tray. It is designed to hold four vials in corner sockets and in addition four evaporating dishes in the larger openings. It also serves as an organizer for rock and mineral samples. The trays nest for storage and stack for space efficiency while soil samples are settling or solutions evaporate during investigations. Vials, with no caps, placed in the corner sockets serve as support columns.

5. Acquire vinegar
You will need about 1 liter (L) of white vinegar for Investigation 1.

▶ **NOTE**
The four soil types are:
River delta (1)
Mountain (2)
Desert (3)
Forest (4)

Soils, Rocks, and Landforms Module—FOSS Next Generation

SOILS, ROCKS, AND LANDFORMS — *Materials*

▶ **NOTE**
The stream-table earth materials are shipped in a small plastic box, not in a FOSS drawer. The bags of white (not gray) clay and fine sand in this plastic box are considered permanent equipment for use in the stream table.

6. **Create new surfaces for the acid test**
   The four rocks subjected to vinegar in Investigation 1, Part 3, might need to have fresh surfaces exposed where vinegar can react with the calcite. You can do this by rubbing a pair of rocks together to scratch the surfaces.

7. **Check earth material (Investigation 2)**
   The total volume of the earth material needed for one stream table is about 1 L. If the kit has been used before, check that each earth-material bag contains about 1 L of sand-clay mixture.

   If the bags seem to have the correct volume, add to each bag one 12-dram vial of white clay. White clay is usually lost during the stream-table investigations.

   If the bags do not have enough earth material, pour all of them into a big bucket. Add a cup of clay to the bucket and mix. Use a 1 L container to measure a volume of the mixture for each of the eight bags. If you still come up short, use the recipe below to make another bag. Additional bags of fine sand (1.35 kilograms [kg]/bag) and white clay (0.45 kg/bag) can be ordered from Delta Education.

   a. Add one plastic cup of white powdered clay (about 100 grams [g]) to a 1.35 kg bag of fine sand. That's about 200 milliliters (mL) of powdered clay to 1 L of sand.

   b. Mix the materials in the zip bag until they are uniformly distributed.

8. **Become familiar with three types of clay**
   There are three kinds of clay in the kit used for different purposes.

   - The gray powdered clay is used in Investigation 1, Part 1, to make the soil samples. This clay is consumable.

   - The white powdered clay is used in Investigation 2, Part 1, as part of the stream table earth material mixture. The earth material can be used over and over again and is considered permanent equipment.

   - The ceramic gray clay is used in Investigation 2, Part 4, to make the fossil models. This is considered consumable.

9. **Collect water bottles**
   Collect a couple recycled gallon plastic jugs or 2 L soft-drink bottles with caps for carrying a water supply outdoors.

10. **Provide magnifying lenses**
    If possible, provide several magnifying lenses for students to use during the module to observe the rocks and the mineral, calcite.

11. Check condition of containers and water sources

A variety of containers are included in the kit. Check that all containers are clean and in good condition. Be sure to use the correct container for each investigation. If you have questions, check the illustrations in the investigations, the kit inventory list, or the online photo cards of the individual pieces of equipment. Check the following items:

- Containers, 1/2 L
- Containers, 1 L
- Plastic cups
- Evaporation dishes
- Vials, 12 dram
- Water source, standard (with small hole in bottom)
- Water source, flood (with larger hole in bottom)
- Vials, 12 dram

All the plastic containers in the module are to be rinsed and reused.

12. Check foam mountains (Investigation 3)

Each foam-mountain set consists of six foam pieces and one dowel. Make sure the sets are complete. If students have used water-based ink pens in tracing the shapes, the edges of the foam pieces may have ink on them. The foam can be washed with soap and water to remove the ink.

13. Check transparencies

The kit contains 8 transparencies of teacher master 13, *Foam-Mountain Topographic Map*. Make sure they are clean and ready for use. Use teacher master 13 to make replacements if needed.

14. Print or photocopy notebook sheets

You will need to print or make copies of science notebook sheets before each investigation. See Getting Ready for Investigation 1, Part 1, for ways to organize the science notebook sheets for this module. If you use a projection system, you can download electronic copies of the sheets from FOSSweb for projection.

15. Plan for the word wall

As the module progresses, you will add new vocabulary words to a word wall or pocket chart and model writing and responding to focus questions. Plan how you will do this in your classroom.

You might also find it beneficial to use a pocket chart to display the equipment photo cards as reference for students as they gather needed items from the materials station for each part. Print the photo cards from FOSSweb.

EL NOTE

You might want to print out the FOSS equipment photo cards (from FOSSweb) to add to the word wall to help students with vocabulary.

Soils, Rocks, and Landforms Module—FOSS Next Generation

SOILS, ROCKS, AND LANDFORMS — *Materials*

16. Review safety issues indoors and outdoors

Two safety posters are included in the kit, *FOSS Science Safety* and *FOSS Outdoor Safety*. You should review the guidelines with students and post the posters in the room as a reminder. Review the Getting Ready for Investigation 1, Part 1, as it offers suggestions for this discussion. Also be aware of any allergies that students might have.

Use the four *Conservation* posters to discuss the importance of conserving natural resources.

17. Plan for letter home and home/school connections

You will need to print or make copies of teacher master 1, *Letter to Family*, for the module and of the Home/School Connection teacher masters for each investigation. Space is left at the top so you can copy the letter on your school letterhead. The *Letter to Family* and *Home/School Connections* are also available electronically on FOSSweb.

TEACHING NOTE

Families can get more information on Home/School Connections from FOSSweb.

18. Check FOSSweb for resources

Go to FOSSweb, register as a FOSS teacher, and review the print and digital resources available for this module, including the eGuide, eBook, Resources by Investigation, and *Teacher Resources*. Be sure to check FOSSweb often for updates and new resources.

Students are fascinated by maps and photos of their local area. If you can, get a topographic map of your local area and an aerial photo that includes your school and neighborhood. Look for Internet resources that provide bird's-eye views of your local area and nearby landforms of interest. Google Earth™ is a terrific resource to introduce your students to at this time.

CARE, *Reuse, and Recycling*

When you finish teaching the module, inventory the kit carefully. Note the items that were used up, lost, or broken, and immediately arrange to replace the items. Use a photocopy of the materials list (the Kit Inventory List), and put your marks in the "Equipment Condition" column. Refill packages and replacement parts are available for FOSS by calling Delta Education at 1-800-258-1302 or by using the online replacement-part catalog (www.DeltaEducation.com).

Standard refill packages of consumable items are available from Delta Education. A refill package for a module includes sufficient quantities of all consumable materials (except those provided by the teacher) to use the kit with three classes of 32 students.

Here are a few tips on storing the equipment after use.

- Inventory and bag the small items.
- Wash vials, cups, and evaporation dishes. Be sure all equipment is dry before returning it to the kit.
- Rocks and minerals that have been in vinegar should be rinsed thoroughly with mild soapy water and dried.
- Sort rocks and minerals into labeled bags.
- Rinse and dry the wood angles before storing them.
- Dry earth material thoroughly and store in a plastic bag or a bucket.
- Secure the lids on all containers of materials and on the food coloring.
- Bag each foam mountain and its dowel in a bag.
- Check the posters and card sets and make sure they are dry and flat in the kit.
- Check quantity of consumables, and order more if necessary.

The items in the kit have been selected for their ease of use and durability. Small items should be inventoried (a good job for students under your supervision) and put into zip bags for storage. Any items that are no longer useful for science should be properly recycled. This is a good opportunity to get students involved in making decisions about what items can be recycled.

Soils, Rocks, and Landforms Module—FOSS Next Generation

SOILS, ROCKS, AND LANDFORMS — Materials

SOILS, ROCKS, AND LANDFORMS — *Technology*

Contents

Introduction	69
Technology for Students	70
Technology for Teachers	76
Requirements for Accessing FOSSweb	80
Troubleshooting and Technical Support	82

INTRODUCTION

Technology is an integral part of the teaching and learning with FOSS Next Generation. FOSSweb is the Internet access to FOSS digital resources. FOSSweb gives students the opportunity to interact with simulations, virtual investigations, tutorials, images, and text—activities that enhance understanding of core ideas. It provides support for teachers, administrators, and families who are actively involved in implementing FOSS.

Different types of online activities are incorporated into investigations where appropriate. Each activity is marked with the technology icon in the *Investigations Guide*. You will sometimes show videos to the class. At other times, individuals or small groups of students will work online to review concepts or reinforce their understanding. Tutorials on specific elements of concepts provide opportunities for review and differentiated instruction.

To use these digital resources, you should have at least one computer with Internet access that can be displayed to the class by an LCD projector with an interactive whiteboard or a large screen. Access to enough devices for students to work in small groups or one-on-one is recommended for other parts.

All FOSS online activities are available at www.FOSSweb.com for teachers, students, and families. We recommend you access FOSSweb well before starting the module, to set up your teacher-user account and to become familiar with the resources.

▶ **NOTE**
To get the most current information, download the latest Technology chapter on FOSSweb.

Full Option Science System

SOILS, ROCKS, AND LANDFORMS — *Technology*

TECHNOLOGY *for Students*

FOSS is committed to providing a rich, accessible technology experience for all FOSS students. Here are brief descriptions of selected resources for students on FOSSweb.

Online activities. The online simulations and activities were designed to support students' learning at all grades. They include virtual investigations and tutorials, grades 3–5, that review the active investigations and support students who have difficulties with the materials or who have been absent. Summaries of some of the online activities are on the next page.

FOSS Science Resources—eBooks. The student book is available as an audio book on FOSSweb, accessible at school or at home. In addition, as premium content, *FOSS Science Resources* is available as an eBook on computer or tablet, either as a read-only PDF or in an interactive format that allows text to be read and provides points of interactivity. The eBook can also be projected for guided reading with the whole class.

Media library. A variety of media enhances students' learning and provides them with opportunities to obtain, evaluate, and communicate information. FOSS has reviewed print books and digital resources that are appropriate for students and prepared a list of these resources with links to content websites. There is also a list of regional resources for virtual and actual field trips for students to use in gathering information for projects, and a database of science and engineering careers. Other resources include vocabulary lists to promote use of academic language.

Home/school connections. Each module includes a letter to families, providing an overview of the goals and objectives of the module. There is also a Module Summary available for families to download. Most investigations have a home/school science activity that connects the classroom experiences with students' lives outside of school. These connections are available as PDFs on FOSSweb.

Class pages. Teachers with a FOSSweb account can easily set up class pages with notes and assignments for each class. Students and families can then access this class information online, using the teacher-assigned class login.

▶ **NOTE**
The following student-facing resources are available in Spanish on FOSSweb using a teacher's class page.

- Vocabulary
- Equipment photo cards
- eBooks
- Select streaming videos
- Home/school connections
- Audio books

Full Option Science System

Technology for Students

Soils, Rocks, and Landforms Online Activities

Here is a sampling of the online activities used in the **Soils, Rocks, and Landforms Module** investigations.

Investigation 1, Extensions

- **"Tutorial—Stream Tables: Weathering"**
  In this tutorial, students review the processes of physical and chemical weathering that break down rocks into smaller pieces.

- **"Virtual Investigation—Water Retention of Soils"**
  This virtual investigation provides an opportunity for students to investigate the water retention properties of different types of soils.

Soils, Rocks, and Landforms Module—FOSS Next Generation

SOILS, ROCKS, AND LANDFORMS — *Technology*

Investigation 2, Part 2: Stream-Table Investigations

- **"Videos: Stream Tables"**
  Students can use stream table videos to compare two stream tables as they continue to test the variables of slope, stream-flow rate, and stream-bed material.

- **"Tutorial—Stream Tables: Slope and Flood"**
  In this tutorial, students review the variables that influence erosion caused by water flowing in rivers. Flood conditions cause faster moving water and more erosion than normal flow. Steeper slopes also cause faster flowing water and more erosion.

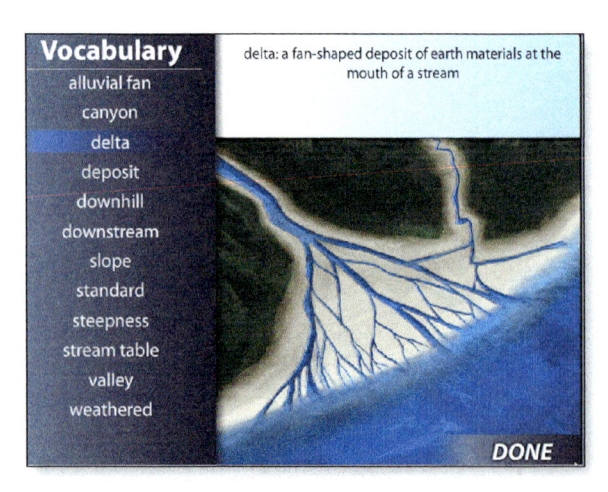

72

Full Option Science System

Technology for Students

Investigation 2, Part 3: Schoolyard Erosion and Deposition

- **"Virtual Investigation—Stream Tables"**
  This virtual investigation provides another opportunity for students to conduct stream-table investigations changing the variables of slope of the land, flood flow, and the addition of a dam in the river. Students can record the simulated results in their notebooks. This can be a substitute experience for students who are absent during the actual classroom stream-table investigations.

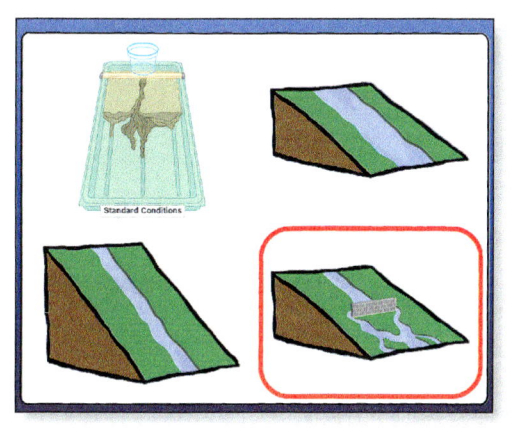

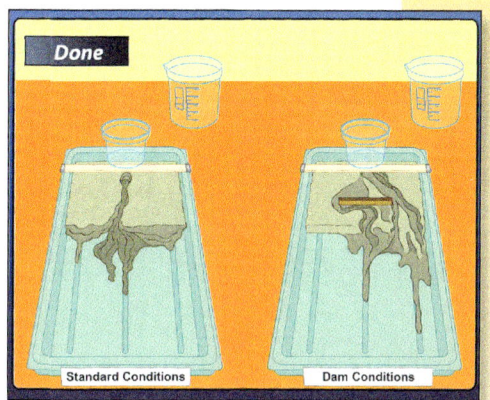

Investigation 2, Extensions

- **"Tutorial—Soil Formation"**
  In this tutorial, students review how soil is formed and how that relates to erosion and deposition.

- **"Tutorial—Fossils"**
  In this tutorial, students learn about different kinds of fossils and how they form.

Soils, Rocks, and Landforms Module—FOSS Next Generation

SOILS, ROCKS, AND LANDFORMS — *Technology*

Investigation 3, Part 2: Drawing a Profile

- **"Topographer"**
  Students can use this interactive to practice making profile maps from topographic maps. Mount Shasta as well as five other mountains are included.

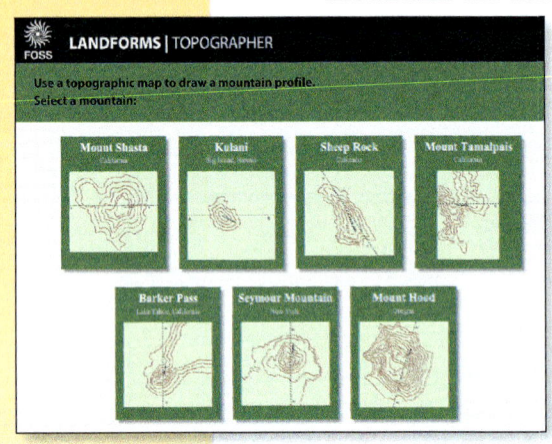

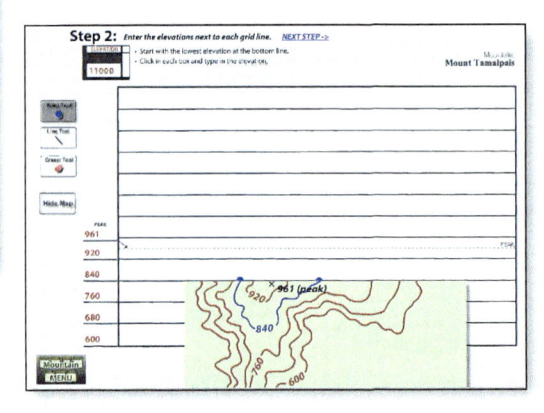

Investigation 3, Extensions

- **"Online Activities"**
  As extensions, students can study the information presented in the online activities on: rock database, sand types, volcanoes, faulting and folding.

74

Full Option Science System

Technology for Students

Investigation 4, Part 1: Introduction to Natural Resources

- **"Resource ID"**
  Students can use this interactive to identify resources that are renewable (plants and animals), nonrenewable (fossil fuels and earth materials such as diamonds), and inexhaustible or perpetual (the Sun).

Investigation 4: Extensions

- **"Virtual Investigation—Natural Resources"**
  In this virtual investigation, students explore what are renewable and nonrenewable resources and how natural resources are used by people.

Soils, Rocks, and Landforms Module—FOSS Next Generation

SOILS, ROCKS, AND LANDFORMS — *Technology*

TECHNOLOGY *for Teachers*

The teacher side of FOSSweb provides access to all the student resources plus those designed for teaching FOSS. By creating a FOSSweb user account and activating your modules, you can personalize FOSSweb for easy access to your instructional materials. You can also set up a class login for students and their families.

Creating a FOSSweb Teacher Account

Setting up an account. Set up a teacher account on FOSSweb before you begin teaching a module. Go to FOSSweb and register for an account with your school e-mail address. Complete registration instructions are available online. If you have a problem, go to the Connecting with FOSS pull-down menu, and look at Technical Help and Access Codes. You can also access online tutorials for getting started with FOSSweb at www.FOSSweb.com/fossweb-walkthrough-videos.

Entering your access code. Once your account is set up, go to FOSSweb and log in. To gain access to all the teacher resources for your module, you will need to enter your access code. Your access code should be printed on the inside cover of your *Investigations Guide*. If you cannot find your FOSSweb access code, contact your school administrator, your district science coordinator, or the purchasing agent for your school or district.

Familiarize yourself with the layout of the site and the additional resources available when you log in to your account. From the module detail page, you will be able to access teacher masters, science notebook masters, assessment masters, the FOSSmap online assessment component, and other digital resources not available to "guests."

Explore the Resources by Investigation, as this will help you plan. This page makes it simple to select the investigation you are teaching, and view all the digital resources organized by part. Resources by Investigation provides immediate access to the streaming videos, online activities, science notebook masters, teacher masters, and other digital resources for each investigation part.

Setting up class pages and student accounts. To enable your students to log in to FOSSweb to see class assignments and student-facing digital resources, set up a class page and generate a username and password for the class. To do this, log in to FOSSweb and go to your teacher page. Under "My Class Pages," follow the instructions to create a new class page and to leave notes for students. Note: student access to the student eBook from your class page requires premium content.

▶ **NOTE**
For more information about FOSS premium content, including pricing and ordering, contact your local Delta sales representative by visiting www.DeltaEducation.com or by calling 1-800-258-1302.

Full Option Science System

Technology for Teachers

Support for Teaching FOSS

FOSSweb is designed to support teachers using FOSS. FOSSweb is your portal to instructional tools to make teaching efficient and effective. Here are some of the tools available to teachers.

- **Grade-level Planning Guide.** The Planning Guide provides strategies for three-dimensional teaching and learning.

- **Resources by Investigation.** The Resources by Investigation organizes in one place all the print and online instructional materials you need for each part of each investigation.

- **Investigations eGuide.** The eGuide is the complete *Investigations Guide* component of the *Teacher Toolkit*, in an electronic web-based format for computers or tablets. If your district rotates modules among several teachers, this option allows all teachers easy access to *Investigations Guide* at all times.

- **Teacher preparation videos.** Videos present information to help you prepare for a module, including detailed investigation information, equipment setup and use, safety, and what students do and learn in each part of the investigation.

- **Interactive whiteboard resources for grades K–5.** You can use these interactive files with or without an interactive whiteboard to facilitate each part of each investigation. You'll need to download the appropriate software to access the files. Links for software downloads are on FOSSweb.

- **Focus questions.** The focus questions address the phenomenon for each part of each investigation, and are formatted for classroom projection and for printing, so that students can glue each focus question into their science notebooks.

- **Module updates.** Important updates cover teacher materials, student equipment, and safety considerations.

- **Module teaching notes.** These notes include teaching suggestions and enhancements to the module, sent in by experienced FOSS users.

- **Home/school connections.** These masters include an introductory letter home (with ideas to reinforce the concepts being taught) and the home/school connection sheets.

- **State and regional resources.** Listings of resources for your geographic region are provided for virtual and actual field trips and for students to use as individual or class projects.

- **Access to FOSS developers.** Through FOSSweb, teachers have a connection to the FOSS developers and expert FOSS teachers.

▶ **NOTE**
There are two versions of the eGuide, a PDF-based eGuide that mimics the hard copy guide, and an HTML interactive eGuide that allows you to write instructional notes and to interface with online resources from the guide.

▶ **NOTE**
The following resources are available on FOSSweb in Spanish.

Teacher-facing resources:
- Notebook masters
- Teacher masters
- Assessment masters
- Focus questions
- Interactive whiteboard files

Student-facing resources:
- Vocabulary
- Equipment photo cards
- eBooks
- Select streaming videos
- Home/school connections
- Audio books

Soils, Rocks, and Landforms Module—FOSS Next Generation

SOILS, ROCKS, AND LANDFORMS — Technology

Technology Components of FOSS Assessment System

FOSSmap for teachers and online assessments for students are the technology components of the FOSS assessment system.

For teachers. FOSSmap is where you set up your class, to schedule online assessments, review/record codes, and run reports. The reports are diagnostic and help you know what students understand and what they still need help with individually and as a class.

The teacher page of FOSSweb has a direct link to FOSSmap. Once you have a login and password for FOSSweb, use the same login and password to access FOSSmap. FOSSmap is a secure site that only you can see. FOSSmap tutorials will get you started with these technology components.

For students. FOSSmap.com/icheck is the URL for students in grades 3–5 who take the assessments online (*Survey/Posttest*, I-Checks, and interim assessments). Students can access this site only when you have scheduled an assessment for them to take. Access codes are generated for each student in the FOSSmap program and can be printed out on mailing labels. Each access code is good for all the assessments taken in one module. When you change modules, students get new access codes.

For more information about the FOSS assessment system and the technology components, see the Assessment chapter.

Technology for Differentiated Instruction

Some resources are for differentiated instruction. They can be used by students at home or by you as part of classroom instruction.

- **Online activities.** The online simulations and activities described earlier in this chapter are designed to support student learning and are often used during instruction. They include virtual investigations and student tutorials for grades 3–5 that you can use to support students who have difficulties with the materials or who have been absent. Virtual investigations require students to record data and answer concluding questions in their notebooks. In some cases, the notebook sheet used in the classroom investigation is also used for the virtual investigation.

- **Vocabulary.** The online word list has science-related vocabulary and definitions used in the module (in both English and Spanish).

- **Equipment photo cards.** Equipment cards provide labeled photos of equipment that students use in the investigations. Cards can be printed and posted on the word wall as part of instruction.

> **NOTE**
> FOSSmap has a number of short online tutorials to get you started. Titles include:
> - Overview
> - Getting Started
> - Module Homepage
> - Embedded Assessment
> - Scheduling Online Assessments
> - Taking an Online Assessment (Teacher Edition)
> - How to Code an Assessment
> - Creating Reports

Technology for Teachers

- **Student eBooks.** Student access to audio-only *FOSS Science Resources* books requires basic access. With premium content, students can access the books from any Internet-enabled device. The eBooks are available in PDF and interactive versions. The PDF version mimics the hard copy book. The interactive eBook reads the text to students—highlighting the text as it is read—and provides students with video clips and online activities.

- **Streaming videos used for extensions.** Some videos are part of the instruction in the investigation and are in Resources by Investigation for each part. Those videos also appear again in the digital resources under "Streaming Videos" along with other videos that extend concepts presented in a module.

- **Recommended books, websites, and careers database.** FOSS-recommended books, websites, and a Science and Engineering Careers Database that introduces students to a variety of career options and diversity of individuals engaged in those careers are provided.

- **Regional resources.** This list provides local resources that can be used to enhance instruction. The list includes website links and PDF documents from local sources.

> **NOTE**
> The eBook is premium content for students.

Support for Classroom Materials Management

- **Materials chapter.** A PDF of the Materials chapter in *Investigations Guide* is available to help you prepare for teaching. A list, organized by drawer, shows the materials included in the FOSS kit for a given module. You can print and use this list for inventory and to monitor equipment condition.

- **Safety Data Sheets (SDS).** A link takes you to the latest safety sheets, with information from materials manufacturers on the safe handling and disposal of materials.

- **Plant and animal care.** This section includes information on caring for organisms used in the investigations.

Professional Learning Connections

FOSSweb provides PDF files of professional development chapters, mostly from *Teacher Resources*, that explain how to integrate instruction to improve learning. Some of them are

- Sense-Making Discussions for Three-Dimensional Learning
- Science-Centered Language Development
- FOSS and Common Core English Language Arts and Math
- Access and Equity
- Taking FOSS Outdoors

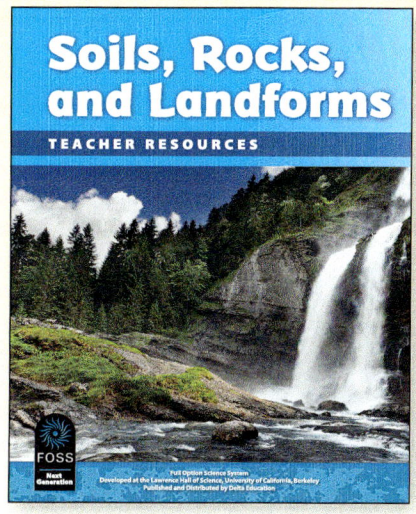

Soils, Rocks, and Landforms Module—FOSS Next Generation

SOILS, ROCKS, AND LANDFORMS — *Technology*

REQUIREMENTS *for Accessing* FOSSweb

FOSSweb Technical Requirements

To use FOSSweb, your computer must meet minimum system requirements and have a compatible browser and recent versions of Flash Player, QuickTime, and Adobe Reader. Many online activities have been updated to an HTML5 version compatible with all devices. (Those designated with "Flash" after the title require Flash Player.) The system requirements are subject to change. It is strongly recommended that you visit FOSSweb to review the most recent minimum system requirements and any plug-in requirements. There, you can access the "Tech Specs and Info" page to confirm that your browser has the minimum requirements to support the online activities.

Preparing your browser. FOSSweb requires a supported browser for Windows or Mac OS with a current version of the Flash Player plug-in, the QuickTime plug-in, and Adobe Reader or an equivalent PDF reader program. You may need administrator privileges on your computer in order to install the required programs and/or help from your school's technology coordinator.

By accessing the "Tech Specs and Info" page on FOSSweb, you can check compatibility for each computer you will use to access FOSSweb, including your classroom computer, computers in a school computer lab, and a home computer. The information on FOSSweb contains the most up-to-date technical requirements for all devices, including tablets and mobile devices.

Support for plug-ins and reader. Flash Player and Adobe Reader are available on www.adobe.com as free downloads. QuickTime is available for free from www.apple.com. FOSS does not support these programs. Please go to the program's website for troubleshooting information.

▶ **NOTE**
It is strongly recommended that you visit FOSSweb to review the most recent minimum system requirements.

Requirements for Accessing FOSSweb

Other FOSSweb Considerations

Firewall or proxy settings. If your school has a firewall or proxy server, contact your IT administrator to add explicit exceptions in your proxy server and firewall for FOSSweb Akamai video servers. For more specific information on servers for firewalls, refer to "Tech Specs and Info" on FOSSweb.

Classroom technology setup. FOSS has a number of digital resources and makes every effort to accommodate users with different levels of access to technology. The digital resources can be used in a variety of ways and can be adapted to a number of classroom setups.

Teachers with classroom computers and an LCD projector, interactive whiteboard, or a large screen will be able to show online materials to the class. If you have access to a computer lab, or enough computers in your classroom for students to work in small groups, you can set up time for students to use the FOSSweb digital resources during the school day. Teachers who have access to only a single computer will find a variety of resources on FOSSweb that can be used to assist with teacher preparation and materials management.

Teachers who have tablets available for student use and have premium content can download the FOSS eBook app onto devices for easy student access to the FOSS eBooks. Instructions for downloading the app can be found on FOSSweb on the Module Detail Page for any module. You'll find them under the Digital-Only Resources section and then under the tab for Student eBooks.

Displaying online content. Throughout each module, you may occasionally want to project online components for instruction through your computer. To do this, you will need a computer with Internet access and either an LCD projector and a large screen, an interactive whiteboard, or a document camera arranged for the class to see.

You might want to display the notebook and teacher masters to the class. In Resources by Investigation, you'll have the option of downloading the masters to project or to copy. Choose "for Display" if you plan on projecting to the class. These masters are optimized for a projection system and allow text entry directly onto the sheet from the computer. The "for Print" versions are sized to minimize paper use when photocopying for the class.

▶ **NOTE**
FOSSweb activities are designed for a minimum screen size of 1024 × 768. It is recommended that you adjust your screen resolution to 1024 × 768 or higher.

Soils, Rocks, and Landforms Module—FOSS Next Generation

SOILS, ROCKS, AND LANDFORMS — Technology

TROUBLESHOOTING and Technical Support

> **NOTE**
> The FOSS digital resources are available online on FOSSweb. You can always access the most up-to-date technology information, including help and troubleshooting, on FOSSweb.

If you experience trouble with FOSSweb, you can troubleshoot in a variety of ways.

1. Test your browser to make sure you have the correct plug-in and browser versions. Even if you have the necessary plug-ins installed on your computer, they may not be recent enough to run FOSSweb correctly. Go to FOSSweb, and select the "Tech Specs and Info" page to review the most recent system requirements and check your browser.
2. Check the FAQs on FOSSweb for additional information that may help resolve the problem.
3. Empty the cache from your browser and/or quit and relaunch.
4. Restart your computer, and make sure all computer hardware turns on and is connected correctly.

If you are still experiencing problems after taking these steps, send FOSS Technical Support an e-mail to support@FOSSweb.com. In addition to describing the problem you are experiencing, include the following information about your computer: Mac or PC, operating system, browser name and version, plug-in names and versions. This will help troubleshoot the problem.

Where to Get Help

For further questions about FOSSweb, please don't hesitate to contact our technical support team.

Account questions/help logging in

School Specialty Online Support
techsupport.science@schoolspecialty.com
loginhelp@schoolspecialty.com

Phone: 1-800-513-2465, 8:30 a.m. –6:00 p.m. ET

General FOSSweb technical questions

FOSSweb Tech Support
support@fossweb.com

INVESTIGATION 1 – Soils and Weathering

Part 1	
Soil Composition	94
Part 2	
Physical Weathering	110
Part 3	
Chemical Weathering	120
Part 4	
Schoolyard Soils	134

Guiding question for phenomenon:
How do soils form?

PURPOSE

Students investigate a familiar phenomenon—the earth material soil. They discover that not all soils are the same and explore the processes of weathering that break down rocks to form soils.

Content

- Soils can be described by their properties.
- Soils are composed of different kinds and amounts of earth materials and humus.
- Weathering is the breakdown of rocks and minerals at or near Earth's surface.
- The physical-weathering processes of abrasion and freezing break rocks and minerals into smaller pieces.
- Chemical weathering occurs when exposure to water and air changes rocks and minerals into something new.

Practices

- Investigate the composition of soils from four different locations.
- Use evidence from investigations to explain the effects of physical and chemical weathering.
- Observe and compare local soils.

Science and Engineering Practices
- Asking questions
- Developing and using models
- Planning and carrying out investigations
- Analyzing and interpreting data
- Constructing explanations
- Engaging in argument from evidence
- Obtaining, evaluating, and communicating information

Disciplinary Core Ideas
ESS2: How and why is Earth constantly changing?
ESS2.A: Earth materials and systems
ESS2.E: Biogeology

Crosscutting Concepts
- Patterns
- Cause and effect
- Systems and system models

FOSS Full Option Science System

INVESTIGATION 1 – *Soils and Weathering*

	Investigation Summary	Time	Focus Question for Phenomenon, Practices
PART 1	**Soil Composition** Students observe and compare four different soils. They learn that soils are composed of essentially the same types of materials (inorganic earth materials and humus), but the amounts of the materials vary. Students speculate on where each of four soils came from: mountain, desert, river delta, or forest.	**Assessment** 1 Session **Active Inv.** 2–3 Sessions * **Reading** 1 Session	**What is soil?** **Practices** Asking questions Planning and carrying out investigations Constructing explanations Engaging in argument from evidence Obtaining, evaluating, and communicating information
PART 2	**Physical Weathering** Students begin to explore how large masses of rock break into smaller pieces. They tumble rocks and freeze water to see how these two types of physical weathering can break rocks. They read about how temperature changes and tree roots (living things) can cause rocks to break down.	**Active Inv.** 2 Sessions	**What causes big rocks to break down into smaller rocks?** **Practices** Developing and using models Planning and carrying out investigations Constructing explanations Engaging in argument from evidence
PART 3	**Chemical Weathering** Students plan and conduct an investigation to test rocks for interaction with "acid rain." They see that some rocks (limestone and marble) are very susceptible to acid rain, one form of chemical weathering, but other rocks (basalt and sandstone) are unaffected.	**Active Inv.** 3 Sessions **Reading** 1 Session	**How are rocks affected by acid rain?** **Practices** Developing and using models Planning and carrying out investigations Analyzing and interpreting data Constructing explanations Obtaining, evaluating, and communicating information
PART 4	**Schoolyard Soils** Students collect and observe different soils from several locations in the schoolyard. They analyze the soil samples to determine how much humus and rock material are in local soils.	**Active Inv.** 2 Sessions **Assessment** 2 Sessions	**What's in our schoolyard soils?** **Practices** Planning and carrying out investigations

* A class session is 45–50 minutes.

At a Glance

Content Related to DCIs	Writing/Reading	Assessment
• Soils can be described by their properties. • Soils are composed of different kinds and amounts of earth materials and humus. • Living things affect the characteristics of soil.	**Science Notebook Entry** *Soil Observations* *Soils in Vials* **Science Resources Book** "What Is Soil?"	**Benchmark Assessment** *Survey* **Embedded Assessment** Science notebook entry
• Weathering is the breakdown of rocks and minerals at or near Earth's surface. • The physical-weathering processes of abrasion, freezing, and pressure of tree roots (living things) break rocks and minerals into smaller pieces.	**Science Notebook Entry** Observations of tumbled rock Answer the focus question	**Embedded Assessment** Response sheet
• Weathering is the breakdown of rocks and minerals at or near Earth's surface. • Chemical weathering occurs when exposure to water and air changes rocks and minerals into something new.	**Science Notebook Entry** *Rock Observations* (optional) *Rocks in Acid Rain* *Acid-Rain Evaporation* **Science Resources Book** "Weathering" **Video** *Weathering and Erosion*	**Embedded Assessment** Performance assessment
• Soils can be described by their properties. • Soils are composed of different kinds and amounts of earth materials and humus. • Weathering is the breakdown of rocks and minerals at or near Earth's surface.	**Science Notebook Entry** *Schoolyard Soil Samples* (optional) **Video** *Soils* **Online Activities** "Tutorial: Weathering" "Virtual Investigation: Water Retention of Soils"	**Benchmark Assessment** *Investigation 1 I-Check* **NGSS Performance Expectation addressed in this investigation** 4-ESS2-1

INVESTIGATION 1 – Soils and Weathering

> **TEACHING NOTE**
>
> Refer to the grade-level Planning Guide chapter in Teacher Resources for a summary explanation of the phenomena students investigate in this module using a three-dimensional learning approach.

BACKGROUND *for the Teacher*

The anchor phenomenon for this module is the surface of Earth's landscape—the shape and the composition of landforms. The driving questions for the module are what are Earth's land surfaces made of? and why are landforms not the same everywhere?

Students start the module by investigating soil as a surface earth-material phenomenon and ask the question how do soils form?

Most of the time when you stroll across Earth's solid surface, you are treading on **soil**. When soil is loose and dry, we observe it as dust. When soil is wet, we observe it as mud. Much of the time we don't see it at all because it is frequently obscured by plants and plant litter (leaves, twigs, and bark). This first investigation is the beginning of an inquiry into the nature of the **earth material** called soil and its composition.

What Is Soil?

Soil is the loose top layer of Earth's surface. It's made of small particles of rock, bits of plant material, and living organisms.

In the beginning there was no soil. Earth's surface was solid **rock**. Over a very long time the rocky surface broke down into smaller and smaller pieces. We have assigned names to rock particles of various sizes. Rocks the size of grapes up to the size of golf balls are called **pebbles**. The next smaller class of rocks, between rice grains and grapes, is called **gravel**. Bits of rock smaller than rice, down to salt grains, are in the class of particles called **sand**. Smaller than sand, but still observable with simple optical magnification and detectable by feel as fine grit, are particles called **silt**. The smallest category of rock pieces, too small to see with a hand lens, is **clay**. When clay is moistened and rubbed between the fingers, it feels slippery. When a significant mass of rock particles of various sizes accumulates, it forms the mineral portion of soil. But to rise to the level of soil, the rock material needs to be infused with organic material (**humus**).

Soil is not a static commodity. Five interwoven factors determine the ever-changing characteristics of soil: (1) type of rock source, (2) climate, which influences **weathering**, (3) topography, (4) actions of living things, and (5) passage of time.

Rocks are composed of minerals. Minerals are chemical compounds. The kinds and concentrations of minerals in a source rock significantly influence the soils they produce. The ways the source rock weathers (breaks down) determines the sizes of the particles in the soil. Topography determines how the weathered particles are distributed—how they get laid down and mixed. Steep gradients encourage smaller particles to travel in flowing water. Flat basins allow silt and clay particles to

"Five interwoven factors determine the ever-changing characteristics of soil: (1) type of rock source, (2) climate, which influences weathering, (3) topography, (4) actions of living things, and (5) passage of time."

Background for the Teacher

accumulate. Organisms create and enrich soil. Rock material without organic matter (humus) is not really soil at all. When insects, worms, and vertebrates dig and burrow in the soil, they stir in leaves and other organic material, and their waste products and decomposing bodies are a critically important contribution to the fertility of soils. Finally, the development of soil requires time. Time for the components to assemble, time for organisms to make the rock material fertile, and time to establish a healthy ecosystem of organisms to perpetuate the vitality of the soil.

Soil provides a medium of support for multitudes of organisms. Animals burrow through and dig into the soil. Plants invade the soil with their roots. The roots extract vital nutrients from the soil—water and minerals. The nutrient minerals come from simple compounds that are the products of decomposition of dead plants and animals as well as from simple chemicals extracted directly from the rock material.

Different soils have different characteristics based on the relative amounts of different-sized rock particles and amount of humus. Mountain soils that are closer to the native rock tend to have higher concentrations of the larger particle sizes. Soils in valleys tend to have higher concentrations of silt and clay particles. Desert soils tend to be heavily infused with sand and gravel, since the wind often carries away the smaller silt and clay particles. Soil can be studied as a **system** composed of different parts that work together.

"Early in its history . . . there was no soil, and there were no small bits of rock—no sand, gravel, or silt."

What Causes Big Rocks to Break Down into Smaller Rocks?

Early in its history, Earth's surface was one continuous expanse of molten rock. In time, it cooled to form a continuous, solid rocky crust. There was no soil, and there were no small bits of rock—no sand, gravel, or silt. As Earth continued to cool, the rock contracted and began to crack. Rock pieces began to break away and tumble downhill. When rocks tumble, they bang into other rocks. The force of impact breaks large rocks into smaller pieces. When rocks break up as a result of physical forces, the process is called **physical weathering**. Rocks don't necessarily require banging into each other to break apart; even friction or **abrasion** causes them to break into increasingly smaller pieces.

When Earth cooled enough for liquid water to condense, water started to flow over Earth's surface. Moving water pushes rocks around and causes more frequent and powerful impacts. If water invades cracks in rocks and then **freezes**, the **expanding** water causes a process called frost wedging. Rocks high in the mountains continually weather because of expanding ice.

Soils, Rocks, and Landforms Module—FOSS Next Generation

INVESTIGATION 1 – Soils and Weathering

"The removal of minerals can weaken a large mass of rock so much that it crumbles into small pieces."

When plants, particularly trees, send their roots into cracks in rocks, pressure from the growing roots can break apart the surrounding rocks. Large pieces of rock dislodged from cliffs fall, and the force of the fall can break a large slab of rock into many pieces. Ice, plants, and gravity all play significant roles in the physical weathering of rock. We can develop **models** to study this natural process.

How Are Rocks Affected by Acid Rain?

Another form of weathering that reduces rock to tiny particles is **chemical weathering**. During chemical weathering, water or acids dissolve the minerals in a rock. The process of reacting removes portions of rock, one molecule at a time. The removal of minerals can weaken a large mass of rock so much that it crumbles into small pieces. The products of the **chemical reaction** can dissolve, creating a solution that eventually precipitates or crystallizes in a new location. Chemical weathering, particularly by **acid rain**, is responsible for a significant amount of the breakdown of rock.

Some rocks and the minerals that make up those rocks are affected by acid rain more than other rocks. Rocks with the mineral **calcite**, such as **limestone** and **marble**, react more in the presence of acid rain than other rocks such as **granite**, **basalt**, **conglomerate**, and **sandstone**.

What's in Our Schoolyard Soils?

Only you and your students will be able to answer this question. Soils vary widely, depending on their source and use. If your school site is relatively pristine, that is, if the soil still represents the soil that was in place before the school was established, your schoolyard soil may be similar to that of an undeveloped location nearby. If the land was previously undeveloped or used for agriculture, you may have a fertile sandy loam composed of a generous mixture of sand, silt, and clay, infused with a significant measure of dark organic humus. If the soil is the aftermath of grading that preceded school construction, the soil may be topsoil (the topmost fertile layer of soil) mixed with subsoil plowed up from deeper in the ground. Such disturbed soils can be heavily biased toward the clay end of the particle-size scale and low in humus. They are easily compacted and can become extremely hard when dry and muddy when wet. If the soil has been "improved," that is, artificially enhanced with topsoil and organic material, it may have a good balance of particle sizes (sand, clay, and silt) and a good measure of organic humus. When students venture into the field to sample the soil in several locations around the school, they will be prepared to answer the question about schoolyard soils.

New Word
Say it · See it · Hear it · Write it

Abrasion
Acid rain
Basalt
Calcite
Chemical reaction
Chemical weathering
Clay
Conglomerate
Earth material
Expand
Freeze
Granite
Gravel
Humus
Limestone
Marble
Model
Pebble
Physical weathering
Rock
Sand
Sandstone
Silt
Soil
System
Weathering

TEACHING CHILDREN about Soils and Weathering

Developing Disciplinary Core Ideas (DCI)

Several of the ideas introduced in this module are pretty abstract. For example, the idea that the sand students find in their schoolyard soil was weathered from a massive rock structure in the mountains a long time ago and carried by erosive forces to its present location is not intuitive. It requires students to think about what might have happened to transform and deliver a tiny piece of rock to its present location. It requires students to infer what happened to deliver those grains of sand to the schoolyard. Inferential thinking is an important element of constructing complex explanations for observed phenomena in the natural world. Helping students build bridges between observations and explanations will take time and lots of experience. One of your roles will be to guide this kind of student thinking.

These experiences will build students' understanding of the core ideas **ESS2.A: Earth materials and systems** and **ESS2.E: Biogeology**, by helping them understand how water, ice, wind, living organisms, and gravity break rocks, soils, and sediments into smaller particles. This introduction to weathering is considered foundational for developing understanding of the processes that shape Earth.

In addition to teaching children foundational ideas about weathering and its importance to Earth processes, the **Soils, Rocks, and Landforms Module** provides students with opportunities to experience two other dimensions of the scientific enterprise—the dimension of science and engineering practices and the dimension of crosscutting concepts.

> **NGSS Foundation Box for DCI**
>
> **ESS2.A: Earth materials and systems**
> - Rainfall helps to shape the land and affects the types of living things found in a region. Water, ice, wind, living organisms, and gravity break rocks, soils, and sediments into smaller particles and move them around. (4-ESS2-1)
>
> **ESS2.E: Biogeology**
> - Living things affect the physical characteristics of their regions. (4-ESS2-1)

INVESTIGATION 1 – Soils and Weathering

NGSS Foundation Box for SEP

- **Ask questions** that can be investigated based on patterns such as cause-and-effect relationships.
- **Identify limitations of models.**
- **Develop models** to describe phenomena.
- **Use a model** to test cause-and-effect relationships or interactions concerning the functioning of a natural system.
- **Plan and conduct an investigation** collaboratively to produce data to serve as the basis for evidence.
- **Make observations and/or measurements to produce data** to serve as the basis for evidence for an explanation of a phenomenon or test a design solution.
- **Analyze and interpret data** to make sense of phenomena, using logical reasoning.
- **Construct an explanation** of observed relationships.
- **Use evidence** (e.g., observations, patterns) to support an explanation.
- **Identify the evidence** that supports particular points in an explanation.
- **Compare and refine arguments** based on an evaluation of the evidence presented.
- **Respectfully provide and receive critiques from peers** about a proposed procedure, explanation, or model by citing relevant evidence and posing specific questions.
- **Construct and/or support an argument** with evidence, data and/or model.
- **Use data to evaluate claims** about cause and effect.

Engaging in Science and Engineering Practices (SEP)

Engaging in the practices of science helps students understand how scientific knowledge develops; such direct involvement gives them an appreciation of the wide range of approaches that are used to investigate, model, and explain the world. Engaging in the practices of engineering likewise helps students understand the work of engineers, as well as the links between engineering and science. Participation in these practices also helps students form an understanding of the crosscutting concepts and disciplinary ideas of science and engineering; moreover, it makes students' knowledge more meaningful and embeds it more deeply into their worldview. (National Research Council, *A Framework for K–12 Science Education,* 2012, page 42)

The focus questions and notebook will provide scaffolds for students in this first investigation, scaffolds that can be carefully removed in later investigations as students become more experienced engaging in practices. In this investigation, students engage in these practices.

- **Asking questions** about soils as they begin an extended study that sets the context for how weathering and erosion change the shape of the land.
- **Developing and using models** to explain why soils are different in different locations and why the rock material in soils is different sizes. Models are used to better understand natural earth systems.
- **Planning and carrying out investigations** dealing with soil properties and aspects of physical and chemical weathering that affect the breakdown of rock material.
- **Analyzing and interpreting data** to draw conclusions about how physical and chemical weathering contribute to the breakdown of rock material.
- **Constructing explanations** using evidence to describe interactions of earth materials with each other and other substances in the environment such as acid rain.
- **Engaging in argument from evidence** about different soils, where they might be from based on properties, and why they are different from one location to another.

Teaching Children about Soils and Weathering

- **Obtaining, evaluating, and communicating information** from books and media and integrating that with their firsthand experiences to construct explanations about soil properties and the weathering of rock material that becomes part of the soil.

Exposing Crosscutting Concepts (CC)

The third dimension of instruction involves the crosscutting concepts, sometimes referred to as unifying principles, themes, or big ideas, that are fundamental to the understanding of science and engineering.

These concepts should become common and familiar touchstones across the disciplines and grade levels. Explicit reference to the concepts, as well as their emergence in multiple disciplinary contexts, can help students develop a cumulative, coherent, and usable understanding of science and engineering. (National Research Council, 2012, page 83)

In this investigation, the focus is on these crosscutting concepts.

- **Patterns:** Patterns in soil composition (of rock and organic material) can help predict where the soil is from and how it formed.
- **Cause and effect:** Water, ice, wind, vegetation, and chemical processes cause rock material to weather, breaking down into smaller pieces.
- **Systems and system models:** Soil can be described in terms of its components. The soil and water can be studied as a system with interacting parts.

Connections to the Nature of Science

- **Scientific knowledge assumes an order and consistency in natural systems.** Science assumes consistent patterns in natural systems.

NGSS Foundation Box for SEP

- **Obtain and combine information** from books and/or other reliable media to explain phenomena.
- **Read and comprehend grade-appropriate complex texts** and/or other reliable media to summarize and obtain scientific and technical ideas and describe how they are supported by evidence.
- **Communicate scientific and/or technical information** orally and/or in written formats, including various forms of media and may include tables, diagrams, and charts.

NGSS Foundation Box for CC

- **Patterns:** Similarities and differences in patterns can be used to sort and classify natural phenomena. Patterns of change can be used to make predictions.
- **Cause and effect:** Events that occur together with regularity might or might not be a cause-and-effect relationship. Cause-and-effect relationships are routinely tested, and used to explain changes.
- **Systems and system models:** A system can be described in terms of its components and their interactions.

INVESTIGATION 1 – Soils and Weathering

Conceptual Flow

The anchor phenomenon for this module is the surface of Earth's landscape—the shape and the composition of landforms. The driving questions for the module are what are Earth's land surfaces made of? and why are landforms not the same everywhere?

In this first investigation, students investigate soil as a surface earth-material phenomenon and ask the question how do soils form?

The **conceptual flow** for this module starts with an introduction to **soil** as an **earth material** (any naturally occurring **solid**, **liquid**, or **gas** matter that makes up Earth). Soil is the top layer of earth material that covers most of Earth's land surface. **Soil is a complex mixture** of rock particles of several sizes (**pebbles**, **gravel**, **sand**, **silt**, **clay**) and decomposing **organic matter** (**humus**). Different environments on Earth tend to have generalizable soil types based on the percentage of the basic soil components. Desert soils tend to have high percentages of sand and gravel and less humus. Mountain soils also tend to have a lot of sand and gravel, but more humus. Forest soils tend to have a fairly even distribution of mineral particle sizes: sand, silt, and clay, as well as high concentrations of organic humus. Delta soils are typically high in clay and silt, with a substantial measure of humus.

In Parts 2 and 3, students grapple with the question of where the small rocks in soil come from. Students are introduced to the idea that small rocks are the broken pieces of large mountain-sized rocks. Those large masses of rock are reduced to boulders by **physical weathering** caused by **frost wedging** and **root wedging** and further reduced to **sand**, **silt**, and **clay** by **abrasion** when rocks fall and tumble downhill in landslides and moving water. Rock is also broken down by **chemical weathering** when **minerals in rocks dissolve in acid rain** and other water sources.

In Part 4, students extend and apply what they have learned in the classroom to soil samples from the schoolyard.

Teaching Children about Soils and Weathering

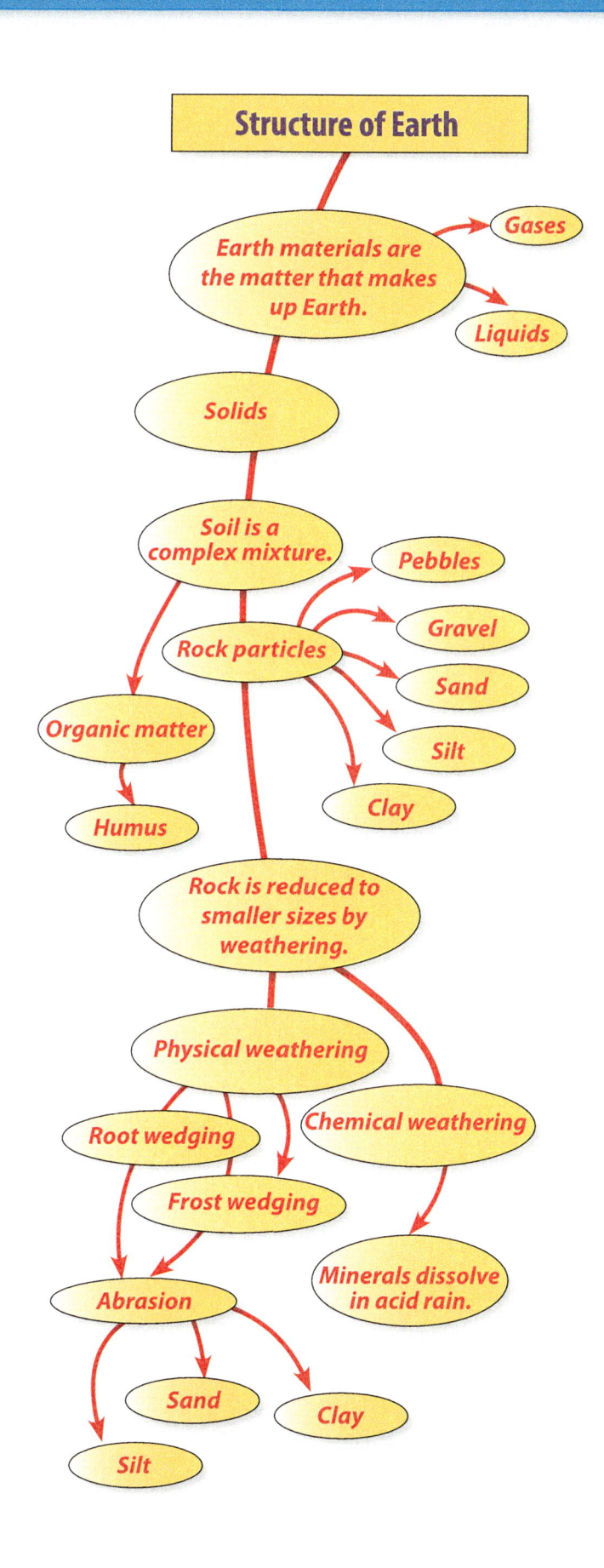

Structure of Earth

- Earth materials are the matter that makes up Earth.
  - Gases
  - Liquids
  - Solids
    - Soil is a complex mixture.
      - Organic matter
        - Humus
      - Rock particles
        - Pebbles
        - Gravel
        - Sand
        - Silt
        - Clay
    - Rock is reduced to smaller sizes by weathering.
      - Physical weathering
        - Root wedging
        - Frost wedging
        - Abrasion
          - Sand
          - Silt
          - Clay
      - Chemical weathering
        - Minerals dissolve in acid rain.

Soils, Rocks, and Landforms Module—FOSS Next Generation

INVESTIGATION 1 – Soils and Weathering

No. 1—Teacher Master

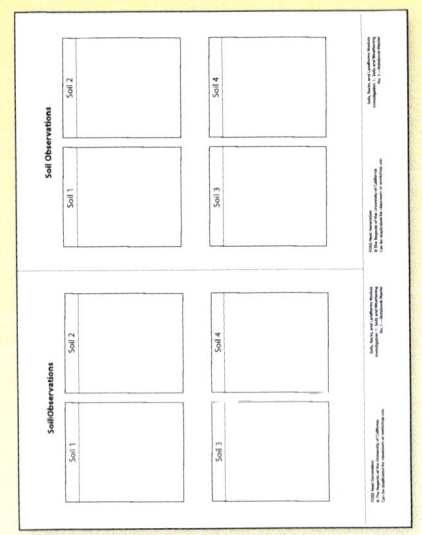

No. 1—Notebook Master

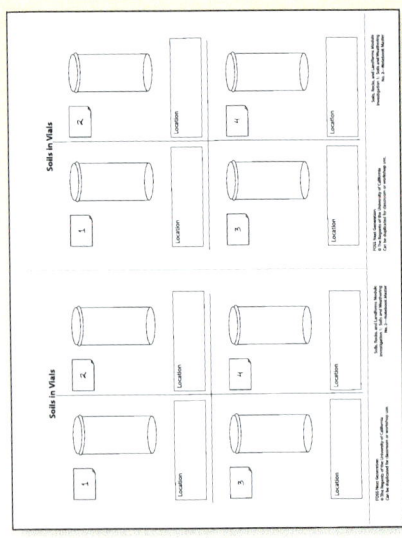

No. 2—Notebook Master

MATERIALS for
Part 1: Soil Composition

For each group

- 1 Basin
- 4 Vials with caps, 12 dram
- 1 FOSS tray
- 4 Paper plates
- 5 Self-stick notes
- 1 Container, 1/2 L
- 1 Syringe
- 4 Science notebooks (composition book) ★
- ❑ 4 *Letter to Family*
- ❑ 4 Notebook sheet 1, *Soil Observations*
- ❑ 4 Notebook sheet 2, *Soils in Vials*
- 4 *FOSS Science Resources: Soils, Rocks, and Landforms*
  - "What Is Soil?"

For the class

- Glue sticks ★
- 1 Vial holder, blue plastic
- 4 Self-stick notes
- 4 Metal spoons
- 4 Zip bags, 1 L
- 1 Beaker, 100 mL
- 4 Containers, with lids, 1 L
- Transparent tape ★
- *FOSS Science Safety* and *FOSS Outdoor Safety* posters
- *Conservation* posters
- ❑ 1 Teacher master 1, *Letter to Family*
- Pebbles
- Gravel
- Sand, coarse
- Clay, gray, powdered
- Potting soil (humus)
- 1 Pitcher
- Water ★

For embedded assessment
- ❑ Embedded Assessment Notes

For benchmark assessment
- ❑ *Survey*
- ❑ *Assessment Record*

★ Supplied by the teacher.   ❑ Use the duplication master to make copies.

Full Option Science System

Part 1: Soil Composition

GETTING READY for
Part 1: Soil Composition

1. **Schedule the investigation**
   This part will take one session for the survey, two to three sessions for the active investigation, and one session for the reading.

2. **Preview Part 1**
   Students observe and compare four different soils. They learn that soils are composed of essentially the same types of materials (inorganic earth materials and humus), but the amounts of the materials vary. Students speculate on where each of four soils came from: mountain, desert, river delta, or forest. The focus question for this part is **What is soil?**

3. **Set up a materials station**
   All of the investigations are written for collaborative groups of four students. Sometimes students will work as individuals or pairs, but there will always be two Getters who are responsible for getting and returning materials for the group.

   Plan to organize the tools for all the investigations in a convenient location where Getters can get the materials efficiently. Provide a basin for Getters to carry materials to and from the materials station.

4. **Prepare water**
   In Step 9 of Guiding the Investigation, students use the 50 mL syringe to measure 25 mL of water into vials with soil. Half fill a 1/2 L container with water for each group. Keep the water containers and syringes out of the reach of students until after the dry-soil observations.

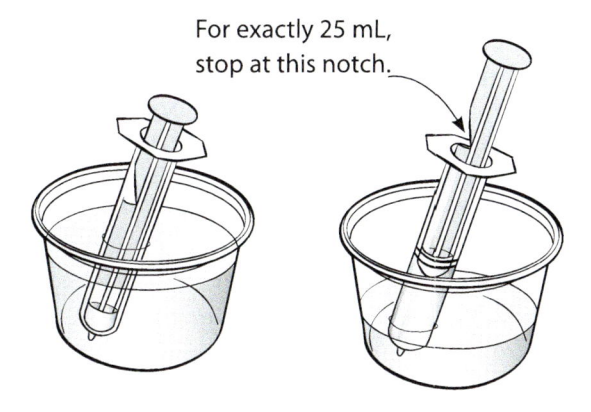

For exactly 25 mL, stop at this notch.

TEACHING NOTE

The Getting Ready section for Part 1 of Investigation 1 is longer than the corresponding section for the other parts. Several of the numbered items appear only here, but may apply to other parts as well, such as the discussions about setting up materials stations (3), preparing student science notebooks (9), photocopying duplication masters (11), and planning a word wall (13).

TEACHING NOTE

Use the collaborative-group role names that work for your classroom. You can use Getter, Starter, and Reporter.

▶ **SAFETY NOTE**
Remind students not to aim syringes at other students.

▶ **NOTE**
To prepare for this investigation, view the teacher preparation video on FOSSweb.

Soils, Rocks, and Landforms Module—FOSS Next Generation

INVESTIGATION 1 – Soils and Weathering

> **NOTE**
> This recipe makes enough "soil" samples for one class use. Each group uses 2/3 of a vial of each soil. At the end of this part, recycle wet and dry soils outside.

> **NOTE**
> The sand used to mix the soils is coarse sand. The container may be labeled "mountain sand." Mountain sand is coarse sand.

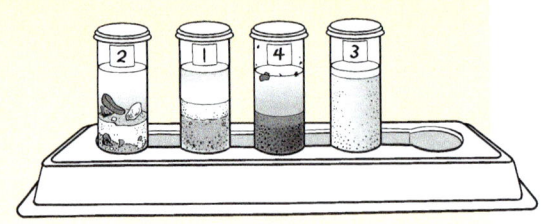

5. **Mix soils**
   Use the materials in the kit and the 100 milliliter (mL) beaker as a measure to make four different soil types. Mix the ingredients thoroughly in a basin before transferring them to the 1 L containers. Label them 1 through 4, but do not include the names of the simulated environments. Students will determine the environments through investigation. (Use the gray clay, not white.)

Soil	Clay (mL)	Sand (mL)	Gravel (mL)	Pebbles (mL)	Humus (mL)	Simulated environment
1	150	110	110	0	30	river delta
2	30	30	150	150	30	mountain
3	30	150	180	0	30	desert
4	15	30	110	15	225	forest

Place the numbered containers of soil and a metal spoon for each at the materials station. Each group needs 2/3 of a vial of each soil.

The soils used in this part will be compared to schoolyard soils in Part 4 of this investigation. Save one set of the four vials for this comparison. Place the four vials in the blue plastic vial holder for the class to use later.

6. **Consider paper plates permanent equipment**
   Don't discard the paper plates after use. You should be able to reuse one set of paper plates for the entire module.

7. **Select your outdoor site (optional)**
   Plan a soil-observation walk around the schoolyard or nearby neighborhood (Step 2 of Guiding the Investigation). Take the students to several locations to discover that soils are everywhere, and that soils may look different from one place to another. If you choose to make this part of the class session, you may want to take along metal spoons and zip bags to collect samples, or wait until

Part 1: Soil Composition

Part 4 to collect soil samples. Refer to the Taking FOSS Outdoors chapter for management strategies for working with students outdoors.

If you do go on an outdoor walk to look for soil, be sure to tour the outdoor site on the morning of your outdoor activity. Do a quick search for potentially distracting, unsafe, or unsightly items.

8. **Plan for safety**
Upper-elementary students must be allowed to demonstrate that they can act responsibly with materials, but they must be given guidelines for safe and appropriate use of materials. Work with students to develop those guidelines so that they participate in making behavior rules and understand the rationale for the rules. Encourage responsible actions toward other students. Display and discuss the *FOSS Science Safety* and *FOSS Outdoor Safety* posters in class. The posters are included in the kit.

Look for the safety-note icon in the Getting Ready and Guiding the Investigation sections, which will point out specific safety alerts throughout the module. Be aware of allergies your students have.

This is a good time to introduce the set of four *Conservation* posters and discuss the importance of natural resources with students.

9. **Plan for student notebooks**
Students will keep a record of their science investigations in individual science notebooks. They will record observations, responses to focus questions, and connections between their classroom learning and the world beyond. This record will be a useful reference document for students and a revealing testament for adults of each student's learning progress.

We recommend that students use bound composition books for their science notebooks. This ensures that student work becomes an organized, sequential record of student learning. Students' notebooks will be a combination of structured sheets printed or photocopied from notebook masters provided in *Teacher Resources* (and glued into the notebook) and unstructured student-generated graphics and writing on blank pages.

> **TEACHING NOTE**
> Refer to the Science Notebooks in Grades 3–5 chapter for more details on how to effectively integrate the use of science notebooks into all FOSS investigations.

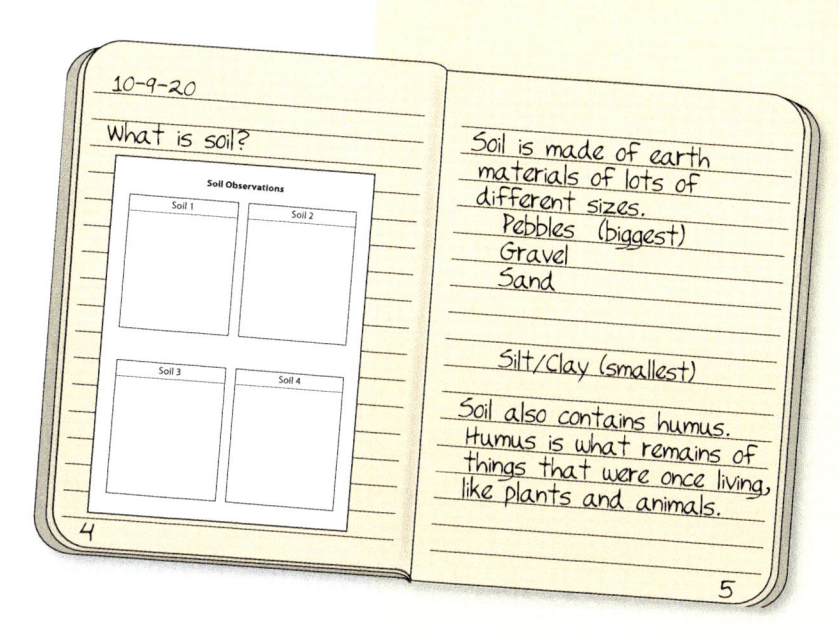

Soils, Rocks, and Landforms Module—FOSS Next Generation

INVESTIGATION 1 – Soils and Weathering

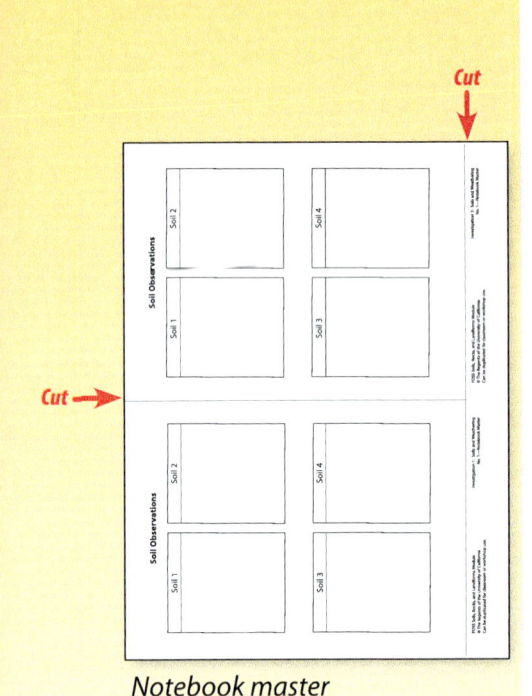

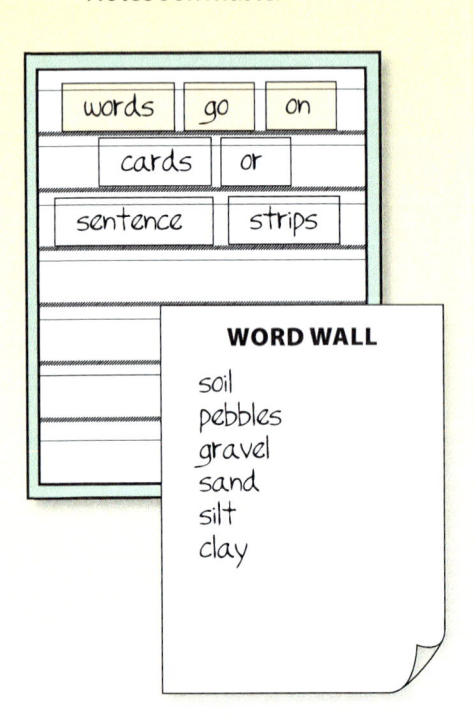

If you are already using an alternative method of organization with your students, such as a sheaf of folded and stapled pages, your method can take the place of the bound composition book.

10. Plan to project images of notebook sheets

Throughout the module, you will see suggestions for projecting images of notebook sheets to orient students to their use or to share data that have been gathered. Use whatever technology is available to you. This may include a document camera, LCD projector and computer, interactive whiteboard, or overhead projector. Digital versions of all duplication masters for duplication or large format for projecting are available at www.FOSSweb.com.

11. Print or photocopy duplication masters

Two sets of duplication masters are used throughout this module: notebook masters and teacher masters. Digital versions of these duplication masters are available on FOSSweb.

Notebook masters are sized to be cut apart and glued into a composition book. Each notebook master has two copies of the sheet, which should be cut apart.

Teacher masters serve various functions such as letter to family, home/school connections, and math extensions.

A notebook sheet or teacher master that requires printing or duplication is flagged with this icon ❏ in the materials list for the part. It is referred to as a notebook sheet in the materials list and a notebook master in the sidebar snapshot.

12. Send a letter home to families

Teacher master 1, *Letter to Family*, is a letter you can use to inform families about this module. The letter states the goals of the module and suggests some home experiences that can contribute to students' learning.

13. Plan to use a word wall

As the module progresses, you will add new vocabulary words to chart paper for posting on a wall and/or to vocabulary cards or sentence strips for use in a pocket chart or a word wall. For additional information, see the Science-Centered Language Development chapter.

Notebook master

WORD WALL
soil
pebbles
gravel
sand
silt
clay

Full Option Science System

Part 1: Soil Composition

14. Plan for working with English learners

At important junctures in an investigation, you'll see an EL note in the sidebar. These notes suggest additional strategies for enhancing access to the science concepts for English learners. You'll also see references to the Science-Centered Language Development chapter in *Teacher Resources*. Refer to this chapter for resources and examples to use when working with science vocabulary and readings in each investigation.

Each time new science vocabulary is introduced, you'll see the new-word icon in the sidebar. This icon lets you know not only that you'll be introducing important vocabulary, but also that you might want to plan to spend more time with those students who need extra help.

15. Plan for storage

You will need a place to store the soil samples while they sit overnight in vials. The FOSS trays are designed to make efficient use of counter space.

If you plan to stack the FOSS trays, have students remove the vial lids. If you have room for each group to place their tray on a counter or table, then have the students keep the lids on the vials of soil and water.

16. Plan for online activities

During instruction, you will need to project online activities or images through a computer for the class to see. The Getting Ready section for each part will indicate what to prepare.

To prepare for the module anchor phenomenon introduction—Earth's surface of landforms—review the images in the Resources by Investigation for Investigation 1, Part 1. These images are selected from the "Landforms Photo Album" article in the *FOSS Science Resources* book. These images are also in the Image Gallery on FOSSweb.

17. Plan to read *Science Resources*: "What Is Soil?"

Plan to read the article "What Is Soil" during a reading period after completing the active investigation for this part.

18. Assess progress throughout the module

Research shows that frequent formative assessment leads to greater student understanding. Embedded (formative) assessments provide a variety of ways to gather information about students' thinking while their ideas are developing. These assessments are meant to be diagnostic. They provide you with information about students' learning so that you know if you need a next step to clarify understanding before going on to the next part of the investigation.

EL NOTE

Display the equipment photo card for each object, and write the object's name on the word wall. Equipment photo cards are available on FOSSweb and can be downloaded and printed.

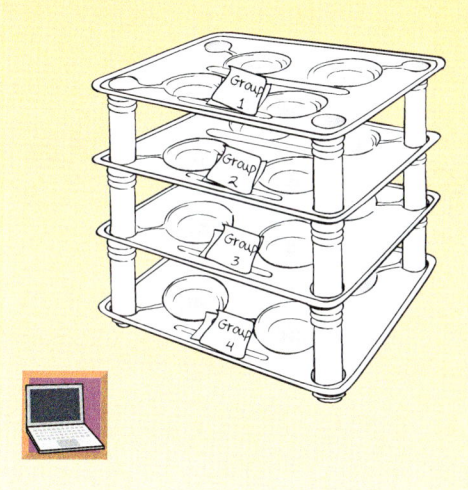

Soils, Rocks, and Landforms Module—FOSS Next Generation

INVESTIGATION 1 – *Soils and Weathering*

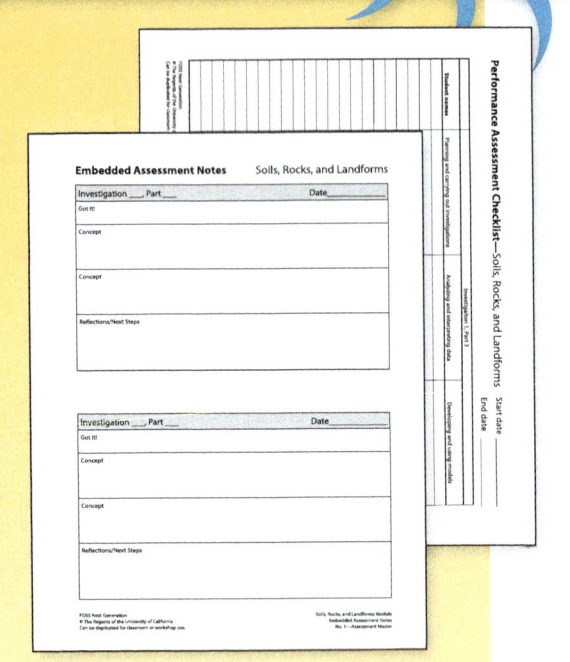

Nos. 1–2—Assessment Masters

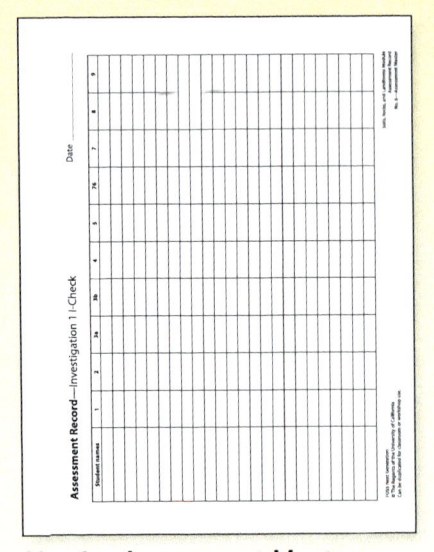

No. 6—Assessment Master

Most Getting Ready sections describe an embedded-assessment strategy that you might find useful in that part. Assessment master 1, *Embedded Assessment Notes,* is a half sheet provided to help you analyze students' embedded-assessment data. (See the Assessment chapter for more on how to use this sheet.) In some parts, the embedded assessment involves a performance assessment. The *Performance Assessment Checklist* is used for recording these observations.

At the end of each investigation (except Investigation 4), there is an I-Check benchmark assessment. The questions on these assessments are summative—they examine the disciplinary core ideas, science and engineering practices, and crosscutting concepts students have learned up to that point in the curriculum. You can find out more about I-Check assessments in the Assessment chapter and in the Getting Ready section for Part 4 of this investigation.

19. Plan benchmark assessment: *Survey*

Before beginning the module, have students take the *Survey* benchmark assessment. This assessment will inform you about students' prior knowledge and give you some insight into which concepts may need more focused instruction. Save the *Surveys* to compare to the identical *Posttest* given after the module instruction has been completed. Masters for the *Survey* can be found in the assessment masters on FOSSweb. Record student performance on the *Assessment Record* (assessment master no. 6). Or plan to administer the *Survey* using FOSSmap. For more information on the assessment system and online tools, see the Assessment chapter.

20. Plan assessment: notebook entry

In Step 17 of Guiding the Investigation, students answer the focus question in their notebooks, and in Step 18 they begin thinking about why soils may differ, based on location. Look for evidence of students' understanding of soil composition and prior knowledge about why different soils have different kinds and sizes of earth materials. Record your observation on a copy of *Embedded Assessment Notes*.

Part 1: Soil Composition

GUIDING the Investigation
Part 1: Soil Composition

1. **Introduce the anchor phenomenon**
   Ask students to observe the projected photos and then talk to their partner about what the images show. Project the images of landforms and introduce the driving questions for the module.

   ➤ *What are Earth's land surfaces made of?*

   ➤ *Why are landforms not the same everywhere?*

2. **Introduce** *earth materials*
   Tell students,

   Today we're going to begin a study of **earth materials** *and how the shape of Earth's surface changes over time. Earth materials are the various solids, liquids, and gases that make up Earth. The earth materials we're going to look at today are in the* **soils** *in these containers.*

   Ask a student to describe to the class what is in one of the containers of soil. Students may call it "dirt," but use the word *soil* to talk about what is in each container (students will develop a specific definition of soil over the course of this activity). Ask students to name places where soils are found.

3. **Go outdoors (optional)**
   Tell students that they are going to go on a quick walk around the schoolyard (or neighborhood). Their task is to look for soil. Where is soil and what is it made of? Explain rules for expected behavior and set boundaries, referring to the *FOSS Outdoor Safety* poster. Have students get coats as needed and head outdoors.

   Ask questions as you tour the area. For example,

   ➤ *Where should we look for soil?*

   ➤ *What kinds of materials do you see in the soil? Is anything growing?*

   ➤ *Are there soil areas where nothing is growing?*

   ➤ *How is this soil the same as the soil we saw over there? How is it different?*

   ➤ *What questions do you have that you might want to investigate?*

   When you see soil near lawn, pavement, or buildings, ask,

   ➤ *Do you think there is soil underneath this lawn, this pavement, or this building?*

   ➤ *Where is there not any soil?* [Where bedrock is exposed; where there is standing or running water, lakes, rivers, the ocean.]

Soils, Rocks, and Landforms Module—FOSS Next Generation

FOCUS QUESTION
What is soil?

TEACHING NOTE
The focus question in each part engages students with the phenomenon to investigate.

Say it — See it — Hear it — Write it — New Word

EL NOTE
Begin a word wall titled, "Soil Composition" and write the words *earth materials* and *soils*. You could show the short video from the first page of the reading "What Is Soil?" in the interactive eBook on FOSSweb.

Materials for Step 3
- *Metal spoons*
- *Zip bags, 1 L*

TEACHING NOTE
Make this trip outside brief, using it simply to get a feel for where soils are found. Most people take soil for granted, so this gives students an opportunity to look at soils from a fresh viewpoint and start asking questions.

SCIENCE AND ENGINEERING PRACTICES
Asking questions

INVESTIGATION 1 – *Soils and Weathering*

CROSSCUTTING CONCEPTS

Patterns

▶ **NOTE**

Glue in the notebook sheet to organize observations or have students divide a blank page into four sections. Choose one method or the other.

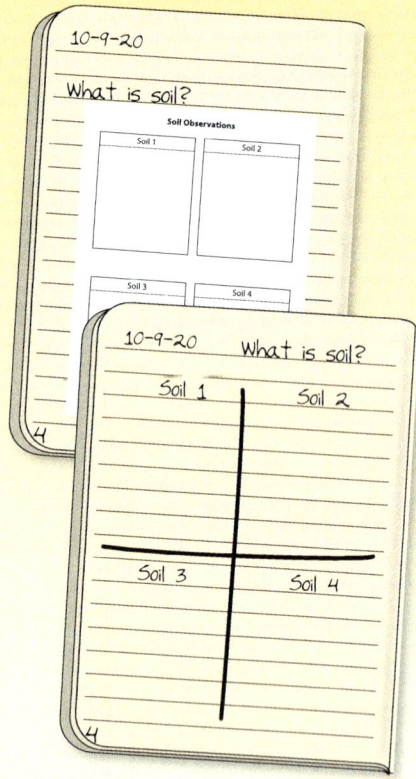

Materials for Step 7
- *Containers of soils, 1–4*
- *Paper plates*
- *Empty vials and caps*
- *Self-stick notes*
- *Transparent tape*

EL NOTE

If necessary, model the procedure for numbering the plate and vial.

4. **Return to class**

   Have students return to the classroom. If they collected soil samples, set those aside for now. Ask students to discuss what they observed.

   ➤ *Where did we find soil?*

   ➤ *What is soil made of?*

   ➤ *Is soil the same in every location? Do you see any patterns?*

   ➤ *Why might geologists (scientists who study Earth) and other people be interested in studying soil?*

5. **Focus question: What is soil?**

   Tell students that they are now going to look more closely at some soil samples. Project or write the focus question on the board.

   ➤ *What is soil?*

6. **Introduce science notebooks**

   Distribute science notebooks to students. Give students a minute to confirm that the pages are all blank. Have students write the numbers 1–10 on the outside corner of the first ten pages, using the front and back of each page. Reserve pages 1–3 for a table of contents.

   Distribute copies of notebook sheet 1, *Soil Observations*. Students should glue the sheet on page 4 of their notebook and record the date.

   Alternatively, if you prefer not to use the notebook sheet, students can write the date at the top and divide the page into four sections, one section for each soil.

7. **Collect soil samples**

   Have Getters go to the materials station and pick up four paper plates, four empty vials and caps, and five self-stick notes. Each person in a group labels one paper plate and its matching vial. Attach a self-stick note with a number ("1," "2," "3," or "4") to the plates. Use one quarter of a numbered self-stick note and a small piece of transparent tape to put labels on vials.

   In tag-team fashion, one person at a time from each group goes to the materials station and fills his or her numbered vial two-thirds full of soil. Be sure students match the number on the vial to the number on the soil container and bring it back to the table. If different groups start with different numbers, it will alleviate the logjam at the materials station.

Full Option Science System

Part 1: Soil Composition

8. **Observe "dry" soils**

   Have students start making observations by pouring soil onto a paper plate, matching the number on the vial with the number on the paper plate. Students should observe and record properties such as color, shape of materials, particle size, and textures.

   As students complete their observations of one soil, have them carefully exchange paper plates so they have the opportunity to observe and record observations for all four soils.

9. **Discuss materials that make up soils**

   Call for attention or gather students together for discussion. (If you have students leave their desks, remind them to bring their notebooks to the discussion.) Ask,

   ➤ *What different kinds of materials did you find in the soils?*

   List the materials that students name on the board.

   When students attempt to name the different sizes of **rock** material, help them by giving each size a name.

   - **Pebbles:** 6–64 mm
   - **Gravel:** 2–6 mm
   - **Sand:** 1–2 mm
   - **Clay** or **silt:** rock particles the size of powder and dust with particles less than 1 mm.

   When students make note of organic material in the mix, introduce the word **humus**. Explain that plants and animals living in or near the soil eventually die and decay. They become part of the soil. Humus is the dead material that becomes part of soil.

   Take a poll. Have students account for the soil that is in front of them. Name each material on the list and ask students to raise their hands if they found that material in the soil they are looking at now. Then ask,

   ➤ *How are the four soils the same and how are they different?*

   Establish the idea that all soils are made of humus (dead plant and animal material) and different sizes of rock material. But the soils may have different amounts of each kind and size of material.

10. **Add water to separate components**

    Tell students that water can be a useful tool to sort the different materials that make up soils. Have Getters go to the materials station and pick up a syringe and 1/2 L container of water for the group. Have students fold their paper plates to create a pathway to convey the soil samples back into their numbered vials.

TEACHING NOTE

The purpose here is for students to notice that soils are made of the same basic materials: different sizes of rock or mineral and humus. Students may also note that water or moisture differs from one soil to the next.

SCIENCE AND ENGINEERING PRACTICES

Planning and carrying out investigations

EL NOTE

Write these rock-particle-size categories on the word wall as reference.

CROSSCUTTING CONCEPTS

Patterns

Materials for Step 10
- *Syringes*
- *Containers of water*

Soils, Rocks, and Landforms Module—FOSS Next Generation

INVESTIGATION 1 – Soils and Weathering

Show students how to use the syringe if this is their first experience using this tool. Remind students not to aim syringes at other students. Have students use syringes to transfer 25 mL of water from the container into their soil-filled vials.

As you go around the room and watch students add the water, point out the bubbles rising from the soil as the water soaks in. Ask them what the bubbles indicate. [As the water seeps in, it pushes air out. The bubbles are evidence that there are air spaces between the particles of soil.]

When everyone has added water, have students cap the vials tightly and shake the contents, holding the cap in place as they shake.

After a minute or two of shaking, ask students to place the vials on their desks and to observe them undisturbed for 3–5 minutes. Students look more closely at the layers of soil materials that form. They should not touch the vials so they can note the order in which the materials settle in the vial.

11. Discuss "wet" observations

Review the earth materials and humus that make up soil. Ask,

➤ *Which materials settled out first after shaking the vials?* [Larger materials settle first. The smaller materials, especially silt or clay, can stay suspended in the water for quite a long time.]

➤ *Where is the humus?* [Students may see pieces of humus floating at the top of the water as well as mixed in with the earth materials, giving it a black color.]

➤ *We observed bubbles when you added water to the soil in the vial. What does that tell you about soil?* [There must be air spaces between the soil particles. When water is added, it pushes the air out from the spaces and we see the air rise as bubbles.]

12. Settle soil vials overnight

Have students put all four vials in the corners of a FOSS tray, label the tray with their group's number or names, and set the trays in a location where they won't be disturbed overnight. Have students return the rest of the materials to the materials station.

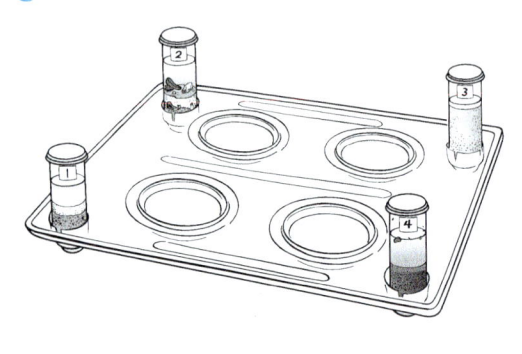

SCIENCE AND ENGINEERING PRACTICES
Analyzing and interpreting data

TEACHING NOTE
Pebbles in the mountain soil may look like they are sitting on top of finer earth material, but they do settle first. The pebbles fall to the bottom immediately, and then the finer materials fill in between the pebbles, so it looks like the layering is different.

EL NOTE
Encourage students to talk about their observations with their partners and/or record them in their notebooks.

Materials for Step 12
- FOSS trays

BREAKPOINT

104

Full Option Science System

Part 1: Soil Composition

13. Observe the settled vials
Have one Getter from each group pick up four copies of notebook sheet 2, *Soils in Vials*, at the materials station. Have another Getter pick up the group's tray with its vials of soil, moving carefully to the table so as not to disturb the settling.

Ask students to record what they see in the vials by completing drawings on the *Soils in Vials* notebook sheet.

14. Have a sense-making discussion
Have students compare the four different soils their group has been observing. Ask,

➤ *Do all the vials contain the same amount of soil?* [All started with the same amount.]

➤ *Do all the vials have the same number of layers?* [They should.]

➤ *Let's look for patterns in the series of vials. Are all the layers of similar earth materials the same size (depth) in all the different soils? For example, do all the soils have the same size sand layer? Do they have the same amount of humus or organic matter?*

Use this discussion to help students define soils. Soils are a mixture of the same kinds of basic rock materials and humus, but the amounts and sizes of the basic materials vary.

15. Clean up
Save one group's set of vials and place the vials in the blue plastic vial holder. Students will compare these vials of soil to local soils in Part 4 of this investigation.

Have students return all equipment to the materials station. The vials (except for the saved set) need to be cleaned and made ready for use in Part 3. The easiest way to do this is to set up a series of basins, some for students to dump the soil and water into, and another filled with water to rinse the vials. All of the soil and water should be recycled into an appropriate outdoor site.

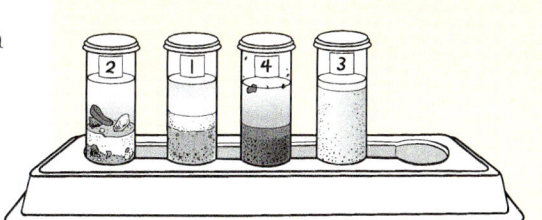

Materials for Step 13
- *Settled vials*
- ***Soils in Vials*** sheets

SCIENCE AND ENGINEERING PRACTICES
Analyzing and interpreting data

CROSSCUTTING CONCEPTS
Patterns

Soils, Rocks, and Landforms Module—FOSS Next Generation

INVESTIGATION 1 – Soils and Weathering

clay
earth material
gravel
humus
pebble
rock
sand
silt
soil

SCIENCE AND ENGINEERING PRACTICES

Constructing explanations

Engaging in argument from evidence

TEACHING NOTE

You might tell students this is a preliminary claim because they do not have much information yet to base their arguments on evidence. After doing more investigations, students will have an opportunity to refine their ideas and engage in argument from evidence in Investigation 2, Part 1, Step 17.

16. Review vocabulary
Review key vocabulary introduced in this part. If those words are not on the word wall, add them now.

17. Answer the focus question
Have students answer the focus question in their notebooks.

➤ *What is soil?*

For students who need scaffolding, provide a sentence frame such as Soil is composed of _____ . In our soil investigation, we found out _____ .

18. Discuss where the soils came from
Tell students that the soil samples they have been observing represent the kinds of soils that might be found in four different locations. The samples represent desert, mountain, river delta, and forest soils. Write these locations on the board or chart paper.

> Where would these soils be found?
>
> desert
> mountain (above the tree line)
> river delta
> forest

Have students take a few minutes to discuss in their groups about why soils might be different from one location to another. Then have students make a preliminary claim about where the soils came from. Students can write the name of the location for each soil in the box below each vial on notebook sheet 2.

Have pairs of students present to each other their preliminary ideas (claims) and why they think so.

Tell students that you are not going to tell them the "right answers." Instead, they are going to learn more about how soils form and other processes that shape Earth in the next few weeks. Then they will come back to these soils with more evidence to argue or change their claims. (This happens in Investigation 2, Part 2.)

19. Assess progress: notebook entry
Have students hand in their notebooks open to the page on which they answered the focus question. Review students' notebooks after class and check to see that they know what makes up soil. You might also want to take a quick look at students' soil location guesses to help you prepare for the next set of lessons.

What to Look For

- Based on direct observation, students write that soil is composed of different sizes of rock (e.g., sand, gravel, pebbles) and humus (decaying plants and animals).

Part 1: Soil Composition

READING *in Science Resources*

20. Read "What Is Soil?"

Give students a few minutes to look at and discuss the cover of *FOSS Science Resources: Soils, Rocks, and Landforms*. Have them examine and discuss the table of contents. Have students locate the glossary and discuss how it is used to determine or clarify the meaning of words. Point out the index as a means for locating specific information.

Tell students that "What Is Soil?" is an expository article that focuses on the various components that make up soil. Have students preview the text by looking at and discussing the photographs and the diagram with a reading partner. Ask students what information they can gather from these illustrations. [Soil comes from the ground; it is composed of different sized rocks that can be categorized by their size.]

Refer students to the title and tell them to look for new information that answers the question. They should also be thinking about how their observations in class compare to what they read in the article. Read the article aloud or have students read independently or in guided-reading groups. For reading strategies to support English learners and below-grade-level readers, see the Science-Centered Language Development chapter.

Pause after the first page and have students take turns summarizing the text. One person can pick out the main idea and the other finds key details that support it. Turn to the next page and ask students to focus on the photograph of the worm. Ask,

➤ *What do you observe?*

➤ *Can you identify any structures?*

➤ *Why do you think the worm is found in humus-enriched soil?*

➤ *How do you think worms affect soil?*

➤ *Do the actions of worms benefit humans in some way?*

Tell students to read the rest of the article to find out how their thinking matches that of the author.

Review the words in bold and then ask students to describe soil in terms of a system. Write the word *system* on the word wall.

➤ *What are the component parts of soil as a **system**?* [Rocks, humus, live organisms, water, and air.]

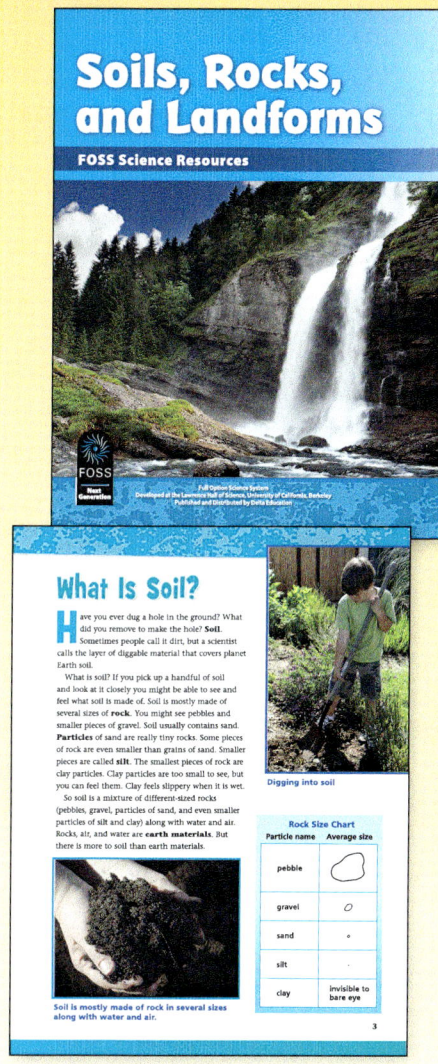

SCIENCE AND ENGINEERING PRACTICES

Obtaining, evaluating, and communicating information

CROSSCUTTING CONCEPTS

Systems and system models

INVESTIGATION 1 – Soils and Weathering

ELA CONNECTION

These suggested reading strategies address the Common Core State Standards for ELA.

RI 1: Refer to details and examples in a text when explaining what the text says explicitly and when drawing inferences from text.

RI 2: Determine the main idea of a text and explain how it is supported by key details; summarize the text.

RI 4: Determine the meaning of general academic domain-specific words or phrases.

RI 6: Compare and contrast a firsthand and secondhand account of the same topic.

RI 7: Interpret information presented visually, and explain how the information contributes to an understanding of the text.

RI 8: Explain how an author uses reasons and evidence to support particular points in a text.

W 5: Strengthen writing by revising.

W 8: Gather relevant information from experiences and print, and categorize the information.

SL 1: Engage in collaborative discussions.

SL 4: Report on a text in an organized manner, using appropriate facts and relevant, descriptive details to support main ideas or themes; speak clearly at an understandable pace.

L 4: Determine or clarify the meaning of unknown and multiple-meaning words.

➤ *How do the soil components work together as a system?* [Living organisms live and die in the soil; they mix up the soil allowing space for air and water to move; humus provides nutrients and helps retain water.]

To support comprehension, make (and/or have students make) a word web for the components of soil.

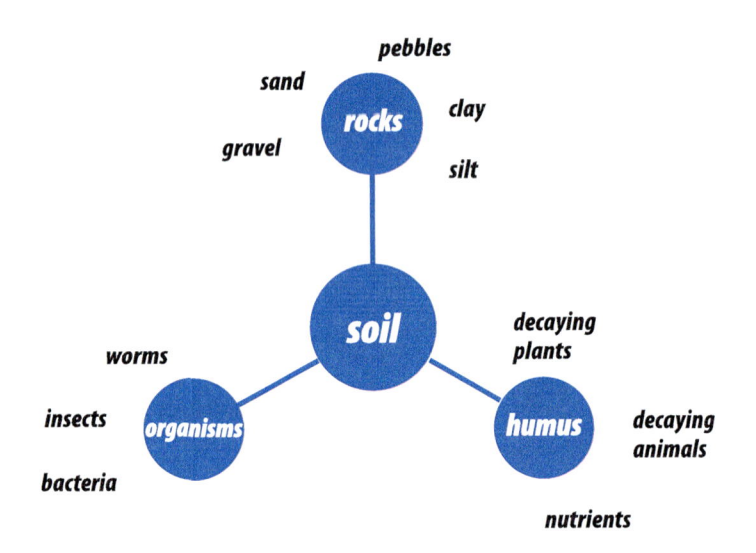

Word web: soil connects to rocks (sand, pebbles, gravel, clay, silt), organisms (worms, insects, bacteria), and humus (decaying plants, decaying animals, nutrients).

21. Discuss the reading

To check for understanding have students discuss these questions in their groups and then ask a Reporter from each group to share the group's responses.

➤ *What did you already know about soil? What was new?*

➤ *What is humus?* [Decayed dead plants and animals.]

➤ *What does humus do for soil?* [Provides nutrients and helps soil retain water.]

Refer students to the statement on page 4, "Worms are good for soil and help plants grow." Ask students to explain how the author uses reasons and evidence to support that statement. Encourage them to cite details and examples from the text.

Ask students to look closely at the soils on page 5 and to describe different kinds of soil. [Some soil has more clay, some more sand, some more pebbles and gravel, some more humus.]

Part 1: Soil Composition

Discuss the questions at the end of the article.

➤ *What differences do you see in the soils shown above?*

➤ *Where do you think these soils are found?*

Call on Reporters to share their group's answers and allow other group Reporters to respond by either agreeing or disagreeing, or offering alternative ideas. To support collaborative discussions, provide sentence frames such as

- I think _____ because _____.
- I agree with _____ because _____.
- I disagree with _____ because _____.
- I would like to add that _____.
- Why do you think _____?
- Do you have evidence that _____?

WRAP-UP/WARM-UP

22. Share notebook entries

Conclude Part 1 or start Part 2 by having students share notebook entries. Ask students to open their science notebooks to the most recent entry. Read the focus question together.

➤ *What is soil?*

Ask students to pair up with a partner to

- share their answers to the focus question;
- discuss the different kinds of materials that were found in the four different soils;
- share their ideas about which soil came from which location—desert, river delta, mountain, forest. Students should support their ideas with reasoning.

Allow time for students to draw a line of learning under their answers and to add any new information about soil they learned from the reading and discussion. See the Science Notebooks in Grades 3–5 chapter for more details.

SCIENCE AND ENGINEERING PRACTICES

Constructing explanations

Engaging in argument from evidence

TEACHING NOTE

Go to FOSSweb for Teacher Resources *and look for the* Science and Engineering Practices—Grade 4 *chapter for details on how to engage students with the practice of engaging in argument from evidence.*

CROSSCUTTING CONCEPTS

Patterns

INVESTIGATION 1 – Soils and Weathering

MATERIALS for
Part 2: Physical Weathering

For each group
- 1 Basin
- 1 Plastic jar with cap, 1/2 L
- 4 Safety goggles ★
- 1 Paper plate

For the class
- 16 Gray granite pieces, tumbled (See Step 3 of Getting Ready.)
- 4 Gray granite pieces, control (See Step 3 of Getting Ready.)
- 16 Conglomerate pieces, tumbled (See Step 3 of Getting Ready.)
- 4 Conglomerate pieces, control (See Step 3 of Getting Ready.)
- 2 Zip bags, 1 L
- 1 Glass bottle with tight-fitting cap (See Step 5 of Getting Ready.) ★
- • Newspaper ★
- 1 Sturdy plastic bag ★ (large enough to contain glass bottle) ★
- 1 Pitcher
- • Water ★
- • Permanent marking pen, black ★

For embedded assessment
- ❑ • Notebook sheet 3, *Response Sheet—Investigation 1*
- ❑ • *Embedded Assessment Notes*

★ Supplied by the teacher.   ❑ Use the duplication master to make copies.

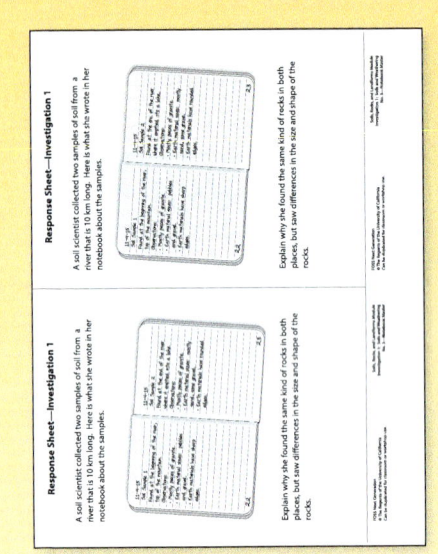

No. 3—Notebook Master

Part 2: Physical Weathering

GETTING READY for
Part 2: Physical Weathering

1. **Schedule the investigation**
   This part will take two active investigation sessions—one session to weather rocks and set up a freezing demonstration, and a second session to look at the results of freezing.

2. **Preview Part 2**
   Students begin to explore how large masses of rock break into smaller pieces. They tumble rocks and freeze water to see how these two types of physical weathering can break rocks. The focus question is **What causes big rocks to break down into smaller rocks?**

3. **Prepare granite and conglomerate rock pieces**
   If you are the first person to use this kit, you will need to separate the pieces of granite and conglomerate into two groups: rocks that students will tumble (four per group) and control rocks that will not be tumbled (one per group). This allows multiple classes to tumble the rocks, and still have unmarred rocks for comparison. Place the rocks to tumble at the materials station, but hold back the control rocks until Step 6 of Guiding the Investigation.

   At the end of the session, put the control rocks into two zip bags that you've labeled "Granite—Control" and "Conglomerate—Control." It is important to keep the control rocks separate from the rocks that are tumbled.

   When you clean up this activity, have students place all the tumbled rocks into the zip bags that the rocks came in.

4. **Think about noise**
   Shaking the rock pieces in the plastic jars creates more than the normal amount of noise in your classroom. Plan to conduct this part when that noise level will be acceptable.

5. **Plan for the freezing demonstration**
   Gather the materials for the freezing demonstration. Get a glass soft-drink or juice bottle with a tight-fitting cap. Narrow-necked bottles work best for this demonstration. The bottle will be filled entirely with water, wrapped in several sheets of newspaper, placed in a strong plastic bag, and left for the day or overnight in a freezer. Because water expands when it freezes, the bottle will break. The effect is memorable.

   To prepare, watch the short video, *Freezing Glass Bottle*, on FOSSweb.

> **NOTE**
> The water will take only a few hours to freeze, so it may be possible to set up the demonstration in the morning and look at the results in the afternoon on the same day.

> **NOTE**
> Put a note in the bag letting the next teacher know how many classes have tumbled the rocks.

> **NOTE**
> This is a teacher demonstration, and students will not get their hands on this material. If you are uncomfortable using glass, a plastic bottle will serve but will not be nearly as effective a demonstration. Or you could show the video.

Soils, Rocks, and Landforms Module—FOSS Next Generation

INVESTIGATION 1 – Soils and Weathering

6. **Plan assessment: response sheet**
   Use notebook sheet 3, *Response Sheet—Investigation 1* for a closer look at students' understanding of why the earth materials in different soils have different properties. Plan to have students spend time reflecting on their responses after you have reviewed them. For more information about next-step and self-assessment strategies, see the Assessment chapter.

Part 2: Physical Weathering

GUIDING *the Investigation*
Part 2: *Physical Weathering*

1. **Review soils**
   Ask students to think about the soils they observed in the last lesson. Review the first focus question and students' answers.

   ➤ *What is soil?*

   Focus on the rock pieces in soils. Tell students,

   Mountains are made of huge masses of rock. The Sierra Nevada, the Appalachian Mountains, and the Rocky Mountains are examples of these huge masses of rock. I'm wondering, if rocks start out as big as mountains, how do they get broken down into pieces small enough to be part of the soil?

2. **Focus question: What causes big rocks to break down into smaller rocks?**
   Project or write the focus question for this part on the board and say it as you write.

   ➤ *What causes big rocks to break down into smaller rocks?*

   Have students date a new page in their science notebooks and write the focus question near the top of the page.

3. **Discuss the focus question**
   Have students discuss the focus question in their groups for a few moments and then call for responses. Student ideas might include lightning, rocks tumbling down hills, floods, and human activity—all contribute to big rocks breaking into smaller rocks.

   Focus on comments that suggest rocks break apart when they bang into each other. For example, rocks tumble over each other in a river or during a flood, or spill down hills in landslides large and small.

4. **Discuss granite and conglomerate rocks**
   Hold up a piece of **granite** and a piece of **conglomerate**. Add these rock names to the word wall. Ask one student to briefly describe the properties of the granite and another student to describe the properties of the conglomerate to the rest of the class. Ask,

   ➤ *How could you break these pieces of granite and conglomerate rock into smaller pieces?*

FOCUS QUESTION
What causes big rocks to break down into smaller rocks?

TEACHING NOTE
The idea here is to help students begin to create a simple model that explains why the rock materials in soils are different from one location to the next.

SCIENCE AND ENGINEERING PRACTICES
Developing and using models

EL NOTE
If necessary for student understanding, show a picture that illustrates these actions.

Materials for Step 4
- *Granite piece, control*
- *Conglomerate piece, control*
- *Plastic jar and lid*

Soils, Rocks, and Landforms Module—FOSS Next Generation

INVESTIGATION 1 – Soils and Weathering

> **Rock Tumble**
> Grps 1–4: granite
> Grps 5–8: conglomerate
>
> 1. Put rocks in jar.
> 2. Screw on lid.
> 3. Shake for 2 minutes.
> 4. Pour rocks onto paper plate.
> 5. Observe and record.

Materials for Step 5
- Basins
- Plastic jars
- Pieces of granite or conglomerate
- Paper plates
- Safety goggles

SCIENCE AND ENGINEERING PRACTICES

Planning and carrying out investigations

CROSSCUTTING CONCEPTS

Cause and effect

Show students the plastic jars with lids. Ask for ideas about how to use the jars to see if the rocks can be broken. Write a plan on chart paper if you think students will need help remembering what to do.

a. Put pieces of granite or conglomerate in a jar. Screw on the lid.

b. Shake the jar enthusiastically for several minutes.

c. Pour the contents of the jar onto a paper plate and observe.

5. **Set up and start shaking**
Half the groups will tumble granite pieces and the other half will tumble conglomerate pieces. Have Getters go to the materials station and use a basin to pick up one jar with a lid, four pieces of one kind of rock—granite or conglomerate, as assigned—one paper plate, and four pairs of safety goggles. Plan on groups shaking for about 2 minutes total, with each student getting a short turn.

6. **Observe results**
Caution students to let the dust settle a bit before opening the jars, then pour the contents of the jar onto a paper plate, being careful not to breathe in the dust. Have students observe and record the results on a new page in their notebooks. Suggest that it might be helpful to include drawings of the rocks. Be sure they include all the different sizes of rock pieces they see, including the dust particles.

Give each group a fresh untumbled control rock to compare with those that have been tumbled together. Have students compare the tumbled rocks with the control piece of rock and draw pictures of what they see.

7. **Trade rocks**
Have groups that tumbled granite switch paper plates and rock materials with a group that tumbled conglomerate. Students should make and record the same observations for the second kind of rock, then compare the two kinds of rocks.

Ask students to describe what happened to the rocks that were tumbled against each other. Ask,

▶ *What is the effect of tumbling the rocks?* [Small sand-size pieces broke off the granite. The conglomerate broke into more and bigger pieces than the granite. The tumbled rocks are more rounded or smoothed than the control rocks.]

TEACHING NOTE

Note that nothing changed except the size and shape of the pieces of rock. They ended up with the same kind of rocks, just smaller pieces.

Part 2: Physical Weathering

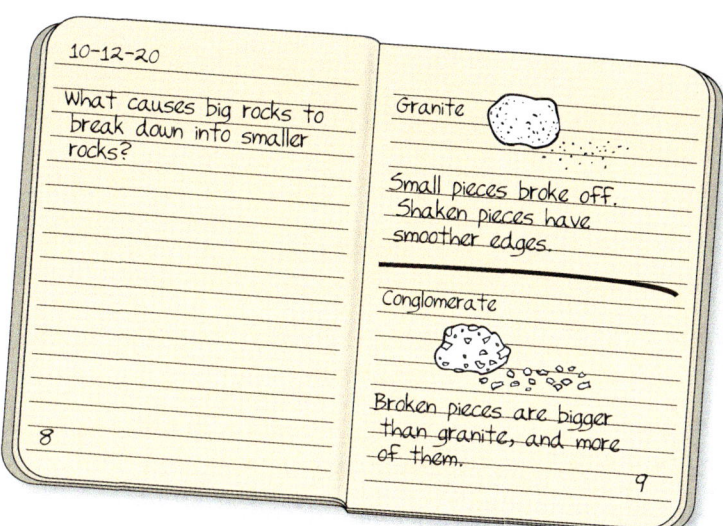

Notebook entry:
> 10-12-20
> What causes big rocks to break down into smaller rocks?
>
> Granite
> Small pieces broke off. Shaken pieces have smoother edges.
>
> Conglomerate
> Broken pieces are bigger than granite, and more of them.

8. **Introduce** *physical weathering*

   Tell students,

   All rocks, including mountains, exposed at Earth's surface break apart over time. Some rocks break apart faster than others. **Weathering** is the geological word used to describe the breakdown of rock into smaller pieces.

   When rocks break into smaller pieces without changing what the rock is made of, it is called **physical weathering**. **Abrasion** (rubbing together) is one kind of physical weathering. Abrasion occurs when rocks fall, when rocks tumble in landslides, and when rocks hit one another when they are pushed around by moving water, waves, and wind.

   Add the new vocabulary to the class word wall. As a class, make a concept definition map for physical weathering. Start with the definition of physical weathering in the What is it? box. Then review what the characteristics are in the What is it like? boxes [rocks breaking apart; rocks rubbing together].

   Next add some examples [rocks falling in landslides, rocks hitting one another in moving water, waves knocking rocks around; wind].

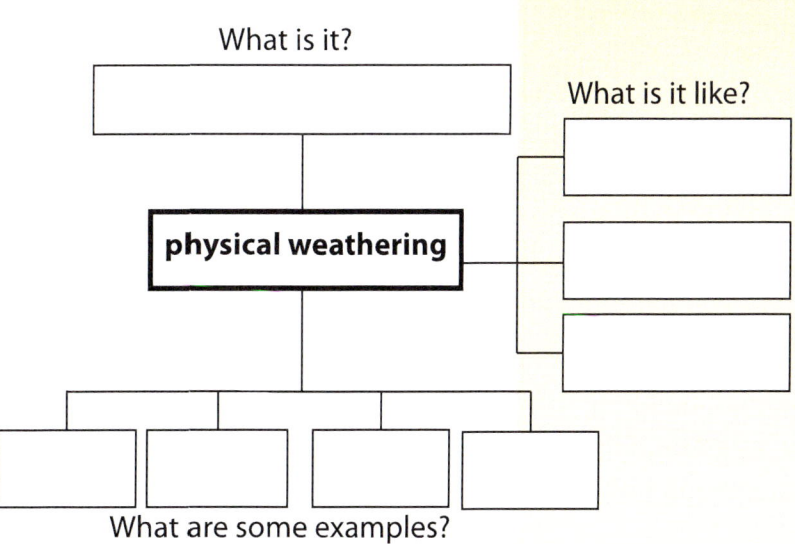

ELA CONNECTION

L 5: Demonstrate understanding of word relationships and nuances in word meanings.

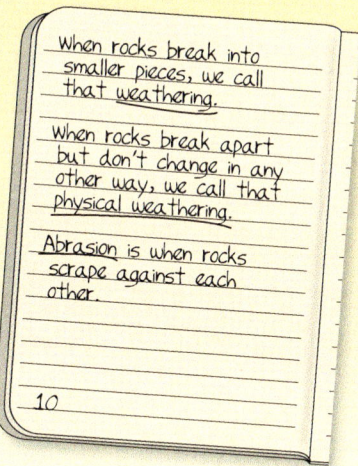

New Word: Say it, See it, Hear it, Write it

Notebook entry:
> When rocks break into smaller pieces, we call that weathering.
>
> When rocks break apart but don't change in any other way, we call that physical weathering.
>
> Abrasion is when rocks scrape against each other.

Concept map:
- What is it?
- physical weathering
- What is it like?
- What are some examples?

Soils, Rocks, and Landforms Module—FOSS Next Generation

INVESTIGATION 1 – Soils and Weathering

Materials for Step 10
- Glass bottle
- Sturdy plastic bag
- Newspaper
- Pitcher of water

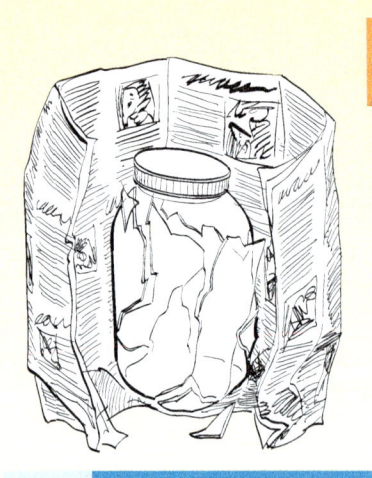

TEACHING NOTE

Students who completed the grade 3 **Water and Climate Module** know water expands when it freezes.

9. **Clean up**

   Have Getters return the jars, safety goggles, and rock materials to the materials station. Return rock pieces to the appropriate storage bags.

10. **Set up the freezing investigation**

    Hold up a glass bottle and tell students,

    Water sometimes flows into cracks in rocks. When the weather gets cold enough, below 0°C, water can **freeze**.

    Let's pretend this bottle is a rock. I'm going to fill it with water. We'll put it into the freezer today and take a look at it tomorrow.

    This bottle-water system will serve as a model of what might happen in the natural world. A **model** *is a representation of the features and actions of a natural system or process, but it isn't exactly the same as the real world—it focuses on the most important features. Scientists use models to help them observe and explain what happens in systems that move too slowly or too quickly, or are too big or too small to observe directly.*

    Fill the bottle completely (absolutely no air space) and screw the cap on tightly. Wrap the bottle in a few sheets of newspaper and then place it in a strong plastic bag (to contain the glass pieces when the bottle breaks). Place the wrapped and bagged bottle in a freezer for a few hours or overnight.

    BREAKPOINT

11. **Observe the frozen bottle**

    Retrieve the bottle from the freezer. Remove the bottle from the plastic bag and carefully unwrap it where students can observe the results, but from a distance. Students should not be near or touch the broken glass. Ask students to offer explanations for the fractured glass using a cause-and-effect statement such as "I think the (cause) made the (effect)."

    After students have offered their ideas, tell them,

    When water freezes, it **expands**. *That means the volume of the water increases—it gets bigger. When it expands, the water pushes on things around it with a lot of force. The force was enough to break this bottle.*

    In the mountains and in other places where it is cold in winter, rocks are exposed to cycles of freezing and thawing.

    ▶ *When the water in a crack in a rock freezes, what might happen to the rock?* [Freezing water can break rocks into smaller pieces.]

116 Full Option Science System

Part 2: Physical Weathering

This cycle of freezing and thawing—water flowing into cracks when it's liquid and expanding when it freezes—is another kind of physical weathering that breaks rock into smaller pieces.

Carefully transfer the ice and broken glass to a basin. After the ice melts, pour off the water and recycle the glass.

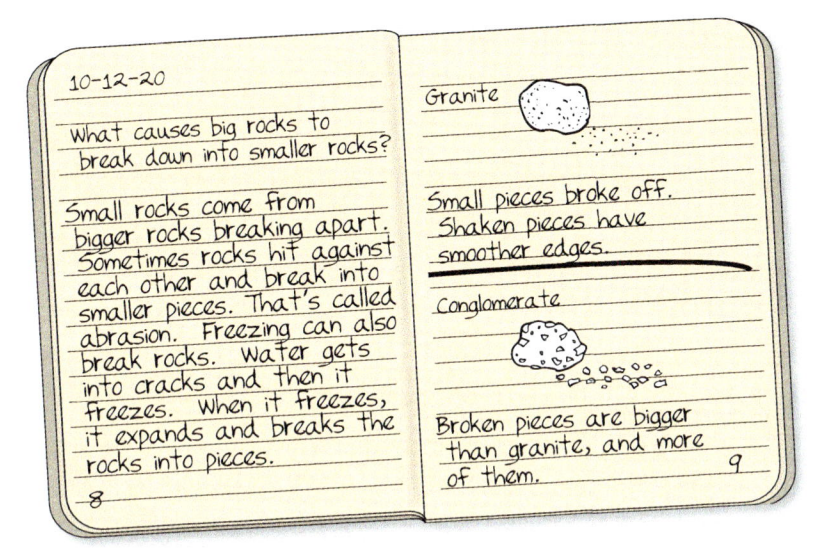

12. Review vocabulary
Review key vocabulary introduced in this part. If these words are not on the word wall, add them now. Add *freezing* and *thawing* to the concept definition map as another example of physical weathering.

13. Answer the focus question
Ask students to work independently to describe one way rocks break down into smaller pieces.

▶ *What causes big rocks to break down into smaller rocks?*

Ask students to think back to what they did with the granite and the conglomerate rocks in the shaking jar and to consider the action of the frozen water in the glass bottle. How do the results of those investigations help us to understand what happens in the natural world with really big rocks and mountains?

For students who need scaffolding, provide a sentence frame such as Big rocks break down into smaller pieces by _____ . We found out that _____ .

---

abrasion
conglomerate
expand
freeze
granite
model
physical weathering
weathering

SCIENCE AND ENGINEERING PRACTICES
Constructing explanations

CROSSCUTTING CONCEPTS
Cause and effect

Soils, Rocks, and Landforms Module—FOSS Next Generation

INVESTIGATION 1 – Soils and Weathering

SCIENCE AND ENGINEERING PRACTICES

Constructing explanations

14. Assess progress: response sheet

Distribute notebook sheet 3, *Response Sheet—Investigation 1*. Have students glue the response sheet to the next blank left-hand page and write their responses on the right-hand page. This allows students to use as many lines as they need to respond.

Ask students to hand in their notebooks open to this page. Review their responses after class and prepare for the next lesson using this information.

What to Look For

- *Both locations have the same kind of rock because the river moves the rock from mountaintop to lake.*

- *Rocks are smaller and smoother near the lake because more abrasion has occurred as the rocks were moved down the river.*

You can use these next-step strategies if you determine one is needed.

Use the key points and revision with color strategies (see the Next-Step Strategies section in the Assessment chapter). Begin by discussing the information on the response sheet. After it is clear that the students understand what is intended, call on individuals or groups to suggest key points that should be included in a complete answer. Write the key points on the board or chart paper as phrases or individual words that will scaffold students' revision. Guide your students to discuss and chart these important key points:

- Both locations have the same kind of rock because the river is moving the same source rock, the difference is location on the river.
- Rocks are smaller farther down the river because more abrasion occurs the farther the rock travels.
- Rocks are more rounded farther down the river because the more abrasion that occurs, the more rough edges are broken off.

After the class discussion, use colored pens or pencils and the three C's to revise answers. As students read over their responses, they confirm correct information by underlining with a green pen; they complete their responses by adding information that is missing, using a blue pen; and they correct wrong information, using a red pen.

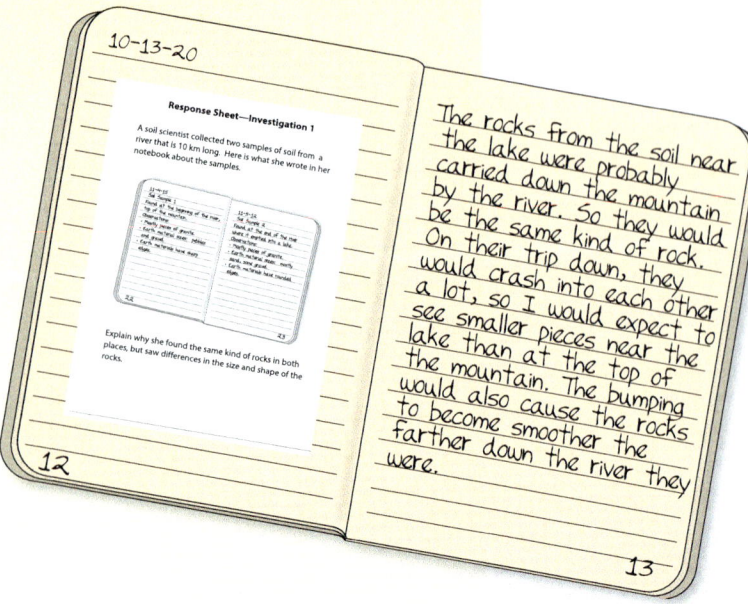

118

Full Option Science System

Part 2: Physical Weathering

WRAP-UP/WARM-UP

15. Share notebook entries

Conclude Part 2 or start Part 3 by having students share notebook entries. Ask students to open their science notebooks to the most recent entry. Read the focus question together.

➤ *What causes big rocks to break down into smaller rocks?*

Ask students to pair up with a partner to

- compare their answers to the focus question;
- share their answers to the response sheet.

Discuss cause-and-effect relationships. Point out that science is about trying to find out the reasons why things happen. Ask students to think about what causes rocks to break into smaller pieces [abrasion, freezing]. With a partner, have them discuss what effects these produce. Encourage students to provide evidence from the investigation and from their own experiences. Here are some sentence frames students can use to build on each other's ideas.

- So what you're saying is _____.
- What caused _____ to happen?
- What was the effect of _____?
- What is your evidence that _____?
- How else could you find evidence for _____?

Another instructional strategy to engage students in argumentation is to provide a statement such as "Freezing causes most of the physical weathering in (location)." Students agree or disagree with that statement and provide some reasoned thoughts about how this might work in the real world. Students will not have enough evidence but they can speculate.

EL NOTE

See the Science-Centered Language Development in Grades 3–5 chapter for strategies on sharing notebook entries.

CROSSCUTTING CONCEPTS

Cause and effect

▶ **NOTE**
Go to FOSSweb for *Teacher Resources* and look for the Crosscutting Concepts—Grade 4 chapter for details on how to engage students with the concept of cause and effect.

SCIENCE AND ENGINEERING PRACTICES

Developing and using models

Engaging in argument from evidence

Soils, Rocks, and Landforms Module—FOSS Next Generation

INVESTIGATION 1 – Soils and Weathering

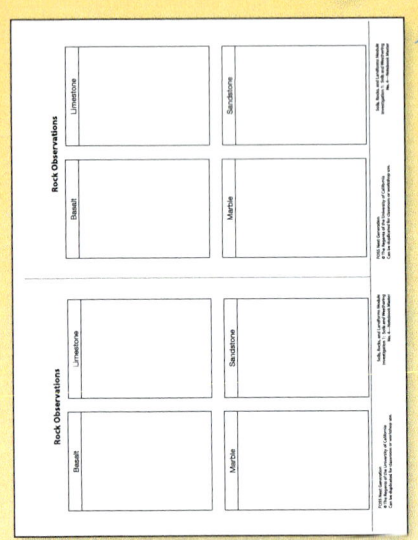

No. 4—Notebook Master

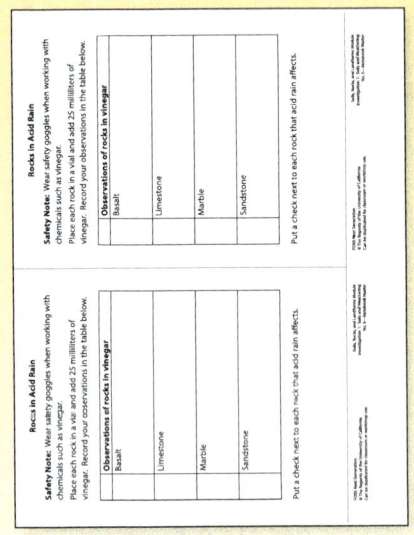

No. 5—Notebook Master

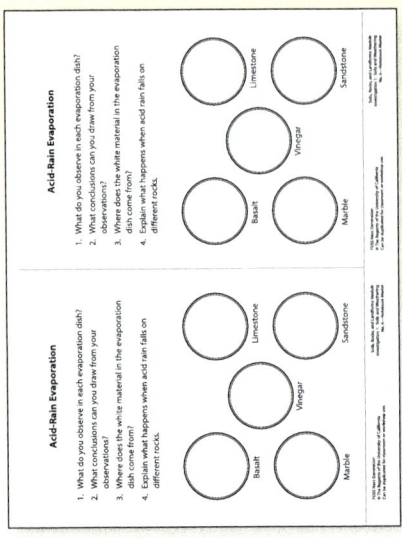

No. 6—Notebook Master

MATERIALS for
Part 3: Chemical Weathering

For each group

- 1 Syringe
- 4 Vials
- 1 Plastic cup
- Rock samples (1 limestone, 1 marble, 1 basalt, 1 sandstone)
- 6 Self-stick notes
- 5 Evaporation dishes
- 1 FOSS tray
- 4 Safety goggles ★
- ❑ 4 Notebook sheet 4, *Rock Observations* (optional)
- ❑ 4 Notebook sheet 5, *Rocks in Acid Rain*
- ❑ 4 Notebook sheet 6, *Acid-Rain Evaporation*
- 4 *FOSS Science Resources: Soils, Rocks, and Landforms*
  - "Weathering"

For the class

- 1 Piece of calcite, large, in a plastic cup (optional)
- 4 Containers, 1/2 L
- 4 Self-stick notes
- 1–2 Basins
- Water ★
- White vinegar, ~1.2 L ★
- Zip bags, 1 L
- Transparent tape ★
- Permanent marking pen, black ★
- 1 Computer with Internet access ★
- ❑ 1 Teacher master 2, *Reference Rocks for Acid-Rain Test*
- ❑ 1 Teacher master 3, *Video Discussion Guide: Weathering* (optional)

For embedded assessment

- ❑ *Performance Assessment Checklist*

★ Supplied by the teacher. ❑ Use the duplication master to make copies.

120 Full Option Science System

Part 3: Chemical Weathering

GETTING READY for
Part 3: Chemical Weathering

1. **Schedule the investigation**
   This part will take three active investigation sessions and one reading session.

2. **Preview Part 3**
   Students plan and conduct an investigation to test rocks for interaction with "acid rain." They see that some rocks (limestone and marble) are very susceptible to acid rain, one form of chemical weathering, but other rocks (basalt and sandstone) are unaffected. The focus question is **How are rocks affected by acid rain?**

3. **Set up a reference sheet of rocks**
   Print or copy teacher master 2, *Reference Rocks for Acid-Rain Test*, and place the sheet in a prominent location (such as the materials station) where students can refer to it when necessary. Place one sample of each rock (from the bags of ten samples) in the labeled circle. After the first use, store these samples in a separate plastic bag to use as "control" rocks. Use a permanent marking pen to label this bag "Acid-Rain Rocks—Control." After soaking in the vinegar, the limestone and marble will show signs of weathering from the acid rain. You can use the control rocks to highlight this change.

4. **Set up the rocks for distribution**
   Use self-stick notes to label four 1/2 L containers with the names of the rocks that will be investigated: basalt, limestone, marble, and sandstone. Place eight samples of each kind of rock in the appropriately labeled containers.

5. **Plan for vinegar distribution**
   Students use the syringes to measure 25 mL of vinegar in Steps 6 and 13 of Guiding the Investigation. To avoid spills, pour a little more than 100 mL of vinegar into cups at the groups' tables; don't have Getters carry full cups from the materials station.

6. **Plan for storage**
   You will need a place to store the rock samples while they sit overnight in vials of vinegar and again for a few days when evaporating the vinegar to look for evidence of chemical weathering. The FOSS trays are designed to make efficient use of counter space.

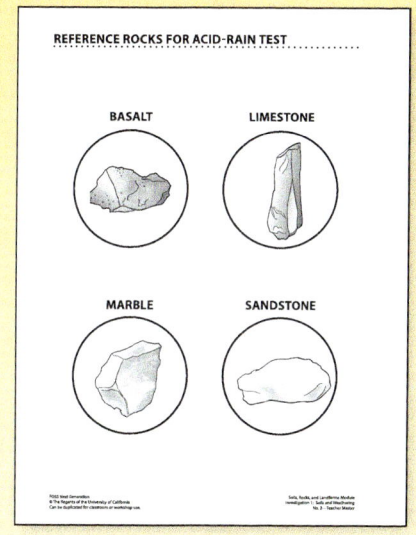

No. 2—Teacher Master

Soils, Rocks, and Landforms Module—FOSS Next Generation

INVESTIGATION 1 – Soils and Weathering

7. Plan to demonstrate calcite in vinegar
In Step 16, you will demonstrate what happens when calcite reacts with vinegar. Have a large piece of calcite and a cup of vinegar ready for that demonstration.

8. Plan for cleanup
Students need to rinse out their vials, evaporation dishes, and plastic cups at various points during this part. Have water in one or two basins strategically placed for students to do this.

9. Plan to read *Science Resources*: "Weathering"
Plan to read the article "Weathering" during a reading period after completing the active investigation for this part.

10. Preview the video
Preview the *Weathering and Erosion* video—chapter 3 on physical weathering (4:11 duration) and chapter 4 on chemical weathering (5:43 duration). The link to the video is found on FOSSweb in the Resources by Investigation section. Have students view the video after the reading. Plan to use teacher master 3, *Video Discussion Guide: Weathering*, to lead a class discussion after watching the video with the class. You can provide each group with a copy of the sheet.

11. Plan assessment: performance assessment
Students work independently (or guided by a student sheet) in a problem-solving environment to determine which rocks are affected by acid rain. This provides an opportunity for you to observe students' science and engineering practices. Carry the *Performance Assessment Checklist* with you as you visit groups while they are working. For more information about what to look for during observation, see the Assessment chapter.

Part 3: Chemical Weathering

GUIDING the Investigation
Part 3: Chemical Weathering

1. **Introduce acid rain**
   Review physical weathering and introduce a new kind of weathering. Say,

   We've seen that rocks break into smaller rocks when they bang and tumble against each other or when ice expands and breaks them apart. That's what we call physical weathering. But there are other ways that rocks can be weathered.

   Sometimes substances in the air dissolve in rain, creating a weak acid. Substances get into the air from natural sources, like volcanoes, and from human-produced sources, such as car exhaust and factories. Let's find out how **acid rain** *affects earth materials.*

2. **Focus question: How are rocks affected by acid rain?**
   Project or write the focus question on the board as you say it aloud.

   ➤ *How are rocks affected by acid rain?*

   Have students write the focus question in their notebooks.

3. **Introduce four rocks**
   Tell students that they are going to investigate the effects of acid rain on four kinds of rock. Write the names of the rocks on the board or the word wall: **basalt**, **limestone**, **marble**, and **sandstone**. Have the Getters get a FOSS tray and four self-stick notes. Have each group member select one of the rock names, write it on *both sides* of a self-stick note, and place the note in one of the tray compartments.

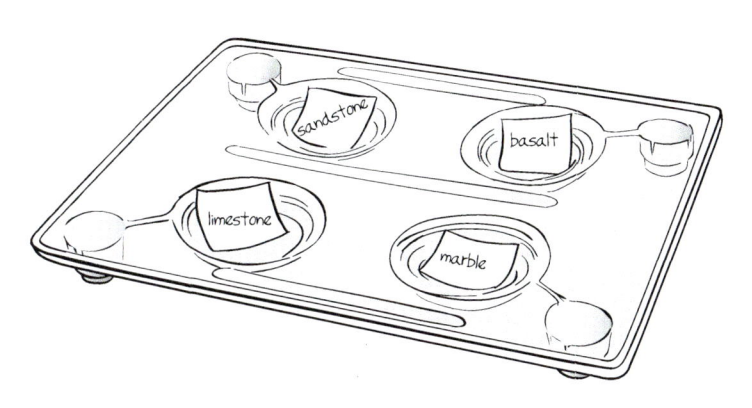

FOCUS QUESTION
How are rocks affected by acid rain?

Materials for Steps 3–4
- *Self-stick notes*
- *Limestone samples*
- *Sandstone samples*
- *Marble samples*
- *Basalt samples*
- *FOSS trays*
- *Reference set of rocks*
- **Rock Observations** *sheets (optional)*

INVESTIGATION 1 – Soils and Weathering

> **NOTE**
> Let students know that a reference set of rocks is available if needed, or they can look at the word-wall picture cards.

SCIENCE AND ENGINEERING PRACTICES

Planning and carrying out investigations

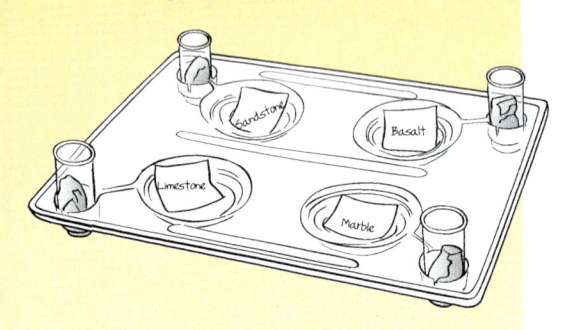

Materials for Step 6
- Plastic cups
- Syringes
- Vials
- Safety goggles
- Vinegar
- **Rocks in Acid Rain** sheets

SCIENCE AND ENGINEERING PRACTICES

Developing and using models

Planning and carrying out investigations

Analyzing and interpreting data

DISCIPLINARY CORE IDEAS

ESS2.A: Earth materials and systems

CROSSCUTTING CONCEPTS

Cause and effect

4. **Distribute the earth materials and observe**
   Have students take turns going to the materials station to pick up the rock samples named on their self-stick notes. When they return, they should put the rock in the compartment of the FOSS tray that is labeled with that rock's name. Have students record their observations of the four rocks on notebook sheet 4, *Rock Observations* (or simply divide a notebook page into quarters and write the name of each rock in one quarter). Allow about 10 minutes for students to complete their observations.

5. **Design the acid-rain investigation**
   Tell students that you have some vinegar they can use to model acid rain. As a class, design an investigation to find out how acid rain affects the four rocks they have been observing.

   a. Use vinegar to simulate acid rain.
   b. Put each rock sample in a vial (be sure labels stay associated with the appropriate rock and vial).
   c. Place the four vials into the corner sockets of the tray.
   d. Carefully put 25 mL of vinegar into each vial, using a syringe to measure the vinegar.
   e. Observe what happens and record those observations on notebook sheet 5, Rocks in Acid Rain.

6. **Conduct the investigation**
   Have Getters pick up materials from the materials station. Go around and give each group enough vinegar to set up the vials. Let the groups start their investigations. Remind students to note observations in their notebooks before placing the rocks in vinegar. Allow about 10 minutes for this part of the investigation.

7. **Assess progress: performance assessment**
   Circulate from group to group throughout the active investigation in this part. Listen to group discussions and observe how students work together. Make notes on the *Performance Assessment Checklist* while students work, and use the same sheet throughout this part.

   What to Look For

   - Students write detailed observations and measure accurately using appropriate tools and techniques. (Planning and carrying out investigations.)
   - Students draw reasonable conclusions based on the data they collect. (Analyzing and interpreting data.)
   - Students can describe a model for how acid rain affects rock. (Developing and using models; ESS2.A: Earth materials and systems; cause and effect.)

Part 3: Chemical Weathering

8. **Report results**

    Ask students to report what they observed. When they mention that they saw bubbles, ask

    ➤ *What do bubbles indicate?*

    Students who have completed the **Motion and Matter Module** will know that bubbles can indicate a chemical reaction. If students can't answer the question, explain,

    *Bubbles can indicate a **chemical reaction**. A chemical reaction is a process in which two or more materials mix in a way that forms new materials. In this case, the limestone reacted with the acid rain and created a gas. Continuous streaming of lots of bubbles is evidence that a gas is being produced as a result of a chemical reaction.*

    Everyone will probably agree that the limestone reacted with the acid rain and basalt did not, but there may be some argument about whether "the right kind of bubbles" were observed for sandstone and marble. A few bubbles may suggest an air pocket in the rock. Suggest to students that another test might provide additional evidence to indicate if a chemical reaction occurred.

9. **Store vials overnight**

    Tell students that they will let the vials rest overnight and then check them for more clues. Hand out one additional self-stick note so that the groups can identify their trays before storing them.

10. **Clean up**

    Have Getters return all of the equipment to the materials station. Return unused vinegar to the bottle. Rinse out and dry cups and syringes.

    B R E A K P O I N T

11. **Retrieve the FOSS trays**

    Have Getters retrieve their group's tray. Give the groups a few minutes to look at the vials and report what they observe. [The rocks are no longer fizzing.]

12. **Discuss evaporating the vinegar**

    Ask,

    ➤ *How could we find out if a chemical reaction occurred, creating a new material that has dissolved in the clear liquid?*

    Suggest to students, if they need help, that evaporating the liquid from each vial would provide evidence of dissolved material. If anything is left behind after evaporation, that is the material that dissolved.

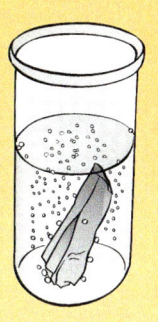

Bubbles that indicate a reaction

Air bubbles

Say it → See it → Hear it → Write it → **New Word**

Materials for Step 9
- Self-stick note

TEACHING NOTE

If students have completed the **Motion and Matter Module**, they know that evaporation will separate the liquid from any dissolved solid material.

Soils, Rocks, and Landforms Module—FOSS Next Generation

125

INVESTIGATION 1 – Soils and Weathering

Discuss the process of setting up four evaporation dishes. Ask students,

> *If something is left in a dish after the liquid has evaporated, how will you know it is evidence of dissolved material in the liquid and not just something in the vinegar itself?*

Have each group plan to set up a control dish that contains only pure vinegar.

13. Set up evaporation dishes

Have each Getter pick up five flat evaporation dishes and one additional self-stick note from the materials station. Students should transfer one self-stick-note label from a tray compartment to the *outside* bottom of a flat evaporation dish and then place that evaporation dish back into the tray compartment next to its corresponding vial.

Have students pour liquid from the vial into the matching evaporation dish, but only enough to cover the bottom. One person in each group can take the four empty vials back to the materials station and rinse them out before putting them back on the trays as pillars for stacking.

Someone in the group should write "vinegar" on both sides of a self-stick note and stick it to the bottom of the fifth evaporation dish. Have each Getter fill a syringe with 25 mL of vinegar, carry it back to their group, carefully squirt the vinegar into the fifth evaporation dish, and place it on the FOSS tray.

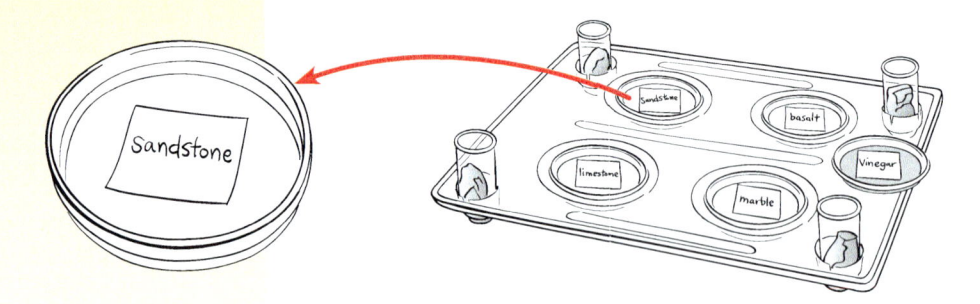

Materials for Step 13
- *Evaporation dishes*
- *FOSS tray with vials*
- *Self-stick notes*
- *Transparent tape*
- *Vinegar*
- *Syringes*
- *Safety goggles*

Have Getters bring the trays of evaporation dishes to the designated evaporation location. Create FOSS tray towers in a convenient place and wait for evaporation to occur over the next few days.

14. Clean up

Have Getters bring the rocks back to the materials station. Assign a few students to rinse the rocks thoroughly.

126   Full Option Science System

Part 3: Chemical Weathering

BREAKPOINT

15. Have a sense-making discussion

Have one Getter pick up their group's FOSS tray and evaporation dishes. Have the other Getter pick up four copies of notebook sheet 6, *Acid-Rain Evaporation*.

Give students 10 minutes to discuss, draw, and write responses to the questions on the notebook sheet. Discuss the results.

➤ *What did you observe in each evaporation dish?* [There are white, needle-shaped crystals in the limestone dish, similar white crystals in the marble dish, but nothing in the basalt and sandstone dishes.]

➤ *Compare your evaporation dishes to the control evaporation dish.* [Plain vinegar didn't leave any white crystals after it evaporated.]

➤ *Where does the white material in the evaporation dishes come from?* [Acid reacts with minerals in the limestone and marble. When the vinegar evaporates, a new material (formed during the chemical reaction) is left behind as white crystals.]

16. Introduce *chemical weathering*

Tell students,

When rocks are exposed at Earth's surface, they come in contact with water and air and living organisms. Chemicals from organisms and in the water and air, like acids, can change rocks. When the minerals in a rock change to new minerals or other materials, it is called **chemical weathering**. *Chemical weathering can change rocks and cause them to break apart.*

Ask,

➤ *Which rocks were weathered by acid rain? What is your evidence?* [The limestone weathered; evidence was bubbling and white crystals. Marble also weathered; evidence was white crystals (even though it didn't produce lots of bubbles). Neither basalt nor sandstone was weathered; evidence was no bubbles formed and no crystals appeared after evaporation.]

Materials for Step 15
- FOSS trays with evaporation dishes
- *Acid-Rain Evaporation* sheets

CROSSCUTTING CONCEPTS
Patterns
Cause and effect

SCIENCE AND ENGINEERING PRACTICES
Analyzing and interpreting data
Constructing explanations

Soils, Rocks, and Landforms Module—FOSS Next Generation

INVESTIGATION 1 – Soils and Weathering

> **TEACHING NOTE**
>
> As a demonstration, put a large piece of calcite in a cup of vinegar and have students observe how it reacts.

Materials for Step 16 (optional)
- Piece of calcite
- Cup of vinegar
- Safety goggles

Tell students,

Rocks are often a mixture of ingredients called minerals. One of the ingredients in both limestone and marble is a mineral called **calcite**. *Calcite reacts with acid, causing some of the rock material to change into a new material. Any rock that includes calcite as a mineral ingredient will be affected by acid rain.*

Add a compare/contrast bubble to the concept definition map. Write *chemical weathering* in the bubble. Discuss how chemical weathering is different from physical weathering [minerals in rocks change to new minerals].

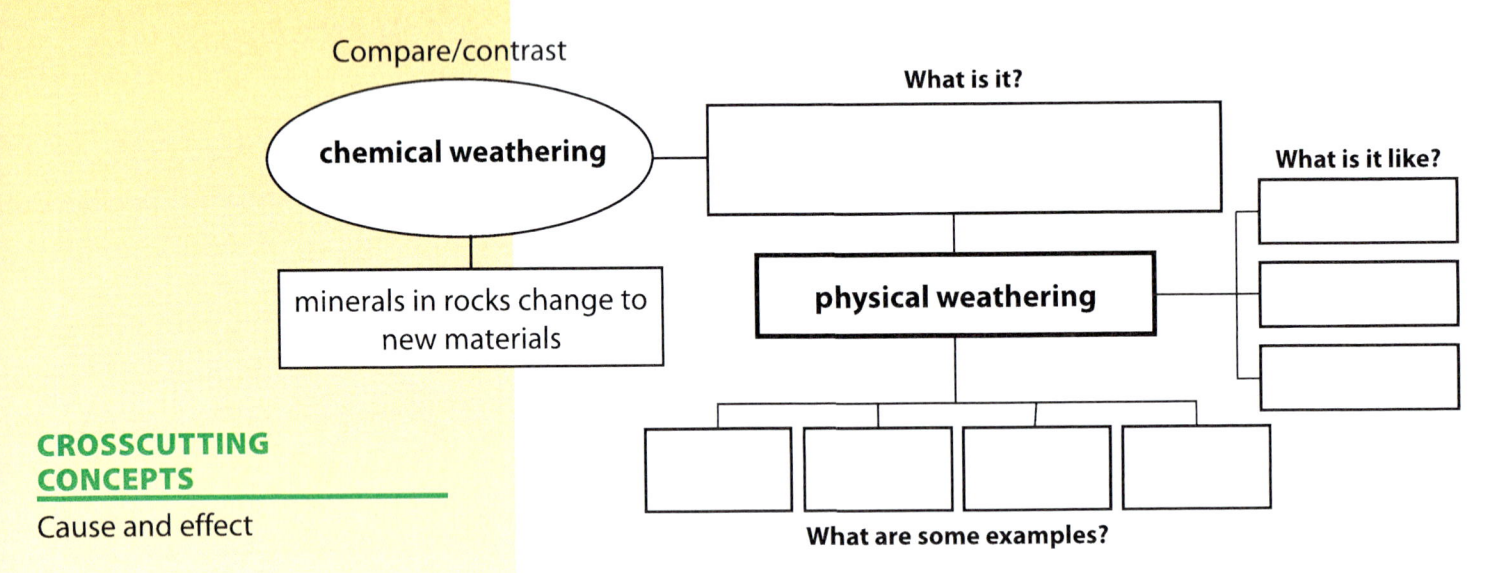

CROSSCUTTING CONCEPTS

Cause and effect

Compare/contrast: **chemical weathering** — minerals in rocks change to new materials

What is it? / **physical weathering** / What is it like? / What are some examples?

17. Clean up

Have Getters return all the materials to the materials station. They should rinse and dry the evaporation dishes.

Part 3: Chemical Weathering

READING *in Science Resources*

18. Read "Weathering"
Read "Weathering" during a reading period after completing the active investigation. The article describes physical and chemical weathering and the landforms that are affected by these processes.

19. Use a photo preview and predict strategy
Before reading, give students a few minutes to look at the photographs. Then tell them to choose a few photos to discuss with their reading partner. Have them take turns explaining what they think is happening in the photos. If needed, provide sentence frames such as

- I think _____ because _____.
- I agree/disagree with you because _____.

Read the article aloud or have students read it independently to look for evidence that supports the ideas they discussed.

Pause after each page to match the information in the text with the photographs. Have students discuss with their partner how the photos help them understand the different ways rocks can be broken down into smaller pieces.

For reading strategies to support English learners and below-grade-level readers, see the Science-Centered Language Development chapter.

20. Discuss the reading
Review any unknown words or phrases. Ask students what they notice about the structure of the text [cause/effect] and how thinking about cause and effect helps them understand weathering. Review the language structures for cause and effect. Have students read the article a second time, looking for cause-and-effect statements.

Ask students to discuss the difference between physical and chemical weathering with a partner based on the information in the text. You might have students use this strategy in pairs to help them listen closely and express their ideas clearly.

First round:

1. *Partner A:* Describe physical weathering to your partner. (Partner B is silent.)
2. *Partner B:* Repeat what your partner said.
3. *Partner A:* Confirm or correct what your partner repeated.

SCIENCE AND ENGINEERING PRACTICES

Obtaining, evaluating, and communicating information

ELA CONNECTION

These suggested reading strategies address the Common Core State Standards for ELA.

RI 3: Explain procedures, ideas, or concepts in a scientific text.

RI 4: Determine the meaning of general academic domain-specific words or phrases.

RI 5: Describe the overall structure of information in a text.

RI 7: Interpret information presented visually, and explain how the information contributes to an understanding of the text.

Soils, Rocks, and Landforms Module—FOSS Next Generation

INVESTIGATION 1 – Soils and Weathering

Second round:

1. *Partner B:* Describe chemical weathering to your partner. (Partner A is silent.)

2. *Partner A:* Repeat what your partner said.

3. *Partner B:* Confirm or correct what your partner repeated.

Students should respond that physical weathering breaks rock into smaller pieces without changing the nature of the rock itself. Chemical weathering breaks down rock by changing minerals into new minerals.

Continue the discussion using these questions to check for understanding. Encourage students to cite details and examples from the text to support their answers.

➤ *What is the effect of acid rain on rocks?* [Acid reacts with the ingredient calcite in limestone and marble to cause weathering. This weakens the rock.]

➤ *How do living organisms contribute to the weathering of rocks?* [Plants send roots (living things) into cracks in rocks. As roots grow, they break rocks apart to cause physical weathering; humans contribute to acid rain.]

➤ *How does ice cause weathering?* [Water expands when it freezes. Water in cracks in rocks acts like a wedge to break rocks apart.]

Conclude by having students connect the idea in the article to what they have investigated in class. Ask,

➤ *What observations have you made that support the idea of physical and chemical weathering in the natural world?*

➤ *What is different about the investigations in class and how things happen in the natural world?*

21. Review vocabulary

Review key vocabulary introduced in this part. If those words are not on the word wall, add them now.

This is a good opportunity to conduct a word sort. Make a set of the vocabulary word cards for each group (or have students make them). Give a set to each group and tell them to work together to group the words into categories. When groups are finished, have a Reporter from each group share their category titles and the common elements the words have in each category.

ELA CONNECTION

These suggested reading strategies address the Common Core State Standards for ELA.

RI 1: Refer to details and examples in a text when explaining what the text says explicitly and when drawing inferences from text.

RI 3: Explain procedures, ideas, or concepts in a scientific text.

RI 6: Compare and contrast a firsthand and secondhand account of the same topic.

SL 1: Engage in collaborative discussions.

L 5: Demonstrate understanding of word relationships.

---

acid rain
basalt
calcite
chemical reaction
chemical weathering
limestone
marble
sandstone

Part 3: Chemical Weathering

22. Answer the focus question

Point to the focus question on the board and review it with students.

➤ *How are rocks affected by acid rain?*

For students who need scaffolding, provide a sentence frame such as "Acid rain affects rocks by ____ . In our investigation we found out that ____ ."

23. View video on weathering

Show students the chapters on physical and chemical weathering from the *Weathering and Erosion* video. The video chapters can be accessed on FOSSweb, Resources by Investigation for this module.

Video topics, Chapter 3: Physical and Mechanical Weathering (4:11)

- There are two types of weathering: physical and chemical.
- Physical weathering, sometimes called mechanical weathering, is the cracking, breaking up, and grinding down of rocks and other objects.
- Temperature changes are some of the most powerful forces of physical weathering.
- Water seeps into cracks in rock, freezes, and expands, causing physical weathering.
- Animals can cause physical weathering by digging holes.
- Plant roots cause physical weathering.
- The force of gravity can cause physical weathering when rocks fall and break apart.
- Sand causes abrasion; physical weathering.
- Moving water can cause abrasion; physical weathering.
- Review of the different types of physical weathering: temperature, ice, organic activity, gravity and abrasion.

Begin by distributing copies of teacher master 3, *Video Discussion Guide: Weathering*. Have students review the questions they aren't sure about. Then have them divide up the questions so that each student has two questions to find answers for in the video.

After watching the video, give students time to share their answers with the rest of their group and to review the questions they weren't sure about before watching the video.

➤ *How do temperature changes and ice weather rocks?* [Heat during the day makes rocks expand, and cold makes them contract.

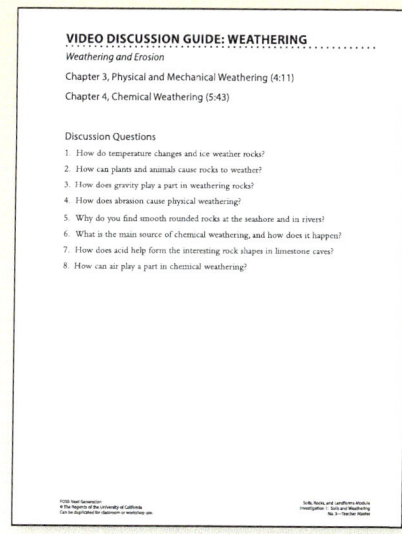

No. 3—Teacher Master

Soils, Rocks, and Landforms Module—FOSS Next Generation

131

INVESTIGATION 1 – Soils and Weathering

Water enters open cracks, and if it freezes, the ice expands and causes the rock to crack further.]

➤ *How can plants and animals cause rocks to weather?* [Tree roots can widen cracks, breaking off pieces of rock. Animals dig around rocks and their digging allows more water and other weathering agents to work on a greater surface area of the rock.]

➤ *How does gravity play a part in weathering rocks?* [When rocks break loose, they tumble downhill due to gravity.]

➤ *How does abrasion cause physical weathering?* [Wind and water carry solid particles of earth materials that break down rocks when they come into contact with each other.]

➤ *Why do you find smooth rounded rocks at the seashore and in rivers?* [The constant abrasion of rocks tumbling against each other and wave action wear pieces away, creating smooth rounded rocks.]

Video topics, Chapter 4: Chemical Weathering (5:23)

- Water is the main cause of chemical weathering on Earth.
- Over many years, exposure to the weather can damage the marble rock used in sculptures.
- Rain combines with certain gases in the air to form acid rain, which weathers rocks.
- When carbon dioxide combines with water, it forms carbonic acid, which dissolves some minerals; carbonation.
- Some plants create weak acids that cause chemical weathering.
- Oxygen can combine with other minerals in rocks and cause chemical weathering; oxidation.
- Review of the causes of chemical weathering.
- The combination of physical and chemical weathering can have dramatic effects on rocks and human-made objects.

Ask these questions about chemical weathering.

➤ *What is the main source of chemical weathering, and how does it happen?* [Water is the main source because it can dissolve some of the minerals that hold rocks together.]

➤ *How does acid help form the interesting rock shapes in limestone caves?* [The water that soaks into the caves has acid in it, which dissolves the limestone over many years.]

➤ *How can air play a part in chemical weathering?* [Oxygen in the air reacts with earth materials like iron and the materials that make bronze, causing them to break down.]

Part 3: Chemical Weathering

WRAP-UP/WARM-UP

24. Share notebook entries

Conclude Part 3 or start Part 4 by having students share notebook entries. Ask students to open their science notebooks to the most recent entry. Read the focus question together.

➤ *How are rocks affected by acid rain?*

Ask students to pair up with a partner to

- discuss how rocks are affected by acid rain;
- discuss how acid rain might affect buildings made of limestone and marble.

Have students pair up with someone from a different group. To reinforce the crosscutting concept, cause and effect, tell students to quiz each other about the causes and effects of weathering. For example, Partner A asks Partner B: What causes physical weathering? Partner B responds: One cause of physical weathering is freezing, then thawing. Partner A responds: Water in the cracks of rocks expands when it freezes making it crack further. Partner B then starts off with another cause question and the process repeats. If needed, make a list on the board of causes and effects for students to choose from.

ELA CONNECTION

This suggested strategy addresses the Common Core State Standards for ELA.

SL 1: Engage in collaborative discussions.

CROSSCUTTING CONCEPTS

Cause and effect

TEACHING NOTE

See the **Home/School Connection** for Investigation 1 at the end of the Interdisciplinary Extensions section. This is a good time to send it home with students.

Soils, Rocks, and Landforms Module—FOSS Next Generation

INVESTIGATION 1 – Soils and Weathering

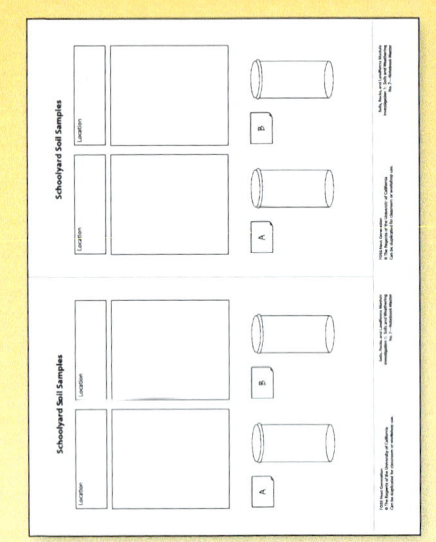

No. 7—Notebook Master

MATERIALS for
Part 4: Schoolyard Soils

For each group
- 1 Large zip bag, 4 L (Optional; see Step 7 of Getting Ready.) ★
- 1 Metal spoon
- 2 Paper plates
- 2 Vials with caps
- 2 Zip bags, 1 L
- 1 Permanent marking pen ★
- 2 Self-stick notes
- 2 Plastic cups (one to take outdoors)
- 1 Syringe
- ❏ 4 Notebook sheet 7, *Schoolyard Soil Samples* (optional)

For the class
- 1 Vial holder with samples from Part 1
- • Water ★
- 1 Pitcher
- 1 Carrying bag (optional) ★
- • String (optional) ★
- • Self-stick notes ★
- • Scissors ★
- 1 Computer with Internet access and projector (See Step 8 of Getting Ready.) ★
- ❏ 1 Teacher master 4, *Video Discussion Guide: Soils*

For benchmark assessment
- ❏ • *Investigation 1 I-Check*
- ❏ • *Assessment Record*

★ Supplied by the teacher.   ❏ Use the duplication master to make copies.

Full Option Science System

Part 4: Schoolyard Soils

GETTING READY *for*
Part 4: *Schoolyard Soils*

1. **Schedule the investigation**
   This part will take two active investigation sessions. Students need to dig in the schoolyard to collect soil samples, so this activity works best if the ground is fairly dry and not frozen. In addition, plan one session for the review and one for the I-Check.

2. **Preview Part 4**
   Students collect and observe different soils from several locations in the schoolyard. They analyze the soil samples to determine how much humus and rock material are in local soils. The focus question for this part is **What's in our schoolyard soils?**

3. **Select an outdoor site**
   Search for a site with a variety of terrain and plant life to ensure multiple soil types. If you are near a wooded area, collect samples under trees. Other places to consider collecting samples include under mulch, at the edge of a grass line, wooded areas, packed soil on a heavily trafficked area, under decaying leaf litter, in garden beds, and at the edge of a building.

   Outdoor activities are always best conducted on a beautiful sunny day, but, if necessary, this one can be conducted in rain or wind. Remind students ahead of time to dress appropriately. If the ground is really wet, you might want to collect the soil samples one day and let some of the moisture evaporate overnight before observing them more closely.

4. **Check the site**
   The morning of the outdoor activity, check the site. Look for unexpected, unsightly and distracting items, or anything else that could affect the activity in the area where students will be working.

5. **Plan to review outdoor safety rules**
   Before going outside, remind students of the rules and expectations for outdoor work. Remind students not to put anything from the ground or from plants in their mouths. Establish the boundaries for their work outdoors.

> **TEACHING NOTE**
>
> *Enlarging your learning space to include the schoolyard and the local neighborhood requires implementing some new teaching methods and learning behaviors. For a more detailed discussion of these methods, see the Taking FOSS Outdoors chapter.*

Soils, Rocks, and Landforms Module—FOSS Next Generation

INVESTIGATION 1 – Soils and Weathering

> **NOTE**
> Simple satchels for carrying materials to the field can be crafted from large zip bags and string. See the Managing Materials section in the Taking FOSS Outdoors chapter.

6. Plan for student notebook use
If the conditions are amenable, students can make their notebook entries outdoors. Have students prepare their notebooks indoors: glue notebook sheet 7, *Schoolyard Soil Samples,* into their notebooks or construct a data table before going outside.

Consider whether you want students to carry their own notebooks or if you would rather transport them *en masse* in one bag to your outdoor base. You may want students to do the soil collecting first and then turn to their notebooks.

7. Set up the materials station
Set up the materials station so that Getters can pick up a large zip bag first, then use it to transport the other materials back to their groups. After labeling the investigation materials, students can use the large zip bags to carry everything outdoors.

Bring a few extra cups, paper plates, vials, and permanent marking pens when you go outdoors.

8. Preview the video
Preview the *Soils* video (chapters 1 and 2). Links to the chapters are found on FOSSweb in the Resources by Investigation section. Have students view these video chapters after answering the focus question. Plan to use teacher master 4, *Video Discussion Guide: Soils,* to lead a class discussion after watching the video with the class. You can provide each group with a copy of the sheet.

9. Plan benchmark assessment: I-Check
When students have completed all the inquiries and readings for this investigation, they will be ready for a review and *Investigation 1 I-Check*. These activities begin with a review session suggested in Step 17 of Guiding the Investigation.

Print or copy the assessment masters or schedule students on FOSSmap to take the I-Check a day or two after the review session. You can read the questions aloud if students need this scaffolding; you want to measure their science understanding, not their reading ability. Students should work independently to mark or write their responses to each item.

The assessment coding guides are on FOSSweb. When you review and code the completed I-Checks, note any concepts students might be struggling with. Plan self-assessment activities to prompt further reflection on those concepts. See the Assessment chapter for guidance and information about online tools.

Part 4: Schoolyard Soils

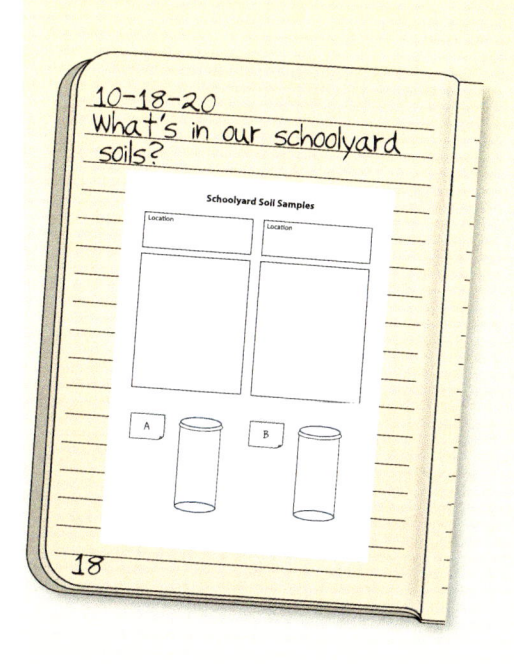

GUIDING *the Investigation*
Part 4: *Schoolyard Soils*

1. **Review soil composition**
   Call for attention and ask,
   ➤ *What is soil?*
   ➤ *What are some of your ideas about how different sizes of rock become part of soil?*

   After completing this discussion, introduce the focus question.

2. **Focus question: What's in our schoolyard soils?**
   Write the focus question on the board and say it aloud.
   ➤ *What's in our schoolyard soils?*

   Have students copy the question into their notebooks.

3. **Discuss the focus question**
   Have students discuss the focus question and what they can do to answer it in their groups. As you visit groups, help students consider collecting and examining schoolyard soils for different properties such as color, content (shape and size of earth materials), smell, texture, moisture (ability to retain water), and so forth. They might also suggest using vials to shake up a small amount of soil and water to see how the ingredients layer.

   Call on groups to summarize their thinking.

4. **Introduce the fieldwork**
   Tell students that today they will go to the schoolyard to collect two different soil samples. Ask students for some suggestions of places that would be good for collecting different kinds of soils. If they are stuck and can only think of one place to collect soil, point out that soil on a hillside might be quite different from soil near a building or in a garden.

   Indicate which door they will exit through and where they will be able to look.

5. **Prepare notebooks**
   Have students set up their notebooks to simplify outdoor record keeping. If students are using notebook sheet 7, *Schoolyard Soil Samples*, have them glue it into their notebooks now. If you're not using the notebook sheet, have students set up an observation table before going outdoors. They should include the date, the focus question, the location of each soil, and a description of both soils. (See example on previous page.)

FOCUS QUESTION
What's in our schoolyard soils?

Soils, Rocks, and Landforms Module—FOSS Next Generation

137

INVESTIGATION 1 – Soils and Weathering

Materials for Steps 6–9
- *Metal spoons*
- *Marking pens*
- *Plastic cups*
- *Paper plates*
- *Vials and caps*
- *Zip bags, 1 L*
- *Self-stick notes*
- *Zip-bag satchels (optional)*

SCIENCE AND ENGINEERING PRACTICES

Planning and carrying out investigations

6. Gather materials

Send Getters to the materials station to pick up the equipment and bring it back to the group. Have Starters organize their groups to label one plate, one vial, and one zip bag for sample A and another set of materials for sample B. Students can write directly on the plates and zip bags with the permanent marking pen and use self-stick labels for the vials.

Have students get their coats (if necessary) and line up.

7. Go outdoors

Head outdoors to your home base near the selected sampling sites. Form a sharing circle. Set the boundaries for the activity. Clearly articulate that students should not harm plants when digging samples but that they can gently dig *near* plants. Review other rules specific to your setting and point out any places where students should *not* dig.

Have students take out the metal spoon, the plastic cup, the permanent marking pen, and two small labeled zip bags. Leave the zip-bag satchels with the rest of the materials, notebooks, and pencils at home base for now.

Review the task. Each group will gather soil samples from two different locations. They should write the name of the location on the small zip bag with the marker, dig enough soil to fill the plastic cup, and then dump the soil into the labeled zip bag. Encourage groups to space themselves out as they collect samples so there will be more variety.

Encourage students to observe and record what is growing in the soil where they are collecting.

8. Collect samples

Send groups off to collect their samples. Monitor the groups. Remind them to remember where they collected their samples.

When the groups have collected their samples, signal them to come back to home base. Have students record in their notebooks where they found the soils.

9. Observe soil samples

Have students pour each sample onto appropriately labeled paper plates. Have them sort the samples into their various components by shaking the plates and pushing things around. Students should observe and record descriptions for each sample in their notebooks: color, shape, size, texture, smell, and moisture level. They should also record what was growing in the soil, if anything.

Part 4: Schoolyard Soils

If students don't address the property of moisture, ask,

▶ *Is there moisture in your soil? How could we find out?* [Place the soil in a zip bag and leave it in the sun for a few minutes. If the soil is moist, condensation will form on the inside surface of the zip bag.]

Help students to make observations of patterns by displaying the plates of soil based on different properties. Have students put their plates in the center of the circle and arrange them from darkest to lightest color. Have a discussion about what we know about the soils. Then sort the plates based on the property of what was growing in the soil—was there a lot growing in the soil or just a little or nothing? Ask,

▶ *What patterns do we see with the soils that had many plants growing in them versus those with no plants growing?*

10. Return to class

After students have had a chance to compare the two soils, have them fold the plates to form a pathway for the soil to slide off the plate back into the labeled zip bag. Fill vials about two-thirds full of soil from the zip bags, matching the labels on the zip bags with the labels on the vials. Cap the vials for transport. Keep the remaining soil in the zip bag.

Have Getters put the rest of the materials into the large zip bags. Have students line up to return to the classroom in the usual orderly manner.

11. Discuss further separation of the soil samples

Ask students how they could further analyze their soil samples. They should remember that adding water helps show the different components. Students can discuss their ideas in their groups.

Then develop a plan with students.

a. Getters pick up a syringe and cup of water.

b. Add 25 mL of water to each vial.

c. Observe what happens when you add the water. [Look for air bubbles.]

d. Cap vials tightly, shake the vials, then let them settle.

When the plan is complete, let students begin.

After the water has been added and vials shaken and observed, ask students to predict what will happen if they let the vials settle overnight. [The soils will settle into layers of the various earth materials and humus.]

CROSSCUTTING CONCEPTS
Patterns

Materials for Step 11
- Syringes
- Cups of water

SCIENCE AND ENGINEERING PRACTICES
Planning and carrying out investigations

CROSSCUTTING CONCEPTS
Patterns

TEACHING NOTE
Students should be able to make this prediction based on previous patterns they have observed.

Soils, Rocks, and Landforms Module—FOSS Next Generation

INVESTIGATION 1 – Soils and Weathering

Materials for Steps 13–14
- Settled vials, schoolyard soils
- Settled vials from Part 1

> **TEACHING NOTE**
>
> Refer to the Sense-Making Discussions for Three-Dimensional Learning chapter in *Teacher Resources* on FOSSweb for more information about how to facilitate this with students.

SCIENCE AND ENGINEERING PRACTICES

Analyzing and interpreting data

Constructing explanations

CROSSCUTTING CONCEPTS

Patterns

12. Clean up
Ask students to check that each vial is labeled A or B and has the group's name on it as well. Have a Getter bring the vials to a designated spot to settle overnight. Have the other Getter return the rest of the materials to the materials station.

BREAKPOINT

13. Observe settled vials and record results
Have one Getter get the group's vials, being careful not to disturb the contents. Have students record what they see in their notebooks for each of the vials.

14. Have a sense-making discussion
Have students look at other groups' vials and the vials of soils you saved from Part 1 to compare the contents.

➤ *Do all vials contain the same amount of soil?* [Approximately.]

➤ *Do all vials have the same number of layers?* [Possibly, but not necessarily.]

➤ *Are the layers the same thickness for each type of earth material?* [Probably not.]

➤ *Can you identify each layer?* [Each soil should include some sand, a significant layer of silt and clay, and some humus. It might also have gravel and pebbles.]

➤ *How are our schoolyard soils alike and different from the mountain, desert, river delta, and forest soils?*

➤ *If we were to dig up soil in (provide a known location), what would you expect to find?*

15. Answer the focus question
Have students answer the focus question in their notebooks.

➤ *What's in our schoolyard soils?*

16. View video on soil
Show students chapters 1 and 2 from the *Soils* video. The video can be accessed on FOSSweb, Resources by Investigation for this module. Before beginning, distribute copies of teacher master 4, *Video Discussion Guide: Soils*, to each group. Have students review the questions and discuss what they already know. Tell them to mark the questions so that each student has one or two questions they will be looking for more information about in the video.

Full Option Science System

Part 4: Schoolyard Soils

Here is a summary of the video.

Chapter 1: Introduction to "Soil" (1:30)

Chapter 2: What is Soil? (4:16)

- Soil is the loose top layer of Earth's surface where plants can grow.
- Because plants are the beginning of all food chains, without soil there would also be very few land animals.
- Soil is composed of many things and ranges in thickness from 1–3 meters.
- Most of the topsoil found on Earth is made up of rocks and minerals mixed with water, air, and organic matter.
- Soil with rounded particles tends to be more fertile, or better for plants.
- The amount of pore space in soil is also affected by the size of the particles in it.
- A mixture of clay, sand, and humus in the topsoil is best for plants to grow.

After watching, give students time to share their answers. Here are the questions to guide discussion.

➤ *What is soil made of?* [Small pieces of rock, bits of plant material or humus.]

➤ *Why is topsoil important for growing plants?* [Topsoil has humus, which has nutrients that plants use to grow.]

➤ *Describe the layers of soil.* [Topsoil with small pieces of rock and humus, water-holding subsoil with more clay, bottom layer with bigger pieces of rock and bedrock.]

➤ *What are the two most important properties of soil?* [How much water it can retain and how much air can move through it.]

➤ *How do rock size and shape affect the quality of soils?* [Size and shape of rock material affect pore space—how much room there is for air and water. Rounded rocks and larger rocks create more space. A balance must be found between drainage and ability to retain water in order for plants to grow best.]

Ask students to discuss this additional question to have them think about the importance of soil to human societies.

➤ *How does soil affect humans?*

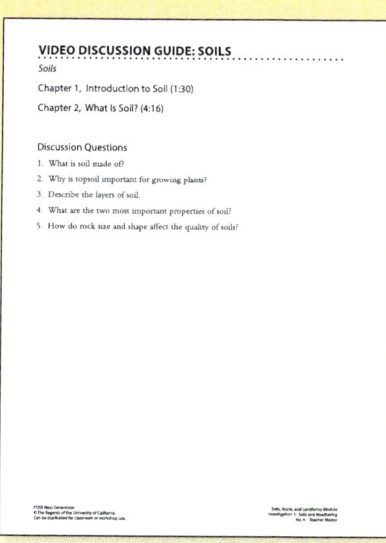

No. 4—Teacher Master

ELA CONNECTION

SL 2: Paraphrase portions of a text presented in diverse media.

Soils, Rocks, and Landforms Module—FOSS Next Generation

INVESTIGATION 1 – Soils and Weathering

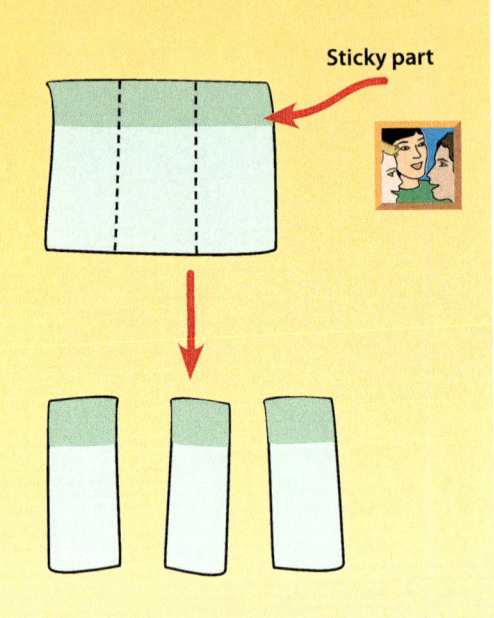

Sticky part

DISCIPLINARY CORE IDEAS

ESS2.A Earth materials and systems

ESS2.E Biogeology

WRAP-UP

17. Review Investigation 1

Distribute one or two self-stick notes to each student. Ask students to cut each note into three pieces, making sure that each piece has a sticky end.

Ask students to take a few minutes to look back through their notebook entries to find the most important things they learned in Investigation 1. Students should include at least one science and engineering practice, one disciplinary core idea, and one crosscutting concept. They should tag those pages with self stick notes.

When students have completed their notebook review and placed their tags, lead a short class discussion to create a list of three-dimensional statements that summarize what students have learned in this investigation.

Here are the big ideas that should come forward in review discussions.

- We conducted investigations to learn about the properties of soils from different places. (Planning and carrying out investigations; systems and system models.)
- From our data analysis, we saw a pattern in soil composition—soils are composed of small rocks (gravel, sand, silt, or clay) and humus. (Analyzing and interpreting data; patterns.)
- We modeled different kinds of weathering: abrasion (physical weathering) and acid rain (chemical weathering). (Developing and using models; cause and effect.)

You can have students record the class-generated list of key points or draw a line of learning under their last entry and compose a short, concise paragraph responding to the investigative guiding question. Students should discuss this question with a partner before responding to it in their notebooks.

➤ How do soils form?

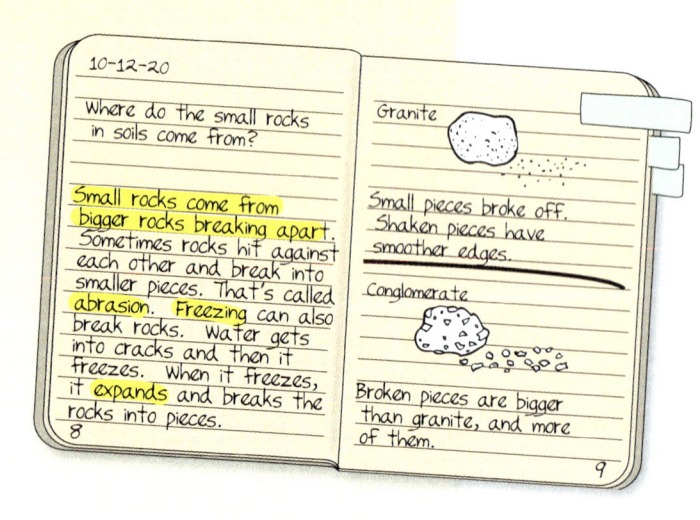

Part 4: Schoolyard Soils

▸ BREAKPOINT

18. **Assess progress: I-Check**

    Give the I-Check assessment at least one day after the wrap-up review. Distribute a copy of *Investigation 1 I-Check* to each student. You can read the items aloud, but students should respond to the items independently in writing. Alternatively, you can schedule the I-Check on FOSSmap for students to take the test and then you can access data through helpful reports to look carefully at students' progress.

    If students take the test on paper, collect the I-Checks. Code the items using the coding guides you downloaded from FOSSweb, and plan for a self-assessment session, identifying disciplinary core ideas, science and engineering practices, and crosscutting concepts students may need additional help with. (If you are using FOSSmap, you will only need to code open-response items and confirm codes for short-answer items as other items are automatically coded.) Refer to the Assessment chapter for next-step/self-assessment strategies.

> **TEACHING NOTE**
>
> During or after these next-steps with the I-Check, you might ask students to make choices for possible derivative products based on their notebooks for inclusion in a summative portfolio. See the Assessment chapter for more information about creating and evaluating portfolios.

INVESTIGATION 1 – Soils and Weathering

INTERDISCIPLINARY EXTENSIONS

Language Extension

- **Write soil stories**
  Have students research more about soils and the organisms that live in soils, then write a story about a day in the life of one of those organisms living in soil.

Math Extension

- **Problem of the week**
  Andy, Bette, Cate, Dustin, Erik, and Franco are rock collectors. Each collector has chosen some rocks from his or her collection to trade. Each collector is going to trade with every other collector. How many different pairs of collectors will trade rocks?

  Notes on the problem. To help students understand the problem, ask three students to come to the front of the class and role-play a simpler problem. If you have three rock collectors and each one is going to trade a rock with another person, how many different pairs of collectors will trade? Once students can visualize the simpler problem, let them work the more complex problem.

  A systematic list is probably the easiest way for students to work out this problem. They might make lists, such as the one shown below. Initials of first names are used in place of the entire name. The important thing for students to remember is that Andy trading with Bette is the same as Bette trading with Andy, and shouldn't be counted twice.

  You might want to provide students with six different-color pencils to help them diagram the trades.

  Andy <—> B, C, D, E, F

  Bette <—> C, D, E, F

  Cate <—> D, E, F

  Dustin <—> E, F

  Erik <—> F

  There will be 15 different pairs of collectors.

No. 5—Teacher Master

144

Full Option Science System

Interdisciplinary Extensions

Science Extensions

- **Contact the U.S. Geological Survey**
  The USGS publishes a number of short publications about geology and soils. Check the website listed on FOSSweb for documents to download.

- **Invite a geologist or soil scientist to class**
  Check with students to see who might have a parent or friend who is a geologist or soil scientist and could come speak to the class. If you have a college or university nearby, check with the earth science department. Have students develop a list of questions to ask the guest speaker before he or she visits.

- **Look for soil profiles**
  Look for a roadside cut or soil pit where you can see several layers in the soil. These layers are called soil horizons. The arrangement of these horizons is known as a soil profile. Soil scientists use soil profiles to help them classify soils for different uses. See the Natural Resources Conservation Service website for more information.

- **Look for soil profiles**
  Look for a roadside cut or soil pit where you can see several layers in the soil. These layers are called soil horizons. The arrangement of these horizons is known as a soil profile. Soil scientists use soil profiles to help them classify soils for different uses. See the Natural Resources Conservation Service website for more information.

- **Engage with the "Tutorial: Weathering"**
  If a student needs more time to work with the concept of weathering, have them go to FOSSweb and engage with this tutorial. The tutorial reviews the process of physical and chemical weathering and how these processes break down rocks.

- **Engage with the "Virtual Investigation: Water Retention of Soils"**
  In this virtual investigation on FOSSweb, students investigate the water retention properties of different types of soils.

> **TEACHING NOTE**
> Refer to the teacher resources on FOSSweb for a list of appropriate trade books that relate to this module.

Soils, Rocks, and Landforms Module—FOSS Next Generation

INVESTIGATION 1 – Soils and Weathering

Art Extensions

- **Soil painting**
  Have students collect different soils and add water to them to make "soil paint." Students can use the paint to create fingerpaintings.

- **Research soil pigments**
  Have students research pigments made from soils and other earth materials.

Environmental Literacy Extensions

- **Invite a geologist or soil scientist to class**
  Check with students to see who might have a parent or friend who is a geologist or soil scientist who could speak to the class. If you have a college or university nearby, check with the earth science department. Have students develop a list of questions to ask the guest speaker before he or she visits. Encourage the soil scientist to emphasize how humans benefit from better understanding of soil composition and formation.

> **TEACHING NOTE**
> Encourage students to use the Science and Engineering Careers Database on FOSSweb.

- **Use the acid-rain test**
  Calcite is the mineral in many rocks that interacts with acid rain. Provide materials to test for calcite, including safety goggles, that students can use on rocks they bring to class. Do not use valuable rocks, such as fancy crystals. The vinegar could change their appearance and lessen their value. Ask students to discuss how acid rain affects things on Earth's surface.

Home/School Connection

Students are encouraged to find and test some rocks from their neighborhoods for interaction with acid rain (vinegar). Students explain to family members that acid rain interacts with a mineral called calcite, an ingredient in many rocks.

Print or make copies of teacher master 6, *Home/School Connection* for Investigation 1, and send it home with students after Part 3.

> **TEACHING NOTE**
> Families can get more information on Home/School Connections from FOSSweb.

No. 6—Teacher Master

146 Full Option Science System

INVESTIGATION 2 – Landforms

Part 1	
Erosion and Deposition	158
Part 2	
Stream-Table Investigations	170
Part 3	
Schoolyard Erosion and Deposition	183
Part 4	
Fossil Evidence	190

Guiding questions for phenomenon: *How do erosion and deposition impact landforms? What do the locations of fossils in rock layers tell us about past life on Earth?*

Science and Engineering Practices
- Asking questions
- Developing and using models
- Planning and carrying out investigations
- Analyzing and interpreting data
- Constructing explanations
- Engaging in argument from evidence
- Obtaining, evaluating, and communicating information

Disciplinary Core Ideas
ESS1: What is the universe and what is Earth's place in it?
ESS1.C: The history of planet Earth
ESS2: How and why is Earth constantly changing?
ESS2.A: Earth materials and systems
ESS2.B: Plate tectonics and large-scale system interactions

Crosscutting Concepts
- Patterns
- Cause and effect
- Scale, proportion, and quantity
- Systems and system models
- Stability and change

PURPOSE

Students create a model stream system to investigate the variables involved with the phenomena of erosion and deposition of weathered earth material by flowing water. Students pursue explanations for the phenomenon of fossils found in layers of sedimentary rock, and what fossils tells us about life on Earth.

Content
- Weathered rock material can be reshaped into new landforms by the slow processes of erosion and deposition.
- Erosion is the transport (movement) of weathered rock material (sediments) by moving water or wind.
- The rate and volume of erosion relate directly to the energy of moving water or wind. The greater the mass and velocity, the greater the energy.
- Deposition is the settling of sediments when the speed of moving water or wind declines.
- Catastrophic events have the potential to change Earth's surface quickly.

Practices
- Use stream tables as models to make predictions about how the slow processes of erosion and deposition by water alter landforms over time.

FOSS Full Option Science System

INVESTIGATION 2 – Landforms

Investigation Summary	Time	Focus Question for Phenomenon, Practices
PART 1 **Erosion and Deposition** Students use stream tables to observe that water moves earth materials from one location to another. After running a volume of water through the stream table, students shake a vial containing a sample of earth material mixed with water to observe the rate at which different particle sizes of earth material settle out.	**Active Inv.** 1 Session * **Reading** 1 Session	**How do weathered rock pieces move from one place to another?** **Practices** Developing and using models Planning and carrying out investigations Analyzing and interpreting data Constructing explanations Obtaining, evaluating, and communicating information
PART 2 **Stream-Table Investigations** Students continue to run stream tables to learn how environmental variables can affect erosion and deposition. They investigate the variables of slope and water volume (flood). Then they plan and conduct their own stream-table investigations.	**Active Inv.** 2–4 Sessions	**How does slope affect erosion and deposition?** **How do floods affect erosion and deposition?** **Practices** Asking questions Developing and using models Planning and carrying out investigations Analyzing and interpreting data Constructing explanations Engaging in argument from evidence Obtaining, evaluating, and communicating information
PART 3 **Schoolyard Erosion and Deposition** Students consider whether erosion and deposition are happening in their own schoolyard. They look for evidence of erosion and for locations where deposition is in evidence. They simulate a rainstorm by pouring water on various outdoor surfaces.	**Active Inv.** 1 Session	**Where are erosion and deposition happening in our schoolyard?** **Practices** Planning and carrying out investigations Constructing explanations
PART 4 **Fossil Evidence** Students think about what happens to and in sediments over long periods of time as sediments layer on top of each other. Students watch a video, make models, and read an article to learn about how sedimentation processes can result in fossils. They learn how fossils provide evidence of life and landscapes from the past.	**Active Inv.** 2 Sessions **Reading** 1 Session **Assessment** 2 Sessions	**How do fossils get in rocks and what can they tell us about the past?** **Practices** Developing and using models Constructing explanations Obtaining, evaluating, and communicating information

* A class session is 45–50 minutes. **Full Option Science System**

At a Glance

Content Related to DCIs	Writing/Reading	Assessment
• Weathered rock material can be reshaped into new landforms by the slow processes of erosion and deposition. • Erosion is the transport (movement) of weathered rock material (sediments) by moving water or wind. • Deposition is the settling of sediments when the speed of moving water or wind declines.	**Science Notebook Entry** *Standard Stream-Table Setup* *Stream-Table Observations* *Landform Vocabulary* **Science Resources Book** "Erosion and Deposition"	**Embedded Assessment** Science notebook entry
• The rate and volume of erosion relate directly to the energy of moving water or wind. • The energy of moving water depends on the mass of water in motion and its velocity. The greater the mass and velocity, the greater the energy.	**Science Notebook Entry** *Stream-Table Observations* *Landform Definitions* *Stream-Table Investigation* **Science Resources Book** "Landforms Photo Album" **Video** *Weathering and Erosion* (optional) **Online Activities** "Videos: Stream Tables" "Tutorial—Stream Tables: Slope and Flood"	**Embedded Assessment** Performance assessment
• Erosion is the transport (movement) of weathered rock material (sediments) by moving water or wind. • Deposition is the settling of sediments when the speed of moving water or wind declines.	**Science Notebook Entry** Record evidence of erosion and deposition in the schoolyard. **Online Activity** "Virtual Investigation—Stream Tables"	**Embedded Assessment** Response sheet
• Fossils provide evidence of organisms that lived long ago as well as clues to changes in the landscape and past environments.	**Science Notebook Entry** *Modeling Fossils* **Science Resources Book** "Fossils Tell a Story" "Pieces of a Dinosaur Puzzle" **Video** *Fossils* **Online Activities** "Tutorial: Soil Formation" "Tutorial: Fossils"	**Embedded Assessment** Science notebook entry **Benchmark Assessment** *Investigation 2 I-Check* **NGSS Performance Expectations addressed in this investigation** 4-ESS1-1 4-ESS2-1 4-ESS2-2

INVESTIGATION 2 – *Landforms*

BACKGROUND *for the Teacher*

Earth is a dynamic place. Great forces continually reshape the face of the planet. Some of these forces bring molten material to the surface or bend and raise the planet's crust, forming new **mountains** or ridges and **valleys**. Opposing forces, such as moving water and wind, wear down the land and fill in irregularities of the planet's surface. If Earth's crust weren't continually being built up, the planet might end up as smooth as a bowling ball.

How Do Weathered Rock Pieces Move from One Place to Another?

We observe the phenomenon of **erosion** as the transport of earth materials from one location to another. Water, working with the force of gravity, is the most powerful agent of erosion on Earth. Flowing water tumbles and transports pieces of rock. Tumbling rocks break into smaller pieces, while simultaneously acting as powerful abrasives, wearing away the rock over which they move. Wind picks up small rock particles and transports them from one place to another. Ice, in the form of glaciers, also picks up and moves rocks, while grinding them into smaller particles. After thousands of years of erosion, mountains become rounded, and **river channels** deepen and widen. In places, curious landforms, such as hoodoos (eroded pillars or columns of rock in bizarre shapes) and hogbacks (sharp, eroded ridges of steeply sloped rocks), decorate the landscape.

Rock particles are classified by size. Boulders are large (256 millimeters [mm] in diameter and larger—from the size of basketballs upward); cobbles are intermediate in size (64–256 mm—the size of potatoes to basketballs); pebbles are small (6–64 mm—the size of peas to potatoes); gravel is yet smaller bits of rock (2–6 mm—the size of rice grains to peas); sand is tiny rock particles (0.06–2 mm—the size of grains of salt to rice grains); silt particles require a hand lens to be seen (0.004–0.06 mm— barely noticeable as fine grit between your fingers); and clay particles are microscopic (less than 0.004 mm—powdery, smooth between the fingers, slippery when wet). The general rule is that the farther a rock travels, the more it bangs and rubs against other rocks and the smaller it becomes. This is how boulders are reduced to silt. In addition to physical weathering, chemical weathering processes are usually involved in the transformation of large rock particles into minute clay particles.

There is a reciprocal phenomenon to erosion. Earth materials, carried by water, wind, and ice, eventually settle out in a process called **deposition**. The deposited rock particles are called **sediments**. Sediments can be huge boulders or motes of silt and clay.

> *"Earth materials, carried by water, wind, and ice, eventually settle out in a process called deposition. The deposited rock particles are called sediments. Sediments can be huge boulders or motes of silt and clay."*

Background for the Teacher

A powerful **flood** might move a boulder a few meters in a riverbed. Sand, silt, and clay may remain suspended in water for long distances. Particles travel with flowing water until the velocity of the current decreases enough for the particles to yield to the pull of gravity and settle on the bottom as sediments. This kind of deposition characteristically occurs where a stream or river pours out into slow-moving or still water in a lake or the ocean, resulting in a **landform** called a **delta** at the **river mouth**. **Alluvial fans** are similar to deltas, except they form on land where the slope of the terrain decreases at the lower end of a steep **canyon**. The amount of erosion of earth materials depends to a large extent on the amount of energy available in the moving water. The greater the mass and speed of water, the greater the energy of the water, and the greater the energy, the larger the particles that can be put into motion and the farther those particles will travel.

© Marli Bryant Miller/Earth Science World Image Bank

Smaller rock particles (sand, silt, and clay) can also be transported by wind. Large deposits of wind-blown sediments form sand dunes in desert and coastal areas.

How Does Slope Affect Erosion and Deposition?

Two variables affect the rate at which water erodes Earth's surface. **Slope** is the angle of the land over which the water flows. Because water is urged along by gravity, the steeper the angle of the surface, the faster the water flows. The energy of a moving object is a factor of both the mass of the object and its velocity. A given mass (volume) of water flowing across a gentle incline will wander across the land slowly, doing little to displace any of the particulate matter in its path. The same mass of water flowing over a steep section of land will find the lowest pathway and surge along that pathway at a much faster pace. The faster flow pushes larger particles of rock around and carries small particles along with the current. The increased energy of the faster-moving water on a steep slope produces significantly more erosion than the same amount of water ambling slowly across a flatter landscape.

How Do Floods Affect Erosion and Deposition?

Flood is the imprecise word used to describe a volume of water in a drainage system that exceeds the normal or typical volume in that system. Floods fill the streambed to overflowing. The mass of water during a flood is more than the environment can handle. During a flood, the energy of the mass of water in the stream channel may be several times greater than the energy of the water in the channel during times of normal flow.

Soils, Rocks, and Landforms Module—FOSS Next Generation

INVESTIGATION 2 – Landforms

Floods propel boulders down stream channels. Boulders in motion gouge the streambed and smash smaller rocks to pieces. Flood energy amplifies both weathering and erosion. Tons of smaller sediments (sand and silt) are carried away. A tremendous amount of reshaping occurs, often not fully appreciated until the flood waters recede. Rivers can take a different course, new sandbars can be created, and vast fields of mud (silt and clay) can be deposited in fields, in towns, and inside homes and other structures.

Where Are Erosion and Deposition Happening in Our Schoolyard?

Schoolyards are rarely absolutely flat and level. Erosion and deposition can occur wherever there is a slope, particularly a slope that is not supporting a dense growth of plants. Plants are quite effective at stabilizing soil and other earth materials under all but the most aggressive flows of water. A modest rain on a bare soil slope results in a small streamlet as water finds a path down the slope. The streamlet produces a small streambed. The bed is the result of material being eroded away to produce the observable path of flowing water. By carefully observing the downhill end of the tiny streambed, you should be able to identify a minute delta or alluvial fan where the eroded sand and silt have been deposited.

Search your schoolyard for slopes where rainwater has established a flow path. This is where you will find evidence of erosion and deposition. Sometimes it is possible to see evidence of erosion and deposition on an asphalt surface. Tiny components of the asphalt are constantly weathered loose by foot traffic, rain, and wind. These particles are transported by water when it rains. By pouring water in strategic locations on your asphalt schoolyard, you may be able locate the paths taken by rainwater as it drains off the hard surface. This will indicate where to look for deposits of weathered asphalt in your hardtop schoolyard.

How Do Fossils Get in Rocks and What Can They Tell Us About the Past?

There are several assumptions scientists make in order to make inferences about the past. One of those is the idea that the processes (phenomena) that we observe happening today also happened in the past. By looking at the environments we see on Earth today and observing conditions conducive to rock formation, we have the data needed to infer that what is now an uplifted plateau in the desert was, at various times in the past, a desert of countless sand dunes, a muddy swamp, and the bottom of the ocean.

New Word — Say it, See it, Hear it, Write it

Alluvial fan
Basin
Canyon
Cast
Delta
Deposition
Erosion
Flood
Floodplain
Fossil
Imprint
Landform
Meander
Mold
Mountain
Petrification
Preserved remains
River channel
River mouth
Sediment
Sedimentary rock
Slope
Superposition
Valley

Background for the Teacher

Another important principle is the idea that when particles of sediment settle out of water or air, they form horizontal layers. Geologists know if layers are no longer horizontal, something happened to tilt or deform them, such as an earthquake, mountain building, or folding.

Finally there is also the idea that for any undisturbed sequence of layered rocks, a particular bed must be older than any bed on top of it. This is the concept of **superposition**. We can infer the relative ages of sedimentary layers based on their relative positions.

We can also look for **fossils** in **sedimentary rocks**. A fossil is defined as any **preserved remains**, trace, or **imprint** of a plant or animal that has been preserved in Earth's crust during prehistoric times. **Molds** are empty spaces in rock formed when an organism decays. A **cast** is formed when minerals are deposited in the empty space where the organism was. Not all fossils are the actual remains of once-living organisms; some fossils are just traces left behind, such as burrows, tracks, and droppings.

Conditions have to be just right for fossilization to occur. When a plant or animal dies, sediments must quickly bury the remains before decomposition erases all evidence of the organism's existence. The hard parts, such as shells, bones and teeth take longer to decompose than the soft parts, so these are the remains most often found as fossils.

Every rock has a story to tell: for sedimentary rocks, that story began when sediments settled in a **basin**. As water slows in the stream table, the larger sediments drop out first. In the case of a river, the larger boulders and cobbles are dropped near their source, then smaller and smaller particles are deposited as the river travels farther, with the silt and clay finally settling in the calm waters of a lake or sea. Can you tell what direction the river flowed millions of years ago? Yes, by reading the rocks and following the trail of sediments from boulders to clay.

By analyzing **banded sandstone**, we see that over time, different colors of sand were deposited and buried under more layers. If we observed a piece of **shale**, we would see that shale is composed of the finest clay and silt, and so must have formed in very still water. Delicate fossils of leaves are often found sandwiched between layers of clay with dark colors, indicating oxygen-deprived environments. The evidence for the environment of shale formation points to a body of still, stagnant water where years of sediments have time to accumulate such as in a bog or swamp.

However, the most powerful tool in analyzing sedimentary rocks comes from looking at the order and thickness of the layers in relation to each other, not in isolation. Layers of sedimentary rocks stacked on top of each other map the changes in the environment of a location. They record in rock the rising and lowering of the ocean, changes in climate, changes in elevation, periods of deposition, and periods of erosion.

Soils, Rocks, and Landforms Module—FOSS Next Generation

INVESTIGATION 2 – *Landforms*

TEACHING CHILDREN about *Landforms*

Developing Disciplinary Core Ideas (DCI)

> **NGSS Foundation Box for DCI**
>
> **ESS1.C: The history of planet Earth**
> - Local, regional, and global patterns of rock formations reveal changes over time due to Earth's forces such as earthquakes. The presence and location of certain fossil types indicate the order in which rock layers were formed. (4-ESS1-1)
>
> **ESS2.A: Earth materials and systems**
> - Rainfall helps to shape the land and affects the types of living things found in a region. Water, ice, wind, living organisms, and gravity break rocks, soils, and sediments into smaller particles and move them around. (4-ESS2-1)
>
> **ESS2.B: Plate tectonics and large-scale system interactions**
> - The locations of mountain ranges, deep ocean trenches, ocean floor structures, earthquakes, and volcanoes occur in patterns. Most earthquakes and volcanoes occur in bands that are often along the boundaries between continents and oceans. Major mountain chains form inside continents or near their edges. Maps can help locate the different land and water features of Earth. (4-ESS2-2)

When young students look at the myriad features of Earth's surface, they have no reason to doubt that what they see is what always has been and what always will be. After all, what is more predictable and permanent than rock?

This investigation is an opportunity for students to reconsider this belief. They will set up model environments in stream tables and observe the effects produced by flowing water. They will discover that water has the power to erode earth materials, moving them from one place to another. The same earth materials are present, but they are redistributed. The process creates new landforms from existing earth material.

The main agent that continuously reshapes Earth's surface is moving water. Rain, streams, rivers, and waves erode the land and carry boulders, sand, gravel, and silt far from their origins. In compressed time, students will observe how sediments deposited in basins of ancient environments can turn into sedimentary rock and turn the remains of living thing into fossils that provide evidence of environments past. They will see how valleys, canyons, deltas, and some other identifiable landforms are destroyed and created. These are big ideas in geology. These introductory experiences will help students develop far-reaching concepts of a dynamic Earth and opportunities to consider the monumental time frame in which they happen. With these basic concepts in place, students will be prepared to look into Earth's deep past for evidence of long-gone landforms and to anticipate distant geological futures.

Three related concepts combine to explain slow, incremental surface reconstruction. Earth's surface is essentially solid rock. Before reshaping can occur, the solid rock must break into bits. This is weathering. After weathering, an agent, such as wind or water, often transports the broken bits. The transportation process is erosion. Finally, the eroded bits settle down in a process called deposition. Breaking, moving, and settling down are all landform modification processes.

These experiences will contribute to students' foundational understanding of the core idea **ESS1.C: The history of planet Earth; ESS2.A: Earth materials and systems;** and **ESS2.B: Plate tectonics and large-scale system interactions**, by helping them model how landforms are shaped and reshaped by natural processes over time. They will also gain experience with the information that the fossil record provides about the landforms and past environments.

Teaching Children about Landforms

Engaging in Science and Engineering Practices (SEP)

In this investigation, students engage in these practices.

- **Asking questions** about the stability of landforms and the factors that affect how erosion moves earth materials from one location to another. These questions are investigated using stream-table models to observe cause-and-effect patterns.

- **Developing and using models** to describe the natural process of landform destruction and construction through the process of erosion. Use the stream-table model to test these relationships and describe the limitations of these models. Use models of fossil formation to understand the process.

- **Planning and carrying out investigations** dealing with erosion using stream tables by changing variables and producing data to serve as the basis for constructing explanations.

- **Analyzing and interpreting data** in the classroom and outdoors in the schoolyard to draw conclusions about how erosion and deposition change the shape of landforms small and large.

- **Constructing explanations** using evidence to describe interactions of solid earth materials and moving water.

- **Engaging in argument from evidence** about different soils, where they might be from based on properties, and why they are different from one location to another.

- **Obtaining, evaluating, and communicating information** from books and media and integrating that with students' firsthand experiences to construct explanations about erosion, deposition, and fossil evidence from past environments.

NGSS Foundation Box for SEP

- **Ask questions** that can be investigated based on patterns such as cause-and-effect relationships.
- **Identify limitations of models.**
- **Develop models** to describe phenomena.
- **Use a model** to test cause-and-effect relationships interactions concerning the functioning of a natural system.
- **Plan and conduct an investigation** collaboratively to produce data to serve as the basis for evidence.
- **Make observations and/or measurements to produce data** to serve as the basis for evidence for an explanation of a phenomenon or test a design solution.
- **Analyze and interpret data** to make sense of phenomena, using logical reasoning.
- **Compare and contrast data** collected by different groups in order to discuss similarities and differences in their findings.
- **Construct an explanation** of observed relationships.
- **Use evidence** (e.g., measurements, observations, patterns) to construct or support an explanation.
- **Compare and refine arguments** based on an evaluation of the evidence presented.
- **Obtain and combine information** from books and other reliable media to explain phenomena.
- **Read and comprehend grade-appropriate complex texts** and/or other reliable media to summarize and obtain scientific and technical ideas and describe how they are supported by evidence.
- **Communicate scientific and/or technical information** orally and/or in written formats, including various forms of media and may include tables, diagrams, and charts.

INVESTIGATION 2 – *Landforms*

> **NGSS Foundation Box for CC**
>
> - **Patterns:** Similarities and differences in patterns can be used to sort and classify natural phenomena. Patterns of change can be used to make predictions.
> - **Cause and effect:** Events that occur together with regularity might or might not be a cause-and-effect relationship. Cause-and-effect relationships are routinely tested, and used to explain changes.
> - **Scale, proportion, and quantity:** Standard units are used to measure and describe physical quantities such as weight, time, temperature, and volume. Observable phenomena exist from very short to very long periods.
> - **Systems and system models:** A system can be described in terms of its components and their interactions.
> - **Stability and change:** Change is measured in terms of differences over time and may occur at different rates. Some systems appear stable, but over long periods of time will eventually change.

Exposing Crosscutting Concepts (CC)

In this investigation, the focus is on these crosscutting concepts.

- **Patterns:** Patterns of erosion and deposition can be predicted from the path and volume of water flowing over earth material.
- **Cause and effect:** Increased slope and flood conditions cause greater weathering, erosion, and deposition.
- **Scale, proportion, and quantity:** Stream tables are a tool to model changes to landforms over time. The models are much smaller than reality and changes occur much faster, but it shows what can happen over long periods of time.
- **Systems and system models:** The process of weathering and erosion can be studied as a system of interacting parts.
- **Stability and change:** Rock is generally stable, but changes can be gradual and occur over long periods of time. The same earth processes that occurred in the past occur now.

Connections to the Nature of Science

- **Scientific knowledge assumes an order and consistency in natural systems.** Science assumes consistent patterns in natural systems.

Teaching Children about Landforms

Conceptual Flow

Students investigate the phenomena of erosion and deposition of weathered earth material by flowing water, and the phenomenon of fossils found in layers of sedimentary rock. The guiding questions for this investigation are how do erosion and deposition impact landforms? and what do the locations of fossils in rock layers tell us about past life on Earth?

The **conceptual flow** for this investigation starts with a model or a simulated **landform**, a plain composed of a mixture of earth material (sand and clay) packed into one end of a stream table (plastic tray). Water enters into the system and flows across the surface of the earth material plain.

As water flows, students observe that the water carves a **canyon** across the plain. The canyon continues to grow as the running water erodes sand and clay, carrying it into the lower portion of the tray where it is deposited. The **erosion and deposition create new landforms**, a river **valley** across the plain, and a **delta** or **alluvial fan** where the sand and clay accumulate in the lower reaches of the tray.

In Part 2, students manipulate the flow of water across the earth material plain, using the stream-table model. When they tilt the stream table to create a steeper slope, the water flows faster across the surface of the plain. By delivering the water in a more robust flow, students simulate a flood flowing over a plain. Students observe that **a steeper slope that induces a faster flow of water increases erosion**. Similarly, the **increased volume of water induces greatly enhanced erosion**. Students learn that fast-moving water has more energy than slow-moving water, and that a greater volume of water has more energy than a lesser volume of water. More energy produces more erosion.

In Part 3, students observe that the processes of erosion and deposition are occurring on a small scale in their schoolyards.

In Part 4, students model the **formation of fossils in sedimentary rock** from the sediments they have been observing. Through modeling, they explore the processes that **take millions of years to transform materials into rock and fossils**. And they consider the ancient environments that provided the conditions for rock to form and turn remains of organisms into fossils.

Earth interactions

- Earth materials form natural structures called landforms.
- Landforms interact with water.
- Erosion and deposition create new landforms.
  - Deposition creates new accumulations of rock particles.
- Erosion removes and carries away rock particles.
  - Valleys
  - Canyons
  - Alluvial fans
  - Deltas
- A steeper slope results in faster flow and more erosion.
- A greater volume of water produces more erosion.
- The energy of a moving object depends on the mass and velocity of the object.
- Landforms change over time due to natural processes.
- Fossils form in sedimentary rock and provide evidence of past environments.

Soils, Rocks, and Landforms Module—FOSS Next Generation

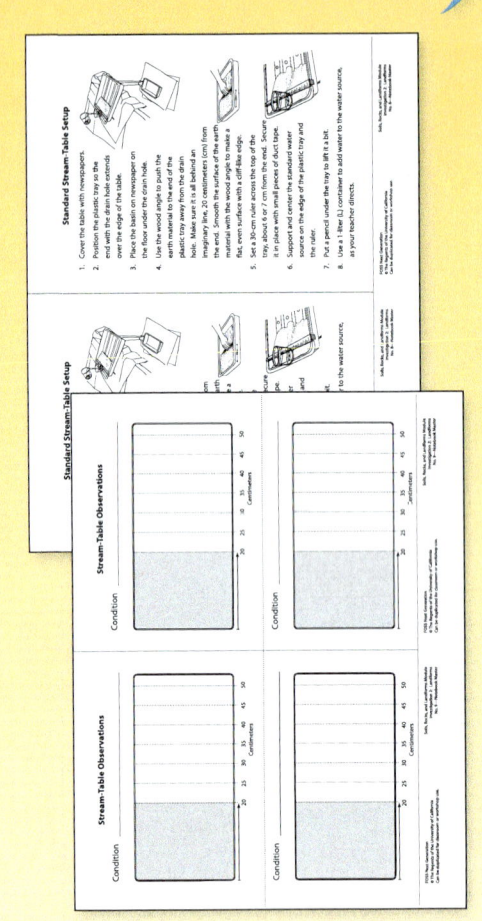

INVESTIGATION 2 – Landforms

MATERIALS for
Part 1: Erosion and Deposition

For each group
- 1 Stream table
- 1 Standard water source, 1/2 L (blue dot)
- 1 Ruler, 30 cm
- 1 Wood angle
- 1 Container, 1 L (without lid)
- 1 Basin
- 1 Meter tape
- 1 Vial with cap
- 1 Syringe
- ❏ 1 Notebook sheet 8, *Standard Stream-Table Setup*
- ❏ 4 Notebook sheet 9, *Stream-Table Observations*
- ❏ 4 Notebook sheet 10, *Landform Vocabulary*
- 4 *FOSS Science Resources: Soils, Rocks, and Landforms*
  - "Erosion and Deposition"

For the class
- 2 Plastic cups
- 2 Bags of white, powdered clay (white, not gray powdered clay)
- 8 Bags of fine sand
- Duct tape
- Water ★
- 1 Pitcher
- Paper towels ★
- Newspaper ★
- 1 Projection system (optional) ★
- 1 Dustpan and broom ★
- Cameras (optional) ★
- Safety goggles and gloves (optional) ★
- ❏ 1 Teacher master 7, *Standard Stream-Table Setup*

For embedded assessment
- ❏ • *Embedded Assessment Notes*

★ Supplied by the teacher.   ❏ Use the duplication master to make copies.

Nos. 8–9—Notebook Masters

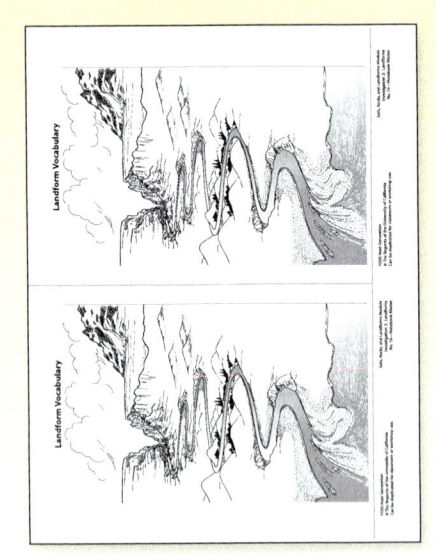

No. 10—Notebook Master

Part 1: Erosion and Deposition

GETTING READY for
Part 1: *Erosion and Deposition*

1. **Schedule the investigation**
   This part will take one active investigation session and one reading session.

2. **Preview Part 1**
   Students use stream tables to observe that water moves earth materials from one location to another. After running a volume of water through the stream table, students shake a vial containing a sample of earth material mixed with water to observe the rate at which different particle sizes of earth material settle out. The focus question is **How do weathered rock pieces move from one place to another?**

3. **Prepare earth material**
   Students flow water over a mixture of sand and clay (earth material) in a stream table to observe erosion and deposition. If you are first to use the kit, you will need to mix the sand and clay. The sand and clay used for the stream table is different from that used to make the soils in Investigation 1. Look for the stream-table earth materials in the small gray box that comes as part of the kit.

   Before you mix them, set aside one plastic cup of sand and another of clay. You will use these to show students the separate materials in Step 4 of Guiding the Investigation.

   The instructions below will make a batch of earth material for one tray. Enlist the help of students, parents, or aides to mix the number of batches needed for your class.
   a. Add about 100 grams (1 plastic cupful) of white powdered clay to a 1.35 kilogram (about 1 liter) bag of fine sand.
   b. Stir the clay and sand until they are uniformly mixed.
   c. Add up to 1 cup of water to the dry materials in the bag and mix thoroughly.

   If the earth material has been used previously, you may need to add a little sand or clay to each bag. The exact amounts are not critical. There should be more sand than clay in the mixture, and the total volume of earth material should be about 1 L.

4. **Consider health and safety**
   You may occasionally have a student with dry skin, which could be aggravated by contact with the earth material. Consider rubber gloves for students with sensitive skin. The sand and clay are not toxic, but students should wash their hands with soap and water after working with them. Check your district guidelines for the use of safety goggles.

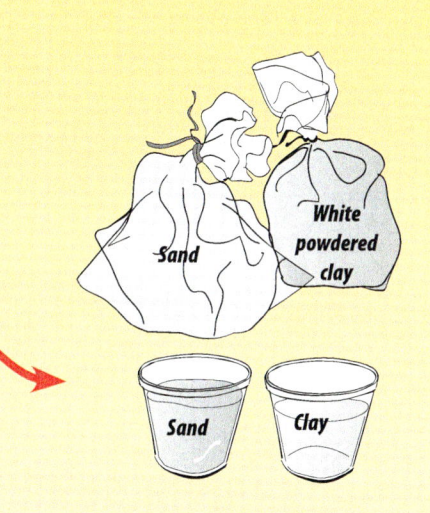

> **NOTE**
> The sand/clay mixture is permanent equipment. Dry it thoroughly after the investigation before returning it to the kit. These materials come in a gray plastic box and should be stored in the same box when the entire stream-table investigation is finished.
>
> Be sure to use the white powdered clay (not the gray clay).

Soils, Rocks, and Landforms Module—FOSS Next Generation

159

INVESTIGATION 2 – Landforms

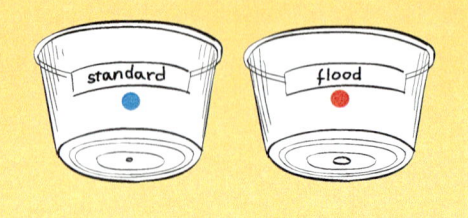

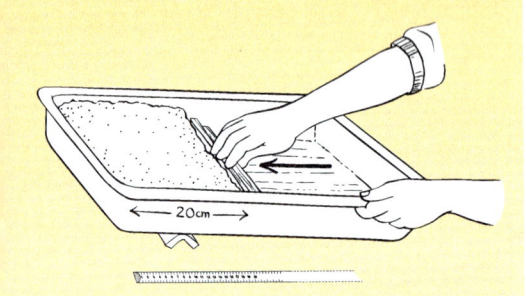

5. **Identify the two water sources**
   The kit contains two different 1/2 L water sources: standard and flood. The flood water source has a larger hole in the bottom. Students only use the standard water sources in this part. Note that the containers have coding dots for easy identification at a distance: blue = standard; red = flood.

   One pitcher is available for a water supply.

6. **Choose a location**
   It is important to have level surfaces for the stream tables. If you are concerned about getting water and earth material on the floor, consider conducting the investigation outdoors.

7. **Plan for multiple trials**
   The earth material in the stream table gets saturated during a run. Before students use the stream table a second time or if you teach two classes back to back, drain excess water from the earth material. Prop the earth-material end of the tray on a wood angle. When most of the water has drained out (2–3 minutes), bulldoze the earth material back to the end of the tray with the wood angle. Absorb excess water by pressing firmly on the earth material with newspaper.

8. **Practice with a stream table**
   Familiarize yourself with the stream table before beginning this investigation. Set one up, as described on notebook sheet 8 and teacher master 7, *Standard Stream-Table Setup*. Conduct several trials, noting the landforms that develop and the texture of the earth material after several uses. Use this stream-table setup for the demonstration in Step 4 of Guiding the Investigation.

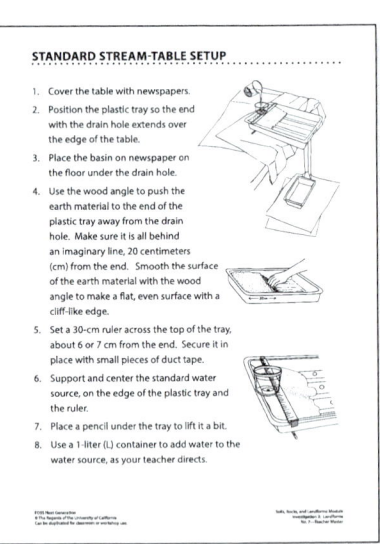

   No. 7—Teacher Master

9. **Prepare for cleanup**
   Plan to put newspaper on desktops and under the catch basins. Some earth material may flow through the drain hole and into the catch basin. After it settles, pour off the water and return the earth material to the stream table. Don't dump the earth material down a drain. Dumping the basins outdoors where there is soil is a better option. Provide paper towels and a broom and dustpan to help clean up spills.

Part 1: Erosion and Deposition

10. ## Store earth material in stream tables
    At the end of a session, drain the trays by tipping them up on the wood angles. After the water has drained, stick a piece of duct tape over the drain hole. Keep the earth material in the trays between sessions. Crisscross the trays for storage.

    Raking the material occasionally as it dries will keep it from clumping and will help the materials dry more quickly and evenly. For subsequent stream-table investigations, students can break up the earth material and add just enough water to make it feel like dry pie-crust dough. If they add too much water, use newspaper to blot up the excess.

11. ## Plan for multiple notebook sheets
    Students will use notebook sheet 9, *Stream-Table Observations,* in this part and the next. You may want to print or make multiple copies of this sheet now. Each student will need up to three copies: one for the standard setup and run, one for slope and flood, and a third if students do an investigation on their own.

12. ## Plan to Read *Science Resources*: "Erosion and Deposition"
    Plan to read "Erosion and Deposition" during a reading period after completing the active investigation for this part.

13. ## Plan assessment: notebook entry
    In Step 14 of Guiding the Investigation, students write answers to the focus question in their notebooks. After class, have students hand in their notebooks open to the page on which they wrote their answers. Students should understand the difference between *erosion* and *deposition* before going on to the next part.

Soils, Rocks, and Landforms Module—FOSS Next Generation

INVESTIGATION 2 – Landforms

FOCUS QUESTION

How do weathered rock pieces move from one place to another?

EL NOTE

This can be done as a cloze review on chart paper or the board. Provide sentences, leaving out the key vocabulary words, for students to fill in such as *soil, humus, physical weathering,* and *chemical weathering.*

SCIENCE AND ENGINEERING PRACTICES

Developing and using models

CROSSCUTTING CONCEPTS

Systems and system models

Say it / See it / Hear it / Write it — **New Word**

Materials for Step 4
- *Stream table*
- ***Standard Stream-Table Setup*** *sheet*
- *Cup of sand*
- *Cup of clay*

GUIDING *the Investigation*
Part 1: *Erosion and Deposition*

1. **Review soils and weathering**
   Ask students about the processes of weathering.

   ➤ *What is soil?* [A mixture of weathered rock and organic humus.]

   ➤ *What is weathering?* [The breaking of rock into smaller pieces.]

   ➤ *What are examples of physical and chemical weathering?* [Frost wedging and roots; minerals reacting with acid rain.]

   ➤ *How does weathering contribute to the formation of soils?* [Weathering reduces rock material to smaller pieces—sand, silt, and clay.]

   Tell students,

   Sometimes the small rock particles we see in soils are found near the parent rock. Parent rock is the original rock from which the smaller pieces broke away. At other times, the parent rock and the sand and silt we see in soils are hundreds of kilometers apart.

2. **Focus question: How do weathered rock pieces move from one place to another?**
   Ask students the focus question as you project or write it on the board, and have students copy it into their notebooks.

   ➤ *How do weathered rock pieces move from one place to another?*

3. **Review *model***
   Explain that one way geologists investigate the movement of earth materials is to develop and study models. Review what a model is.

   A **model** is a representation of the features and actions of a natural system or process, but it isn't exactly the same as the real world—it focuses on the most important features. Scientists use models to help them observe what happens in systems that move too slowly or too quickly, or are too big or too small to observe directly.

4. **Describe stream tables**
   Show students a stream table. Tell them,

   This stream table is a model of a piece of land. In this model, the land is composed of soil made of two particle sizes: sand and clay.

   Show students the samples of sand and clay in two plastic cups.

   Give each student a copy of notebook sheet 8, *Standard Stream-Table Setup.* Project the *Standard Stream-Table Setup* teacher master as you preview the procedure for preparing and using the stream table.

Full Option Science System

Part 1: Erosion and Deposition

a. Place the stream table at the edge of the table with a basin (on newspaper) on the floor or on a chair to catch water that flows out of the stream table.

b. Use a wood angle like a bulldozer to push the earth material into the top 20 centimeters (cm) of the tray (away from the hole). Use the meter tape to measure the 20 cm.

c. Pat the earth material into a smooth, even slope. It is important to have it the same depth all the way across, or barely V-shaped from side to side so the mass of earth material is slightly lower in the middle.

d. Position and tape the ruler over the earth-material slope. The ruler will help support the water-source container.

5. **Set up stream tables**
Ask Getters to pick up materials for their groups. Have the Starters supervise the stream-table setup. As groups complete the stream-table setups, give them notebook sheet 9, *Stream-Table Observations*, to glue into their notebooks. Students should label the condition for the first stream table on the sheet "standard—before run" and the second stream table on the sheet "standard—after run." They should also draw in the stream-table setup—before run.

As an alternative to notebook sheet 9, have cameras available. Students can take pictures of the stream tables after setup and after running the water through. They'll need to download the pictures, print them out, glue them into their notebooks, label the parts, and write notes to go with the pictures.

6. **Start the investigation**
Remind students that everyone will run exactly 1 L of water through the stream tables. Explain that it is important for everyone to use the same amount of water since this is one of the variables to be controlled in later runs. While the water is flowing, no one should touch the stream-table tray or the earth material in the tray, or bump or shake the table.

Send Getters to fill the 1 L containers with water. (If another class previously set up the stream tables, students may need to remove duct tape from the hole in the stream tables before using the water.) Let them begin the investigation by pouring the liter of water into the standard water-source container positioned over the earth material in the stream table.

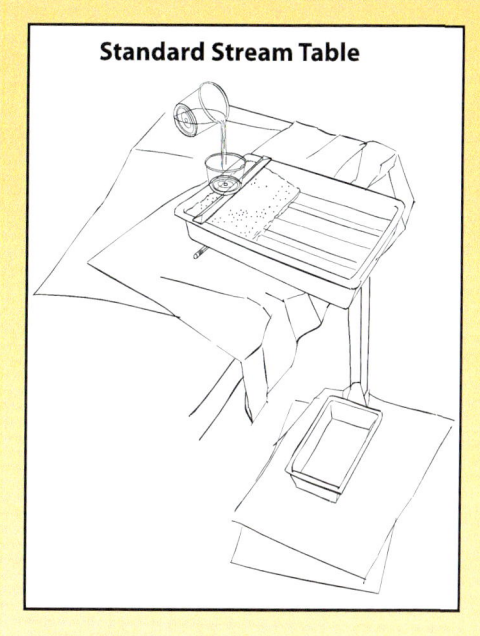

Standard Stream Table

Materials for Step 5
- Stream tables
- Bags of earth material
- Standard water sources
- Rulers
- Wood angles
- Containers, 1 L
- Basins
- Newspapers
- Duct tape
- Water
- Meter tapes
- Cameras (optional)
- Safety goggles and gloves (optional)
- **Stream-Table Observations** sheets

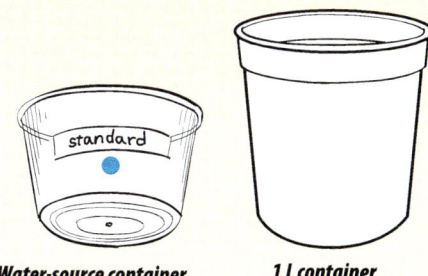

Water-source container   1 L container

SCIENCE AND ENGINEERING PRACTICES
Planning and carrying out investigations

Soils, Rocks, and Landforms Module—FOSS Next Generation

INVESTIGATION 2 – *Landforms*

> **TEACHING NOTE**
>
> *Go to FOSSweb for Teacher Resources and look for the Science and Engineering Practices—Grade 4 chapter for details on how to engage students with the practice of developing and using models.*

SCIENCE AND ENGINEERING PRACTICES

Developing and using models

Analyzing and interpreting data

CROSSCUTTING CONCEPTS

Patterns

Cause and effect

Scale, proportion, and quantity

Stability and change

Say it • Write it • See it • Hear it
New Word

7. **Guide the analysis of data**

   When all the water has run through the stream tables, hold a 2-minute discussion to review what students observed. Ask them to think about these questions.

   ➤ *Where did the earth material start in the tray, and where did it go? the sand? the clay?*

   ➤ *How does this model show what happens to earth materials in nature?*

   ➤ *How do the size of these landforms compare to those we are modeling?*

   ➤ *What does this tell us about the effect of flowing water on earth materials?*

   Have students draw a second picture of the stream table on notebook sheet 9, showing how it looks after the water has run.

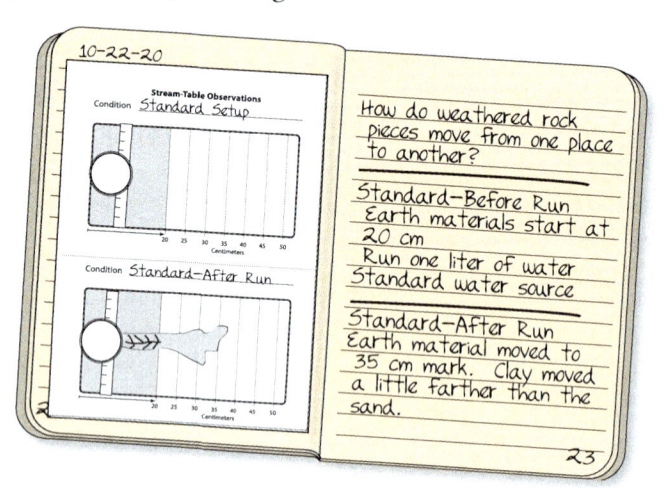

8. **Share observations and build vocabulary**

   Have the Reporter from each group report one or two of the group's observations to the class. As students report the various formations they have seen, develop landform vocabulary.

   For example, when students mention the depression in the earth through which the stream of water flowed, identify that space as a **valley** or **canyon**. When they mention the fan-shaped materials deposited at the stream or **river mouth**, identify it as an **alluvial fan** or **delta**.

9. **Distribute the *Landform Vocabulary* sheets**

   Have Getters pick up copies of notebook sheet 10, *Landform Vocabulary*, for their groups. Have students glue this sheet onto the left-hand side of a two-page spread. (Leave the right-hand page blank for definitions that will be added later.) Project the sheet to help students identify the valley and write the name in the appropriate location. Have students label the canyon and delta.

Part 1: Erosion and Deposition

Explain that the **landform** that was created in the stream tables is called an *alluvial fan* because it formed on "dry land." When this landform is created in a body of water it is called a *delta*. The landform shown on the sheet is a delta. Ask,

➤ *What patterns do you notice when you look at the results of other groups' stream-table investigation?*

10. Introduce *erosion* and *deposition*

Tell students,

One thing you observed in your stream tables is called **erosion**. Erosion is the movement of pieces of weathered rock by water or wind. Erosion changes the shape of the land. In your stream tables, water eroded the earth material away to form a valley or a canyon. A valley is the broad area through which a river flows. A canyon is a narrow, very steep-sided valley.

Earth material carried by moving water eventually stops moving and settles out. The settling process is called **deposition**. The eroded materials that settle are called **sediments**. When the water slows down, sediments are deposited in new locations, somewhere along the path of the moving water. The fan-shaped deposits of sediments you see at the mouth of your stream-table river form an alluvial fan.

Emphasize that *erosion* means to take away (from), and *deposition* means to place or settle out. The eroded materials end up being deposited materials in a new location.

11. Conduct the shake test

Remind students that many of them noticed that some of the eroded earth materials were carried farther by the flow of water than others. Ask,

➤ *Why do you think the clay was deposited farther downstream?*

Have students conduct the earth-materials-shake test (like they did in Investigation 1, Part 2, with the soils) to give them additional information that will help them think about why some eroded materials travel farther than others.

Give each group a 12-dram vial with cap. Provide these instructions.

a. Fill a vial half full with earth material from the stream table.

b. Use the syringe to transfer water into the vial until it is almost full.

c. Put the cap on the vial. Shake it for 5 seconds.

d. Set the vial on the table and observe.

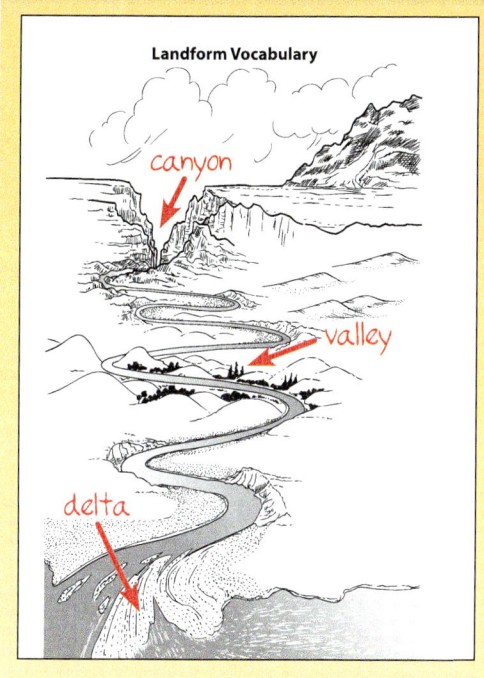

Landform Vocabulary

TEACHING NOTE

This is important to emphasize. Many students think erosion and deposition are the same thing.

Materials for Step 11
- *Vials with caps*
- *Syringes*
- *Water*

Soils, Rocks, and Landforms Module—FOSS Next Generation

INVESTIGATION 2 – Landforms

Let students conduct the shake test. Remind them that their task is to observe the rate at which different materials settle. Students will notice that the larger-grained sand particles settle to the bottom of the vial as soon as it is placed on the table. The smaller clay particles settle more slowly, forming a distinct clay layer on top of the sand.

12. Have a sense-making discussion
Ask,

➤ *What happened when you shook the earth material with water?* [The sand immediately settled to the bottom of the vial; clay settled on top of the sand.]

➤ *Why do you think the sand was deposited on the bottom of the vial and the clay on top of the sand?* [Sand is heavier (larger particles) and settled to the bottom first when the shaking stopped.]

➤ *What caused the sand and clay to deposit in different places in the stream table?* [Larger (heavier) earth materials settled as soon as the water flow started to slow down. Smaller (lighter) earth materials, such as clay particles, stayed suspended in the water longer, so they traveled farther and didn't settle until the flow stopped.]

➤ *Can you relate this to soils? How was the settling like that in the soil vials? How might the size of the rock particles in soils at different locations be determined by erosion and deposition?* [The settling process was just like that in the soil vials. Smaller particles stayed suspended longer so they are deposited farther downstream.]

13. Review vocabulary
Review the vocabulary words that have been introduced in this part so far. Make sure that these words are on the word wall. Add any other words students might need.

14. Answer the focus question
Have students consider the focus question and write a response in their notebooks.

➤ *How do weathered rock pieces move from one place to another?*

Encourage students to use new vocabulary words in their answers.

SCIENCE AND ENGINEERING PRACTICES
Developing and using models
Constructing explanations

CROSSCUTTING CONCEPTS
Cause and effect
Systems and system models

alluvial fan
canyon
delta
deposition
erosion
landform
river mouth
sediment
valley

EL NOTE
For students who need scaffolding, provide a sentence frame such as: Weathered rock pieces move from one place to another by _____. In our stream table we noticed _____.

Part 1: Erosion and Deposition

15. Assess progress: notebook entry
Have students hand in their notebooks open to the page on which they answered the focus question. After class, review students' entries to see if students clearly understand the difference between *erosion* and *deposition*.

What to Look For
- *Erosion is part of the sediment-moving process: earth materials are carried away by moving water that flows into rivers, forming valleys.*
- *Deposition is another part of the process: the eroded earth materials eventually are deposited or settled out somewhere downstream, for example, in an alluvial fan or delta—smaller particles usually move farther than larger particles.*

16. Clean up
Have students return everything but the stream table with the earth material to the materials station. Ask the Starters to recycle the water in the basins outdoors and clean the tabletops.

17. Store the trays
Have students place a small piece of duct tape over the drain hole in their trays and stack the trays crisscross in an appropriate location.

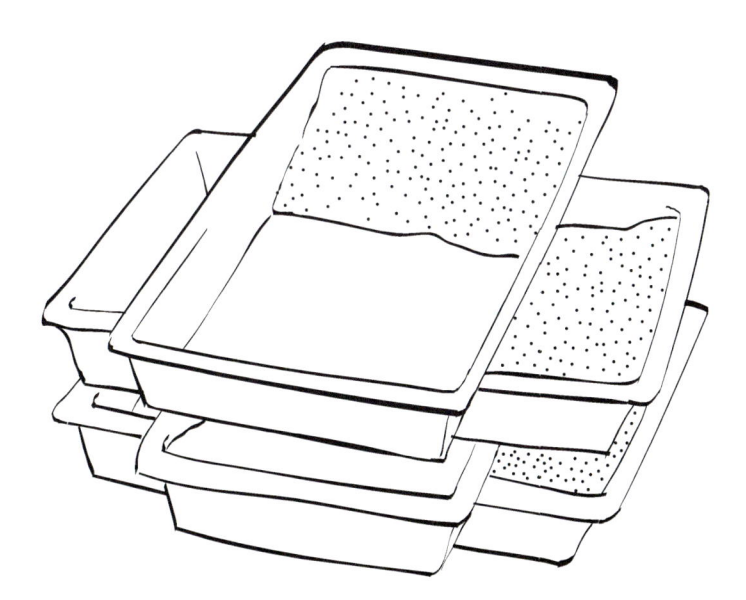

▶ **NOTE**
Allow the basins of water to settle after running the stream tables. Decant most of the water. Blot off as much water as possible. Stir the clay back into the stream-table earth material.

Soils, Rocks, and Landforms Module—FOSS Next Generation

INVESTIGATION 2 – Landforms

READING in Science Resources

18. Read "Erosion and Deposition"

Read the article "Erosion and Deposition" during a reading period. In this article, students learn more about the processes that move and deposit earth materials, creating landforms.

Before reading, give students time to preview the text by reading the subtitles, looking at and discussing the photographs and captions, and reviewing the words in bold. Review that the stream-table investigation is a way to present a model of erosion and deposition. Explain that this article will give them more information about how these processes work in the natural world.

19. Use a summarizing reading strategy

During the reading, tell students to read each section independently and to think about how their observations in class compare to the ideas presented in the text. To support reading comprehension, have students pause after each section and then take turns summarizing with a partner. Model by reading and summarizing the first page aloud. Ask students to discuss why they think the author put the photo of the beach here.

Have students read the next section on erosion. Pause and have Partner A give the main idea; Partner B gives the details. Repeat the process for deposition starting with Partner B. Have students read the last section independently and then take turns summarizing.

For reading strategies to support English learners and below-grade-level readers, see the Science-Centered Language Development chapter.

20. Discuss the reading

After reading, use these review questions to discuss the article. Have students discuss the first two questions in their groups. Call on a Reporter from each group to share one example. Encourage students to cite details and examples from the text to support their answers.

➤ *Describe and give examples of erosion.* [Rock is eroded by flowing water, wind, glaciers, and waves.]

➤ *Describe and give examples of deposition.* [Rock is deposited by slowing water, slowing wind, and melting glaciers.]

SCIENCE AND ENGINEERING PRACTICES

Obtaining, evaluating, and communicating information

ELA CONNECTION

These suggested reading strategies address the Common Core State Standards for ELA.

RI 1: Refer to details and examples in a text when explaining what the text says explicitly and when drawing inferences from text.

RI 6: Compare and contrast a firsthand and secondhand account of the same topic.

RI 7: Interpret information presented visually, and explain how the information contributes to an understanding of the text.

SL 1: Engage in collaborative discussions.

Part 1: Erosion and Deposition

Have students answer the next two questions by drawing a model either in their notebooks individually or as a group task on chart paper. Project a few of the notebook pages or post the posters on the wall for students to compare and critique.

➤ *Describe how rocks in the mountains become sand on a beach along the coast.* [Rock is weathered by roots, ice, abrasion, chemicals, and gravity. It is then eroded (transported) by streams and rivers until it reaches the sea. It is deposited at the shore when the river flows into the sea.]

➤ *What do you think will happen to the Sierra Nevada Mountains in California in the next hundred million years?* [They will probably weather and erode, ending up as sand in the sea.]

WRAP-UP/WARM-UP

21. Share notebook entries

Conclude Part 1 or start Part 2 by having students share notebook entries. Ask students to open their science notebooks to the most recent entry. Read the focus question together.

➤ *How do weathered rock pieces move from one place to another?*

Have students pair up with a student from another group to
- share their stream-table drawings;
- discuss how rock materials get from mountains to a new location.

Encourage students to describe the cause-and-effect relationships involved in the processes.

Explain to students that their task is to help their partner improve their notebook entry. Have them take turns telling each other one thing they like about the entry and one thing they could do to make it better. Allow time for students to add to or revise their entries based on their partner's feedback.

ELA CONNECTION

These suggested reading strategies address the Common Core State Standards for ELA.

RI 3: Explain procedures, ideas, or concepts in a scientific text.

SL 5: Add visual displays to presentations.

EL NOTE

See the Science-Centered Language Development chapter for strategies on sharing notebook entries.

ELA CONNECTION

This suggested reading strategy addresses the Common Core State Standards for ELA.

W 5: Strengthen writing by revising.

CROSSCUTTING CONCEPTS

Cause and effect

Systems and system models

Soils, Rocks, and Landforms Module—FOSS Next Generation

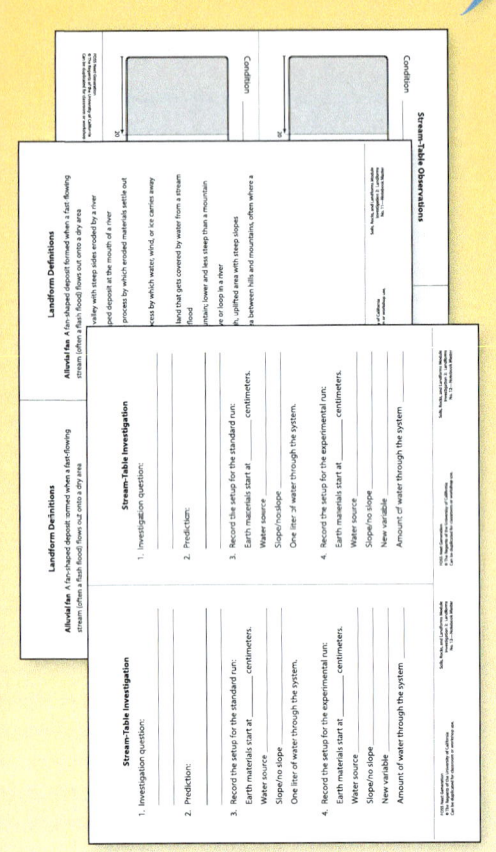

INVESTIGATION 2 – Landforms

MATERIALS for
Part 2: *Stream-Table Investigations*

For each group
- 1 Stream table with earth material
- 1 Water source, standard or flood
- 1 Container, 1 L
- 1 Ruler
- 1–2 Wood angles
- 1 Meter tape
- 1 Basin
- ❑ 4–8 Notebook sheet 9, *Stream-Table Observations*
- ❑ 4 Notebook sheet 11 *Landform Definitions*
- ❑ 4 Notebook sheet 12, *Stream-Table Investigation*
- 4 FOSS Science Resources: *Soils, Rocks, and Landforms*
  - "Landforms Photo Album"

For the class
- Duct tape
- Cotton swabs
- Food coloring
- 3 Plastic cups
- 1 Vial, 12 dr
- White, powdered clay (See Step 3 of Getting Ready.)
- Additional materials for experiments (See Step 5 of Getting Ready.) ★
- Paper towels and newspaper ★
- Water ★
- 1 Pitcher
- 1 Dustpan and broom ★
- 1 Computer with Internet access ★
- ❑ 1 Teacher master 8, *Stream-Table Ideas*
- ❑ 1 Teacher master 9, *Video Discussion Guide: Erosion and Deposition*

For embedded assessment
- ❑ • *Performance Assessment Checklist*

★ Supplied by the teacher.   ❑ Use the duplication master to make copies.

Nos. 9, 11, 12—Notebook Masters

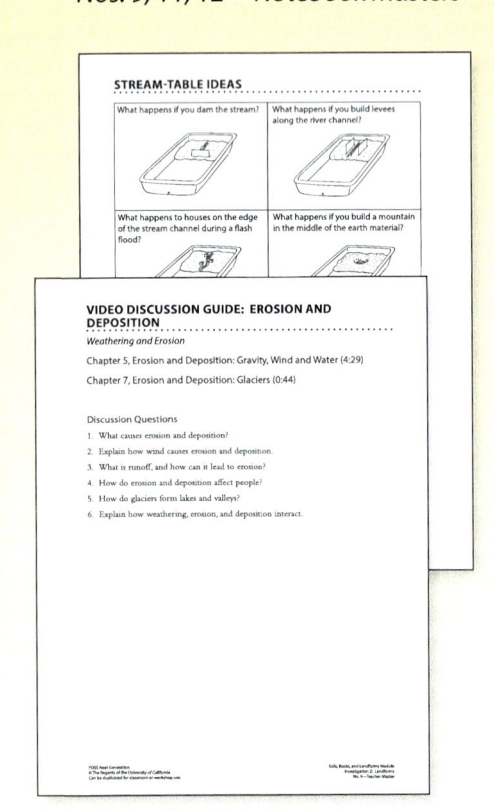

Nos. 8–9—Teacher Masters

170   Full Option Science System

Part 2: Stream-Table Investigations

GETTING READY for
Part 2: Stream-Table Investigations

1. **Schedule the investigation**
   This part will take from two to four active investigation sessions. The article in *FOSS Science Resources* is integrated as a photo reference at the beginning of the active investigation.

2. **Preview Part 2**
   Students continue to run stream tables to learn how variables can affect erosion and deposition. They investigate the variables of slope and water volume (flood). Then they plan and conduct their own stream-table investigations. The focus questions are **How does slope affect erosion and deposition?** and **How do floods affect erosion and deposition?**

3. **Check earth materials**
   After three or four uses of the earth materials in the stream tables, you may want to replenish the clay sediments. Work a 12-dram vial of clay into the earth-material mixture of each stream table.

4. **Prepare food coloring**
   Put 5–6 drops of red, blue, or green food coloring in cups to distribute during the stream-table investigations. Place one cotton swab in each cup. When you visit groups as they are running water through their stream tables, suggest that adding food coloring to the water will allow them to observe flow patterns more effectively. Do this by plunging a color-saturated swab into the water source for a few seconds, just above the hole. This will add an intense stream of color to the water flow.

5. **Plan for additional materials**
   In Step 19 of Guiding the Investigation, students plan their own stream-table investigations. Think about extra materials you can provide for them, such as gram pieces, cardboard, and craft sticks for dams and levees. If students need some starter ideas for an investigation, plan to distribute copies of teacher master 8, *Stream-Table Ideas*.

6. **Plan to watch video (optional)**
   Preview the *Weathering and Erosion* video—chapter 5 (4:29) and chapter 7 (0:44). Links to the video are found on FOSSweb in the Resources by Investigation section. Have students view the video at the end of this part. Plan to use teacher master 9, *Video Discussion Guide: Erosion and Deposition*, to lead a class discussion after watching the video with the class.

Soils, Rocks, and Landforms Module—FOSS Next Generation

INVESTIGATION 2 – Landforms

7. **Preview the online activities**

    Preview the online activities and videos by going to the Resources by Investigation section on FOSSweb. Students can use videos of the stream tables in the to compare two stream tables as they continue to test the variables of slope, stream-flow rate, and stream-bed material.

    Also preview the tutorial "Stream Tables—Slope and Flood." This is a more structured opportunity for students to work with an animation dealing with changing stream-table variables.

    Students can engage with these activities after the reading.

8. **Plan assessment: performance assessment**

    Students can work independently (or guided by notebook sheet 12, *Stream-Table Investigation*) to plan and conduct their own stream-table investigations as a third session of this part. This provides an opportunity for you to observe their science and engineering practices. Carry the *Performance Assessment Checklist* with you as you visit the groups.

Part 2: Stream-Table Investigations

GUIDING *the Investigation*
Part 2: *Stream-Table Investigations*

1. **Introduce variables to investigate**
   To begin this investigation, review the landforms vocabulary sheet as a class and make a concept grid on chart paper. Add the landforms in the first column and title the other columns as shown below. Tell students that they will find out more about how landforms are formed and add that information to the concept grid. They can also speculate about how human actions might impact the landforms.

Landform	Characteristics	How it was formed	Human impact
Alluvial fan			
Canyon			
Delta			
Floodplain			
Meander			
Valley			

   Have students open their *FOSS Science Resources* books to "Landforms Photo Album" and focus on the pictures of landforms caused by weathering, erosion, and deposition. Ask students to brainstorm in their groups a list of factors that explain how the valleys, alluvial fans or deltas, and other land features they see in the photos might have formed. Make sure students consider how deep the valleys are and how much water is flowing in the bodies of water. You might assign each group a page to read and analyze.

2. **Focus question: How does slope affect erosion and deposition?**
   Tell students that today they get to test two new conditions: **slope** of the land and **flood**. Have students open their science notebooks, date a clean page, and copy the first focus question while you project or write it on the board.

   ➤ *How does slope affect erosion and deposition?*

3. **Plan the slope investigation**
   Pair up groups so they can run their stream tables side by side. One group will run a standard stream table (as they did in Part 1) and the other will test the slope variable during this first run of the day. (Let students know that groups will switch in the next run and those doing the standard run will get to test the flood variable.) By working in group pairs, students will be able to easily compare their results.

Soils, Rocks, and Landforms Module—FOSS Next Generation

FOCUS QUESTION
How does slope affect erosion and deposition?
How do floods affect erosion and deposition?

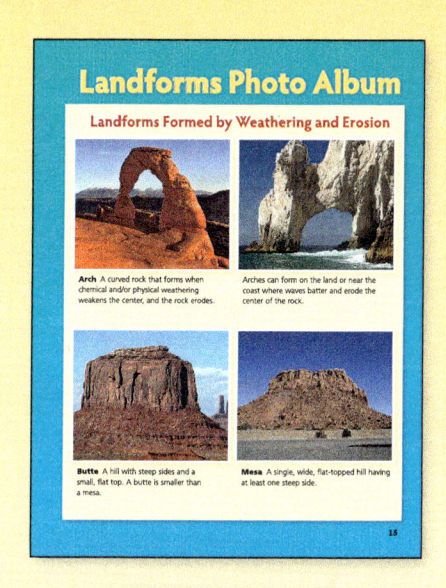

ELA CONNECTION
This suggested strategy addresses the Common Core State Standards for ELA.

RI 7: Interpret information presented visually, and explain how the information contributes to an understanding of the text.

TEACHING NOTE
This part is written for you to guide the investigation planning, given that most elementary students have little experience. Keep these guidelines in mind, but let students lead the investigation-planning discussion. If your students have experience planning investigations, let them go ahead on their own, but check plans and setup before they start the water flowing.

INVESTIGATION 2 – Landforms

SCIENCE AND ENGINEERING PRACTICES
Planning and carrying out investigations

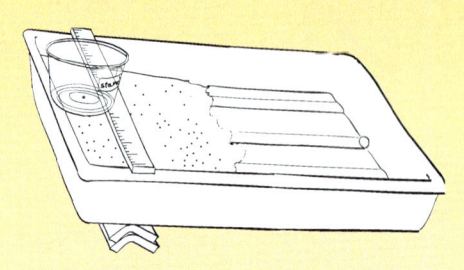

Note two wood angles are used by the "slope" groups.

Materials for Step 4
- Stream tables, with earth materials
- Water sources, standard
- Rulers
- Wood angles
- Containers, 1 L
- Basins
- Meter tapes
- Newspapers
- Duct tape
- Water

Plan the investigation together. These points should be included.

- The focus question: How does slope affect erosion and deposition?
- Prediction: How do you think increasing the slope will affect erosion and deposition? (Use a sentence frame if needed: I think _____ because _____ .)
- Variables to control: Both stream tables have earth materials pushed to 20 cm, both will run exactly 1 L of water through the system, and the water flow in both stream tables will start at the same time.
- Variable to change: One group (standard stream) will leave their stream table flat on the table (as they did in Part 1). The other group (slope) will prop up the water-source end of the stream table, using wood angles.

Have copies of notebook sheet 9, *Stream-Table Observations,* available for students to pick up and record results of both the standard tray and the slope tray after the water has run. (Students should already have a drawing of the standard stream-table setup, so they don't need to redraw that here.)

4. Set up stream-table systems
Have Getters get the materials they will need to run the stream tables. Give students a few minutes to get the earth material ready. They might need to break up the material if it has solidified. Adding a little water will speed up the process.

When the groups have their earth material conditioned, all the parts of their systems assembled, and newspaper strategically deployed, make sure that the slope groups have placed two wood angles under their stream tables to prop up the water-source end of the tray.

5. Run the stream tables
Send Getters back to the materials station for 1 L of water and let the trials begin. Remind students that both groups should start the water running at the same time for the stream tables that are running side by side. When all of the water has flowed through the table, they should begin to draw and write their observations.

6. Guide the recording of the stream-table run
As you visit the groups, have students compare the slope and standard results. Ask students a few questions to help guide their recording and to continue to build vocabulary.

➤ *Where did erosion occur? Did the slope make a difference in how much earth material eroded?* [Erosion occurred in the canyon; more material eroded in the steeper tray.]

Part 2: Stream-Table Investigations

➤ *Did you notice any difference in the time it took for erosion or deposition?* [Things usually happen faster in the slope tray.]

➤ *Did the slope make a difference in the shape of the canyon that was left after the earth materials were carried away?* [The canyon is usually deeper and has steeper sides.]

➤ *How far did the eroded materials travel in the two trays?* [The earth material traveled farther in the slope tray.]

➤ *Why do you think there was more erosion in the steeper tray?* [When the land was steeper, the water flowed faster. Faster flowing water has more energy to move sand and clay farther along the stream or **river channel**.]

➤ *How does this help you explain what happened in the photos in the "Landforms Photo Album?"*

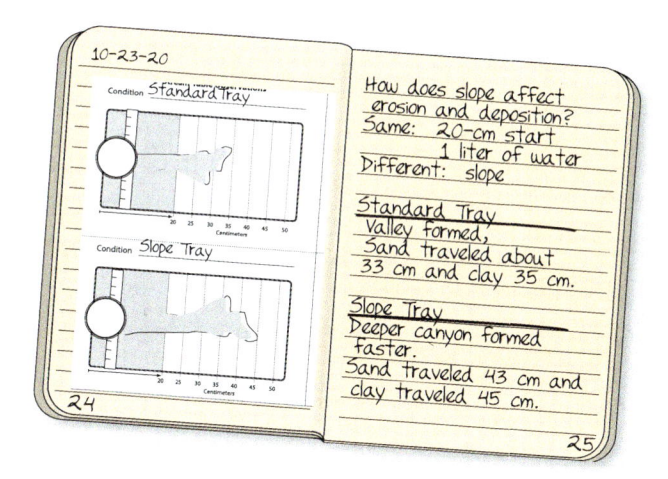

Have students complete drawings and descriptions of the two stream tables.

7. **Answer the focus question**
   Have students consider the focus question and write a response in their notebooks.

   ➤ *How does slope affect erosion and deposition?*

   Encourage students to use new vocabulary words in their answers.

8. **Store trays (optional)**
   If you decide to stop here, remove excess water from the earth materials and basins as described in Part 1. Stack the trays crisscross so they take up less counter space but still have airflow between trays.

   POSSIBLE BREAKPOINT

SCIENCE AND ENGINEERING PRACTICES

Developing and using models

Planning and carrying out investigations

Constructing explanations

TEACHING NOTE

This important idea may not be obvious to students. You may want to ask questions to help students think about a ball rolling on a gentle slope compared to one rolling on a steep slope. Any object moving in response to gravity down a steeper slope will move faster and have more energy than the same object moving down a gentler slope. Faster = more energy.

CROSSCUTTING CONCEPTS

Cause and effect

EL NOTE

For students who need scaffolding, provide a sentence frame such as:
We found out that _____.
When we compared _____ to _____, we noticed _____.

Soils, Rocks, and Landforms Module—FOSS Next Generation

INVESTIGATION 2 – Landforms

Materials for Step 11
- Stream tables, with earth materials
- Water sources, flood
- Rulers
- Wood angles
- Containers, 1 L
- Basins
- Meter tapes
- Newspapers
- Duct tape
- Water

SCIENCE AND ENGINEERING PRACTICES

Planning and carrying out investigations

Analyzing and interpreting data

Constructing explanations

9. Introduce flood conditions
Ask students if they know what a flood is. If they need help, here are some key points to help guide the discussion.
- Floods are caused by excessive runoff from rain or snowmelt.
- If soil is frozen or saturated with water, water can't soak in, so runoff is greater.
- Because there is so much water in a flood, land that is normally dry gets covered with water.

10. Focus question: How do floods affect erosion and deposition?
Have students copy the second focus question while you project or write it on the board.

➤ How do floods affect erosion and deposition?

11. Plan the flood investigation
Have groups switch roles. Those who tested slope will now run a standard stream table. Those who ran the standard will now run a flood. Point out the flood water-source containers half the groups will use, then let them plan the next investigation on their own.

If this is the second run of the day, students may need to soak up some of the water in the earth materials with newspaper in order to stabilize the mass of earth material behind the 20 cm mark.

Have students check their plans with you before they start to run the water through the stream tables. Be sure each pair of groups has prepared a standard tray and a flood tray and is controlling all variables except for the water source. (Both stream tables should be resting "flat" on the tables.) Students should glue a second recording sheet (notebook sheet 9, *Stream-Table Observations*) in their notebooks.

12. Test flood conditions
When students have a complete plan and their stream tables are set up and ready to go, send Getters to the materials station for 1 L of water. Remind students that they should start the water running in both trays at the same time. Let the investigation begin.

13. Have a sense-making discussion
As students finish their investigations, remind them to draw pictures and write descriptions of the results of their stream-table runs, as they did when they were testing slope.

When the data are recorded, discuss the results. Ask students to make some statements about how slope and flood affect erosion and deposition. Chart the statements.

Part 2: Stream-Table Investigations

Students should find that in the tray set at a steeper slope, the earth material eroded faster, moved farther, and formed a deeper valley—a canyon. Flood conditions also result in more erosion; broader, deeper, straighter canyons, and more extensive deposition that reached farther into the basin. The flood eroded more earth material than the standard and sloped stream tables because a larger mass of moving water has more energy than a smaller mass of moving water.

14. Review vocabulary

Review the new vocabulary introduced in this part. Have students turn to notebook sheet 10, *Landform Vocabulary*. Have them glue notebook sheet 11, *Landform Definitions,* on the opposite page. In their groups, have students review the landform definitions and place new landform terms in the appropriate place on the graphic.

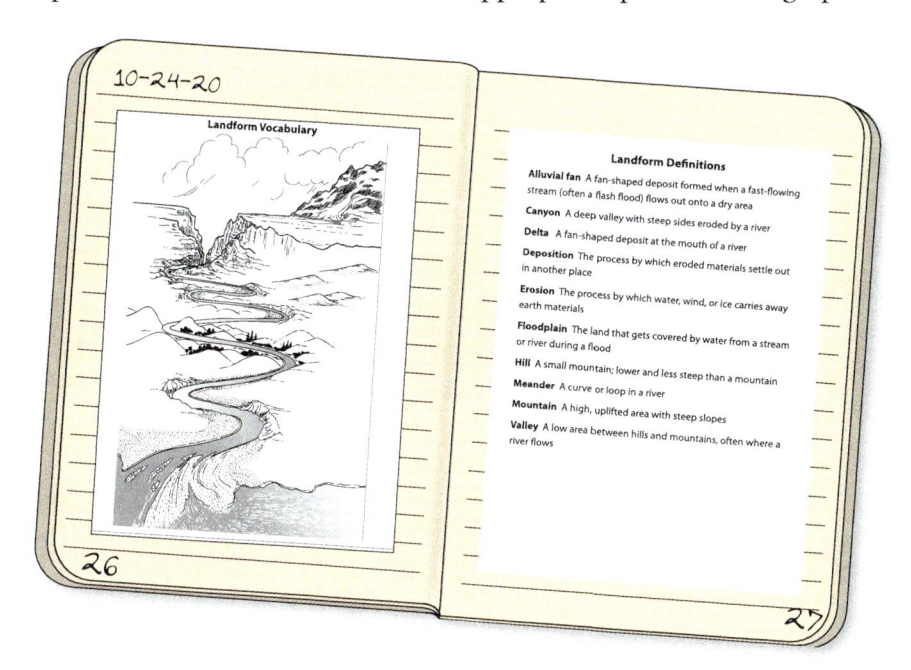

15. Answer the focus question

Have students consider the focus question and write a response in their notebooks.

➤ *How do floods affect erosion and deposition?*

CROSSCUTTING CONCEPTS
Cause and effect

flood
floodplain
meander
mountain
river channel
slope

EL NOTE

For students who need scaffolding, provide a sentence frame such as: Floods cause _____. When we compared _____ to _____, we noticed _____.

Soils, Rocks, and Landforms Module—FOSS Next Generation

177

INVESTIGATION 2 – Landforms

16. Discuss erosion, deposition, and soils
Tell students,

Normal weathering and erosion change Earth's surface a tiny bit at a time. It can take thousands of years to create a valley or alluvial fan.

A flood is an event that can change Earth's surface in a short period of time. Floods can destroy existing landforms and create new landforms in hours and days. A flood can erode and redeposit more rock material in 10 hours than normal water flow can erode and redeposit in 10 years.

Ask students,

➤ *How do erosion and deposition contribute to soil formation?* [Earth materials are carried from one location to another and, especially during floods, deposit many more new layers of rock material.]

➤ *How might a flood be detrimental to soils?* [Erosion can take away the fertile topsoil that is good for growing crops.]

➤ *How might erosion and deposition contribute to the kinds and sizes of rock material that you find in soils?* [The kinds of rock you find in a soil depends on the kind of parent rock (the source rock) and how the rock has weathered and moved. As rocks erode, they tumble against one another; for instance in a river or flood, more pieces break off and they get smaller and smaller. When rock particles are deposited, they can become part of the soil.]

17. Identify soils from Investigation 1
Have students turn back to the entries in their science notebooks from Investigation 1 where they first attempted to identify where each soil studied comes from: desert, delta, forest, or mountaintop.

Have them discuss these soils in their groups, providing additional evidence for their arguments from what they have learned about erosion and deposition. Have them share their final decisions with the class, then reveal the soil sources (locations).

Soil 1—River delta (Evidence: the smallest earth materials, carried a long way when deposited in a delta)

Soil 2—Mountaintop (Evidence: large pieces of earth material, not a lot of humus)

Soil 3—Desert (Evidence: mostly sand and gravel, not a lot of humus)

Soil 4—Forest (Evidence: lots of humus and different sizes of earth materials)

TEACHING NOTE

Make sure students understand that soil formation takes many years, in fact so many that some people consider soil a nonrenewable resource.

CROSSCUTTING CONCEPTS

Cause and effect
Stability and change

TEACHING NOTE

Go to FOSSweb for *Teacher Resources* and look for the *Science and Engineering Practices—Grade 4* chapter for details on how to engage students with the practice of engaging in argument from evidence.

SCIENCE AND ENGINEERING PRACTICES

Constructing explanations
Engaging in argument from evidence

▶ **POSSIBLE BREAKPOINT**

Part 2: Stream-Table Investigations

18. **Design a new investigation**
    Tell students that now they will get to plan and conduct their own investigations. Two groups will work together to come up with a new question to investigate.

19. **Discuss investigation criteria**
    Remind students that one group will set up a "standard" run and the other group will set up the "experimental" run—two runs to compare to each other with only one variable (condition) changed. They can use the flat (standard) setup, the slope setup, or the flood setup for the new "standard run." Then they need to identify one change for the experimental run.

20. **Consider scaffolding the process**
    If students need scaffolding for this investigation, you can distribute copies of teacher master 8, *Stream-Table Ideas* to stimulate their thinking. If you want to leave it more open-ended, have students devise a plan on their own, but they should check with you before they get materials to make sure they have controlled all necessary variables. Groups can use a copy of notebook sheet 12, *Stream-Table Investigation*, to help them design their investigations. The plan should include

    - a question to investigate (a focus question);
    - a prediction;
    - a description of the stream-table setup they plan to use as a standard;
    - a description of the experimental setup.

21. **Plan student investigations**
    Let students start their planning. Encourage them to talk through the process of setup and investigation before they start writing their plans on paper. Students often revise and modify their plans as they consider the process, step by step.

    Students should have complete plans and a clear vision of their investigation before they go to the materials station to pick up their equipment.

22. **Conduct the investigations**
    Let students get the materials they need to conduct their investigations. Circulate among the groups as they work. Have copies of notebook sheet 9, *Stream-Table Observations*, available for students to use when recording their results.

TEACHING NOTE

Go to FOSSweb for *Teacher Resources* and look for the *Science and Engineering Practices—Grade 4* chapter for details on how to engage students with the practice of asking questions.

SCIENCE AND ENGINEERING PRACTICES

Asking questions

Planning and carrying out investigations

EL NOTE

If needed, provide sentence frames for planning an investigation such as:
First, we will _____. Then we will _____. The variable we are changing is _____. The variable that is not changing is _____. We will measure _____ using _____.

Materials for Step 22
- *Additional materials for experiments*

INVESTIGATION 2 – Landforms

SCIENCE AND ENGINEERING PRACTICES

Asking questions

Planning and carrying out investigations

Analyzing and interpreting data

Constructing explanations

DISCIPLINARY CORE IDEAS

ESS2.A: Earth materials and systems

CROSSCUTTING CONCEPTS

Patterns

Cause and effect

EL NOTE

For students who need scaffolding, provide sentence frames such as:
We expected the water to _____ because _____. We found out _____.

We expected the earth material to erode at the _____ because _____. We found out _____.

We expected the earth material to be deposited at the _____ because _____. We found out _____.

---

23. Assess progress: performance assessment

Circulate from group to group throughout the active investigation in this session. Conduct 30-second interviews, listen to group discussions and observe how students work together to learn more about erosion and deposition. Make notes on the *Performance Assessment Checklist* while students work.

What to Look For

- *Students write a focus question to guide their investigations. (Asking questions.)*
- *Students plan and carry out investigations, controlling variables and recording observations as needed. (Planning and carrying out investigations; ESS2.A: Earth materials and systems.)*
- *Students analyze and interpret the data they collect in order to compare the new conditions they test. (Analyzing and interpreting data.)*
- *Students construct explanations to account for the changes they see in the stream table when new variables are tested. (Constructing explanations.)*
- *Students note trends and patterns as compared to other stream-table runs they have conducted and think about how a new variable affects the formation of various landforms. (Patterns; cause and effect.)*

24. Answer the focus question

Remind students to answer their focus question after they've recorded their observations.

25. Present results

Call the class to attention. Have each group briefly describe the question they were investigating, what happened when they ran the water through the stream table, and how they answered their focus question. You can do this while the stream tables are still set up, or use a document camera to have students share their notebook entries for planning the investigations and the drawings of their results.

Students should answer these questions during their presentations.

➤ *Did the water flow where you expected it to? Explain.*

➤ *Did the earth material erode where you expected it to? Explain.*

➤ *Was the earth material deposited where you expected it? Explain.*

➤ *What was the effect of the flowing water over the earth material?*

Part 2: Stream-Table Investigations

26. Clean up
Ask students to drain and stack the stream tables. Crisscross the stream tables for storage. Gritty materials should be rinsed in a basin of water (never down the sink) and set out to dry.

27. View video (optional)
Before showing the video, distribute teacher master 9, *Video Discussion Guide: Erosion and Deposition*. Have students review the questions and discuss what they already know. Tell them to mark the questions they aren't sure about. Then, have them divide up the questions so that each student has one or two questions they will be looking for more information about in the video.

Show students the videos on erosion and deposition from the *Weathering and Erosion* video (chapter 5, duration 4:29 and chapter 7, duration 0:44).

After watching the video, give students time to share their answers with the rest of their group and review the questions they weren't sure about before watching the video. Here is what students should discuss.

➤ *What causes erosion and deposition?* [Gravity, wind, running water, waves, and glaciers.]

➤ *Explain how wind causes erosion and deposition.* [Wind blows sand and soil away from weathered rock. When the wind stops, it drops the sand and soil in a new place.]

➤ *What is runoff, and how can it lead to erosion?* [When soil has soaked up all the water it can, the excess runs over the top of the soil and can carry it away.]

➤ *How do erosion and deposition affect people?* [Soil can erode away, but rich new soil can be deposited in deltas. Blowing sand causes weathering, but it can form beaches and sand dunes.]

➤ *How do glaciers form lakes and valleys?* [They move downhill, crushing rock and moving it to new places. Glaciers can dig out valleys, and when they melt, the water can form new lakes.]

➤ *Explain how weathering, erosion, and deposition interact.* [Weathering breaks down landscapes, erosion carries away landscapes, and deposition creates new landscapes elsewhere.]

Have students return to the class concept guide created in Step 1 of Guiding the Investigation to add information gathered from the video.

ELA CONNECTION

This suggested reading strategy addresses the Common Core State Standards for ELA.

SL 2: Paraphrase portions of a text presented in diverse media.

SCIENCE AND ENGINEERING PRACTICES

Obtaining, evaluating, and communicating information

Soils, Rocks, and Landforms Module—FOSS Next Generation

INVESTIGATION 2 – Landforms

28. View online activities

Have students engage with the stream table online activities and videos as they continue to test the variables of slope, stream-flow rate, and stream-bed materials. For a more structured experience, have students use the tutorial, "Stream Tables: Slope and Flood."

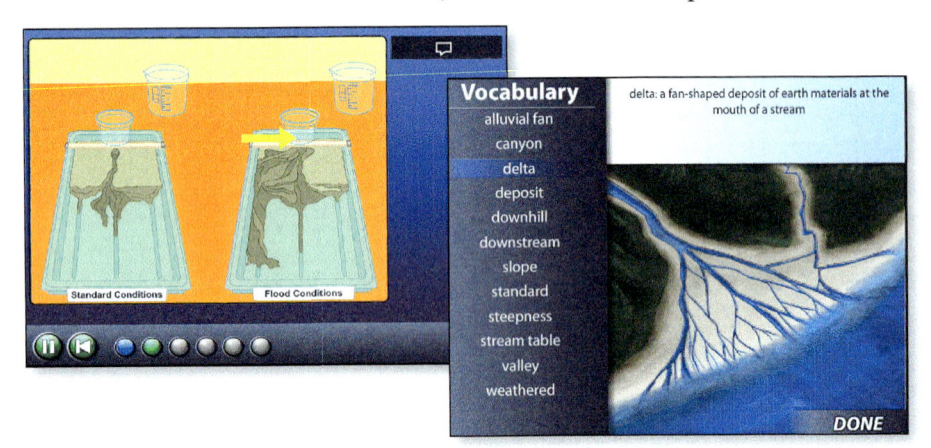

WRAP-UP/WARM-UP

29. Share notebook entries

Conclude Part 2 or start Part 3 by having students share notebook entries. Ask students to open their science notebooks to the most recent entry. Ask students to pair up with a partner with whom they did not conduct investigations to

- share the plans and results of their own investigations;
- review landform vocabulary on notebook sheet 10.

Encourage students to critique each other's work.

➤ *What was clear? What could be improved?*

Have students discuss how their investigations helped them answer the original focus question.

➤ *What evidence did your results provide to support your answer?*

Allow time for students to revise their entries and to add new questions they could investigate.

ELA CONNECTION

This suggested reading strategy addresses the Common Core State Standards for ELA.

SL 1: Engage in collaborative discussions.

SCIENCE AND ENGINEERING PRACTICES

Constructing explanations

---

182   Full Option Science System

Part 3: Schoolyard Erosion and Deposition

MATERIALS for
Part 3: Schoolyard Erosion and Deposition

For each student
- 1 Clipboard (optional) ★
- 1 Pencil ★

For the class
- 1 Large outdoor watering can (optional) ★
- 4 Empty recycled soft-drink bottles with caps, 2 L ★
- Water ★
- Extra pencils ★
- 1 Computer with Internet access ★

For embedded assessment
- ❏ Notebook sheet 13, *Response Sheet—Investigation 2*
- ❏ *Embedded Assessment Notes*

★ Supplied by the teacher.   ❏ Use the duplication master to make copies.

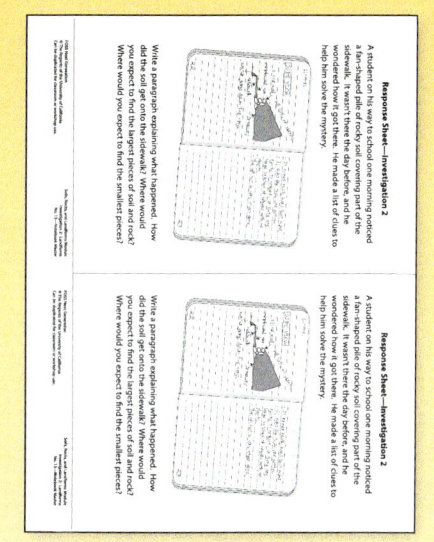

No. 13—Notebook Master

Soils, Rocks, and Landforms Module—FOSS Next Generation

INVESTIGATION 2 – Landforms

GETTING READY for
Part 3: Schoolyard Erosion and Deposition

1. **Schedule the investigation**
   This part will take one active investigation session. Ideally, the ground will be dry, but you can conduct the outdoor session on a day when the ground is wet, even when it is sprinkling or snowing.

2. **Preview Part 3**
   Students consider whether erosion and deposition are happening in their own schoolyard. They look for evidence of erosion and for locations where deposition is in evidence. They simulate a rainstorm by pouring water on various outdoor surfaces. The focus question for this part is **Where are erosion and deposition happening in our schoolyard?**

3. **Select your outdoor site**
   Scout out the most visible evidence of erosion and deposition in the schoolyard or nearby neighborhood. If you have recently had a rainstorm, this may be obvious.

   For Steps 7 and 8 of Guiding the Investigation, you will need four locations for erosion and deposition simulations. Select areas that provide a nice variety and contrast—a grassy slope versus a slope with little vegetation, a steep slope versus a more moderate slope, a place with loose topsoil versus compacted earth, and so forth. Also think about different surfaces such as pavement, gravel, and soil. By pouring water on your asphalt schoolyard, you may be able to locate the paths taken by rainwater as it drains off the hard surface.

   If you have an extremely flat schoolyard, consider a walking field trip to a nearby site that has the conditions you are looking for. Try to find another adult to join you on this walking trip. You should be able to find evidence of erosion, even on fairly flat surfaces. Look for these conditions:

   - Depressions in the ground
   - Areas around drain grates or spouts
   - Areas around a building that are often exposed to wind and sunlight
   - High-foot-traffic areas versus low-foot-traffic areas

4. **Check the site**
   Check the outdoor site on the morning of the outdoor activity. Look for unexpected, unsightly, and distracting items, or anything else that could affect the activity in the area where students will be working.

> **TEACHING NOTE**
>
> For a more detailed discussion about methods for working with students outdoors, see the Taking FOSS Outdoors chapter.

Part 3: Schoolyard Erosion and Deposition

If doing this activity in the winter, be careful not to pour the water on areas that could become hazardous if the water freezes to form a patch of ice.

5. **Plan to review outdoor safety rules**
   Before going outside, remind students of the rules and expectations for the outdoor work. Remind students not to put anything from the ground or from plants in their mouths. Establish the boundaries for their work outdoors.

6. **Prepare materials**
   If possible, get a large (at least 2 L) watering can with a sprinkling head that can be used to simulate a rain shower. Be sure the sprinkling spout is securely attached. If you can't get a watering can, use a 2 L soda bottle and make it sprinkle by piercing multiple holes in the cap, using a drill or hot nail. Squeezing the capped bottle will deliver a forced spray and simulate a downpour. Transport water outdoors using four recycled 2 L soda bottles filled with equal amounts of water. Screw the caps on tightly and have students carry them.

7. **Preview the effects**
   The results of this outdoor experience will vary depending on the moisture level in the earth materials and how hard packed they are. Consider doing a little experimenting yourself before taking students outside so you will know what to expect.

8. **Preview the online activity**
   Preview the virtual investigation called "Stream Tables" by going to the Resources by Investigation section on FOSSweb. This virtual investigation provides another opportunity for students to conduct stream-table investigations changing the variables of slope of the land, flood flow, and the addition of a dam in the river. Students can record the simulated results in their notebooks. This can be a substitute experience for students who are absent during the actual classroom stream-table investigations.

9. **Plan assessment: response sheet**
   In Step 11 of Guiding the Investigation, use notebook sheet 13, *Response Sheet—Investigation 2*, for a closer look at students' understanding of the processes of erosion and deposition. Plan to have students spend time reflecting on their responses after you have reviewed them. For more information about next-step and self-assessment strategies, see the Assessment chapter.

2-liter bottle cap

Soils, Rocks, and Landforms Module—FOSS Next Generation

INVESTIGATION 2 – Landforms

FOCUS QUESTION

Where are erosion and deposition happening in our schoolyard?

> **TEACHING NOTE**
>
> Wind erosion was not investigated firsthand, but it is important to erosion. Ask students to use information from the readings and videos as evidence about wind erosion.

GUIDING the Investigation
Part 3: Schoolyard Erosion and Deposition

1. **Review erosion and deposition**
   Review the processes of erosion and deposition.

   ➤ *What is erosion?* [The transportation of earth materials by water or wind.]

   ➤ *What does it look like when erosion has happened?* [Valleys and stream channels are carved into the surface of the soil.]

   ➤ *What is deposition?* [The settling of earth materials (sediments) in a new location by water or wind.]

   ➤ *What does deposition look like?* [Sand and mud accumulate.]

   Students should recall the investigations they did with the stream tables and know that erosion is the process that carries away sediments and deposition is the settling of the materials in a new place. Students may also mention that slope and the amount of water that rushes over the land affects how much earth material is moved.

2. **Focus question: Where are erosion and deposition happening in our schoolyard?**
   Ask students the focus question as you write it on the board and have students copy it into their notebooks.

   ➤ *Where are erosion and deposition happening in our schoolyard?*

   Have students turn and talk to someone in their group about where they might find evidence of erosion and deposition in the schoolyard. Have students share a few of their ideas.

3. **Describe the investigation**
   Tell students,

   Today we will go outdoors to look for evidence of erosion and deposition. We will walk to many of the places you mentioned to check our predictions. The second part of our work will include pouring water on four surfaces to simulate a heavy rainstorm and see if erosion and deposition happen.

4. **Set up notebooks**
   To make recording easier when the class is outdoors, have students set up a data-table page in their notebooks now. Draw an example on the board for students to copy that will help them organize their data and give them room to record their observations.

	Date
Where are erosion and deposition happening in our schoolyard?	
Evidence of erosion and deposition in the schoolyard.	
Surface 1	Surface 2
Surface 3	Surface 4

186 Full Option Science System

Part 3: Schoolyard Erosion and Deposition

5. **Go outdoors**

   Have students line up with their notebooks and pencils. Distribute water bottles and the watering can for students to carry outside. Head out to your outdoor home base and form a sharing circle. Have students look around the schoolyard and decide if there are any other locations that you should consider searching, such as cracks in pavement.

6. **Look for evidence of erosion and deposition**

   Walk together as a class to the various locations to search for evidence of erosion and deposition. If students see clearly eroded areas, ask

   ➤ *Where did erosion occur and where did the eroded materials get deposited?*

   Analyze the erosion–deposition relationship. For example, encourage students to think about the direction of the slope.

   Or vice versa, if they find deposited material ask,

   ➤ *Where did the deposited material come from? What eroded?*

   ➤ *What particle sizes can you observe in the deposited materials?*

7. **Simulate erosion and deposition**

   Gather students together to begin the erosion and deposition simulations. Begin with one of the four locations that students are familiar with and where they can predict what will happen (e.g., a bare soil slope). Then move on to locations and surfaces that students may not have considered as places where erosion or deposition occurs.

   Remind students that you want them to watch while the entire 2 liters of water sprinkles the surface. Tell them to step out of the way of the water if it comes toward them, and keep hands off any earth materials carried by the water. When the water stops moving, they will record what they've seen in their notebooks.

   Pour the water from one 2 L bottle into the watering can and have a student pour the water over one spot on the surface of the first location you have chosen. When the water starts flowing, ask students questions such as

   ➤ *Where is the water moving? What do you notice about the water? What do you notice about the surface it is moving over?*

   ➤ *Is the water moving anything?*

   ➤ *Is the surface (soil/gravel/pavement) soaking up any of the water?*

   ➤ *Is anything left behind after the water runs through?*

Materials for Step 5
- *Bottles of water*
- *Watering can*
- *Clipboards (optional)*
- *Pencils*

SCIENCE AND ENGINEERING PRACTICES

Planning and carrying out investigations

CROSSCUTTING CONCEPTS

Cause and effect

INVESTIGATION 2 – Landforms

When the water stops moving, students should record their observations: where the water started, where it flowed, what was eroded, where it moved, if the ground soaked up any of the water, where earth material was deposited, if any organic material was moved by the water.

8. Pour water on other surfaces
Repeat the above procedure at the other locations and surfaces you have chosen. Ask students to describe the location you've gathered them around. Then repeat the "rainstorm" with a different student pouring the water. Repeat the questions, compare locations as they are added to the repertoire, and write observations in notebooks after each 2 L water run.

9. Have a sense-making discussion
Decide if you want to hold a sense-making discussion while you are still outdoors, or return to class for the discussion. Have students think about what they've seen in the stream tables and in the schoolyard. Have them make a list of general principles to think about when considering erosion and deposition. Ask,

➤ *What is the difference between erosion and deposition?*

➤ *How would you describe the general flow direction of water?* [High area to low: water flows downhill.]

➤ *What is the general distribution of earth materials in deposited areas?* [Bigger pieces settle first. Organic material is often carried farthest.]

➤ *How might these processes contribute to soil formation?* [Water carries earth materials from one place to another, and soil builds as earth materials settle in layers. Soil in a particular location can include earth materials that have been carried from far away. The size and shape of the particles is determined by how much weathering has taken place and how far water has carried them. Soil formation takes a very long time.]

➤ *How stable is the shape of the landform? What causes the shape of the land to change?*

10. Answer the focus question
Review the focus question and ask students to record an answer using data from the schoolyard to support their answers.

➤ *Where are erosion and deposition happening in our schoolyard?*

SCIENCE AND ENGINEERING PRACTICES
Constructing explanations

CROSSCUTTING CONCEPTS
Patterns
Stability and change

EL NOTE
For students who need scaffolding, provide a sentence frame such as: In our schoolyard we discovered that water _____. We also noticed that earth materials _____.

Part 3: Schoolyard Erosion and Deposition

11. Assess progress: response sheet
Distribute a copy of *Response Sheet—Investigation 2* to each student. This assessment should be completed during class. Collect the response sheets after class and check students' ability to interpret the information on the sheet and craft a response to the question.

What to Look For
- Soil on the sidewalk is a result of erosion caused by rainwater flowing on the hillside and deposition on the sidewalk.
- The larger pieces of soil will be deposited near the rock wall and smaller pieces will travel farther along the water flow.

12. View online activity
Have students engage with the virtual investigation "Stream Tables" to gain more experience with observing and recording the effects of changing variables of slope of the land, amount of river flow (flood conditions), and the addition of a dam in the river.

Then ask students to discuss the guiding question for the investigation with a partner.

➤ *How do erosion and deposition impact landforms?*

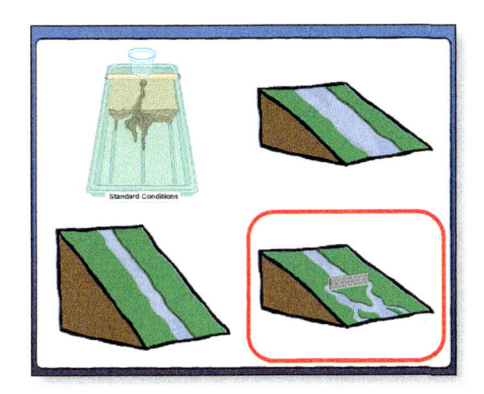

WRAP-UP/WARM-UP

13. Share notebook entries
Conclude Part 3 or start Part 4 by having students share notebook entries. Ask students to open their science notebooks to the most recent entry. Read the focus question together.

➤ *Where are erosion and deposition happening in our schoolyard?*

Have students discuss the possible causes of the erosion and deposition examples they identified and their effects. Ask students to identify any potential problems erosion and deposition might cause on their schoolyard and possible solutions.

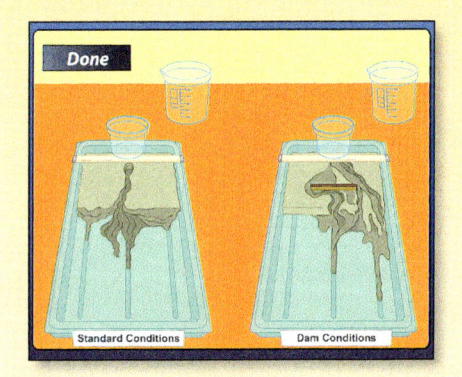

SCIENCE AND ENGINEERING PRACTICES
Constructing explanations

CROSSCUTTING CONCEPTS
Cause and effect
Stability and change

ELA CONNECTION
This suggested reading strategy addresses the Common Core State Standards for ELA.

SL 1: Engage in collaborative discussions.

Soils, Rocks, and Landforms Module—FOSS Next Generation

INVESTIGATION 2 – *Landforms*

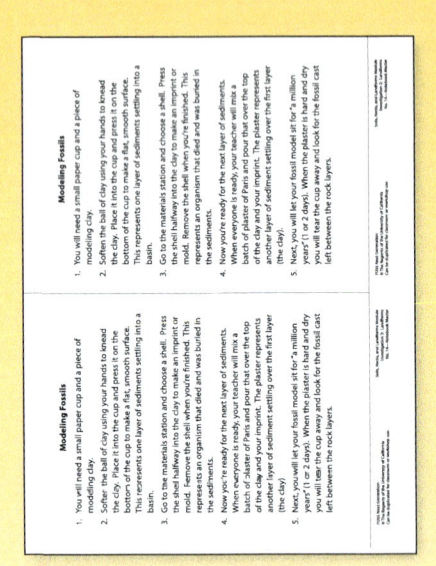

No. 14—Notebook Master

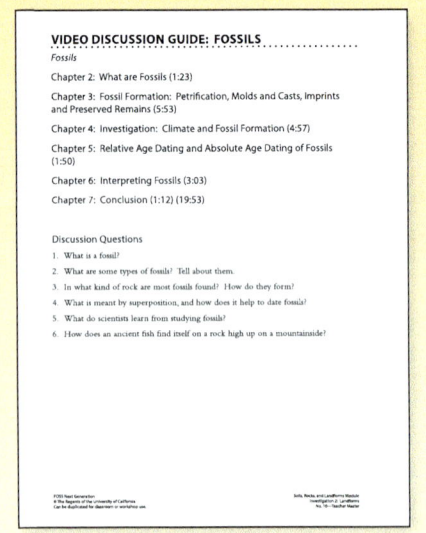

No. 10—Teacher Master

MATERIALS *for*
Part 4: *Fossil Evidence*

For each student
- ❏ 4 Notebook sheet 14, *Modeling Fossils*
- 4 *FOSS Science Resources: Soils, Rocks, and Landforms*
  - "Fossils Tell a Story"
  - "Pieces of a Dinosaur Puzzle"

For each group
- 1 Basin
- 1 Cups, plastic
- 2 Banded sandstone
- 4 Paper cups, small (3 oz.)
- 4 Pieces of ceramic gray clay, 2.5 cm cube

For the class
- Small shells, assorted (See Getting Ready, Step 4.)
- Plaster of paris (See Getting Ready, Step 5) ★
- Container for mixing plaster (old plastic bowl) ★
- Spoon or stick for mixing plaster ★
- Water ★
- 1 Computer with Internet access ★ (See Step 6 of Getting Ready.) ★
- ❏ 1 Teacher master 10, *Video Discussion Guide: Fossils*

For embedded assessment
- ❏ • *Embedded Assessment Notes*

For benchmark assessment
- ❏ • *Investigation 2 I-Check*
- ❏ • *Assessment Record*

★ Supplied by the teacher.   ❏ Use the duplication master to make copies.

Full Option Science System

Part 4: Fossil Evidence

GETTING READY for
Part 4: *Fossil Evidence*

1. **Schedule the investigation**
   This part will take a total of five sessions: two sessions for active investigation, one for reading, one for review, and one for the I-Check.

2. **Preview Part 4**
   Students think about what happens to and in sediments over long periods of time as sediments layer on top of each other. Students watch a video, make models, and read an article to learn about how sedimentation processes can result in fossils. They learn how fossils provide evidence of life and landscapes from the past. The focus question is **How do fossils get in rocks and what can they tell us about the past?**

3. **Cut pieces of clay**
   Use strong monofilament fishing line or a large knife to slice the ceramic clay into 2.5 cm cubes (about 1" cubes). You'll need one cube for each student in the class.

4. **Set up rocks in cups**
   Place two pieces of banded sandstone in a cup for each group. Place these at the materials station to pick up in Step 2 of Guiding the Investigation.

   In Step 8 of Guiding the Investigation, you will distribute another set of materials. Place sets of four small paper cups and four cubes of potters clay in a basin for each group to acquire at the appropriate time. Set up a separate area for the shells that students will use to make imprints.

5. **Prepare for plaster of paris**
   When you mix the plaster of paris, use a ratio of 2 parts plaster to 1 part water and use an old plastic bowl or disposable paper paint bucket. Each student needs enough plaster mix to cover the modeling clay imprint in a small cup. Plan to prepare about 1 cup of dry plaster for every eight students.

   Have the plaster and water premeasured in separate containers and ready to go, but DO NOT premix the plaster. It will take some time for students to soften, press, and flatten their clay balls into the bottom of a small paper cup. They also need to go to the materials station and use a shell to make a mold in the clay. After all students

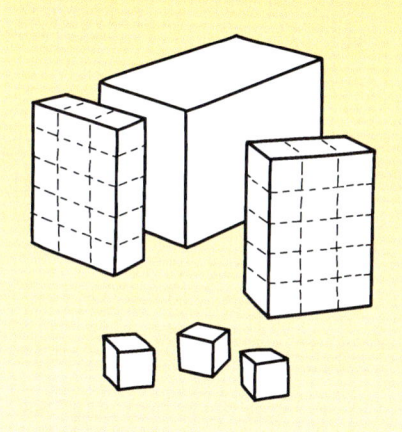

▶ **SAFETY NOTE**
Be sure to read the directions on the plaster of paris package you purchase for this activity. Study the safety considerations and the suggestions for cleanup. Try to mix the plaster in a paper bucket that can be recycled.

Soils, Rocks, and Landforms Module—FOSS Next Generation

INVESTIGATION 2 – Landforms

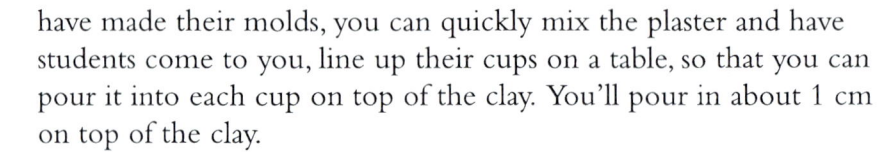

have made their molds, you can quickly mix the plaster and have students come to you, line up their cups on a table, so that you can pour it into each cup on top of the clay. You'll pour in about 1 cm on top of the clay.

6. **Plan to view *Fossils* video**
   Preview the *Fossils* video (chapters 2–7, duration about 20:00). Links to the video is found on FOSSweb in the Resources by Investigation section. Have students view the video in Step 5. Plan to use teacher master 10, *Video Discussion Guide: Fossils*, to lead class discussion after watching the video with the class. Make 8 copies, one for each group.

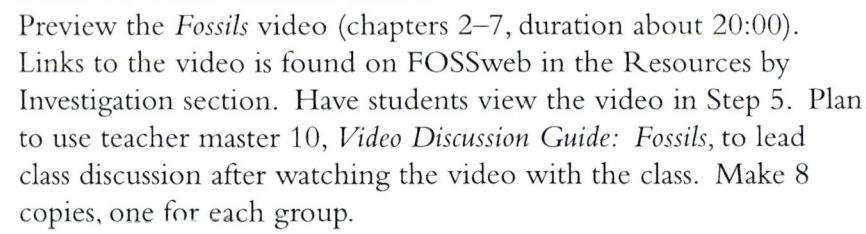

7. **Plan to read *Science Resources*: "Fossils Tell a Story" and "Pieces of a Dinosaur Puzzle"**
   Plan to read "Fossils Tell a Story" and "Pieces of a Dinosaur Puzzle" during a reading period after completing the active investigation for this part.

8. **Plan assessment: notebook entry**
   In Step 14 of Guiding the Investigation, students answer the focus question. After class, have students hand in their notebooks open to the page on which they wrote their answers. Check students' understanding of how fossils provide evidence for changes in the landscape over long periods of time.

9. **Plan benchmark assessment: I-Check**
   When students have completed all the activities and readings for this investigation, plan a review session, and then give them *Investigation 2 I-Check*. Students should complete their responses to the items independently. After you review and code students' work, plan which items you will bring back for additional discussion and self-assessment. See the Assessment chapter for self-assessment guidance and information about online tools. The assessment coding guides are available on FOSSweb.

Part 4: Fossil Evidence

GUIDING the Investigation
Part 4: Fossil Evidence

1. **Review investigations**
   Tell students,

   In the first investigation, we learned how rocks are continually breaking down into smaller and smaller pieces, called sediments. In this investigation we have been using the stream tables to model how water can move sediments to new locations.

   Now let's think about how sediments can become beds of rock.

2. **Introduce basin**
   Remind students that they used stream tables to model flowing water in a river or stream to study how water carries sediments (weathered and eroded materials). The materials move downhill to lower areas where the water slows down and the sediments settle out. Tell students that the low areas where sediments deposit are called **basins**.

   ➤ *What happens to sediments that water is carrying when the water enters a basin?* [If the water is calm, the sediments will settle out. Bigger particles are deposited first; smaller sediments get deposited later.]

   Have Getters pick up a cup containing two rocks (banded sandstone). Have students examine the rocks and write their observations in their notebooks.

   Ask students to share their observations.

   Introduce the rock as **banded sandstone**. Tell students that sandstone forms when a layer of very small sediments (sand) is compressed over millions of years. Rocks made out of sediments are called **sedimentary rocks**.

3. **Discuss fossils**
   Ask students to look at the *FOSS Science Resources* book pages 69–70 for examples of fossils in sedimentary rocks. Have a brief discussion to elicit students' prior knowledge about fossils, what they are, and how they form.

FOCUS QUESTION
How do fossils get in rocks and what can they tell us about the past?

Materials for Step 2
- Half liter containers with rocks

Soils, Rocks, and Landforms Module—FOSS Next Generation

INVESTIGATION 2 – *Landforms*

4. **Focus question: How do fossils get in rocks and what can they tell us about the past?**

   Tell students that the question they will be focusing on today is,

   ➤ *How do fossils get in rocks and what can they tell us about the past?*

   Have students write the focus question in their notebooks as you say it and project or write it on the board.

5. **View video about fossils**

   Distribute a copy of teacher master 10, *Video Discussion Guide: Fossils* to each group and have them discuss the questions. Have the groups assign the questions so each student has one or two to look for more information about while watching the video.

   Show students the *Fossils* video (chapters 2–7, duration about 20:00). The video can be accessed on FOSSweb Resources by Investigation for this module. The topics in the video include:

   - What are fossils?
   - Fossil formation: petrification, molds and casts, imprints, and preserved remains.
   - Climate and fossil formation investigation.
   - Relative age dating and absolute age dating of fossils.
   - Interpreting fossils.

   After watching the video, give students time to share their answers to the questions in the discussion guide with the rest of the group.

6. **Discuss the video with the class**

   Use the questions on the teacher master to guide the discussion.

   ➤ *What is a fossil?* [The preserved remains or other evidence of living things that lived on Earth long ago.]

   ➤ *What are some types of fossils?* [Petrified forests are an example of **petrification**. Trees were quickly buried in sediments (volcanic ash and mud) preventing decomposition. Minerals in the sediments replaced the wood tissue and when the fossils were uncovered millions of years later, they were rock. **Molds** are the hollow spaces where objects such as shells were buried after the shell decomposed and a space remained. Minerals fill the hollow spaces forming a **cast**. Other types of fossils are **imprints** left by leaves or footprints providing evidence that something existed. Frozen **preserved remains** of ancient animals can occur in very cold environments.]

   ➤ *In what kind of rock are most fossils found? How do they form?* [Most fossils form in sedimentary rock when an animal or plant dies in a water environment and is buried in mud and layers of

EL NOTE

Add the new words to the word wall. Make a pictorial (labeled diagram) on chart paper to go over these new words that come up in the discussion.

194   Full Option Science System

Part 4: Fossil Evidence

sediments. Over millions of years, under great pressure, these layers turn into sedimentary rocks.]

➤ *What is meant by* **superposition** *and how does it help to date fossils?* [This is the principle that rock layers and the fossils they contain are arranged in layers and that lower sedimentary layers are older than higher layers of rock.]

➤ *What do scientists learn from studying fossils?* [They learn about past climates and environments, how environments have changed (for example, what is now a desert may have once been an ocean), what organisms lived together in ancient times, and how organisms have changed over time.]

➤ *How does an ancient fish find itself on a rock high up on a mountainside?* [The fish fossil tells us that rock layer was once underwater. The water has dried up or the land was pushed up over hundreds of millions of years to form the mountain.]

7. Demonstrate modeling fossil formation

Tell students that each student will now make a model of a fossil. Distribute notebook sheet 14, *Modeling Fossils*, and have students follow along as you demonstrate the procedure.

a. Each student gets a small paper cup, and a piece of clay.

b. Soften the clay, roll it into a ball, then place it into the cup, pressing it to make it a flat, smooth surface on the top. This represents one layer of sediment deposited in a basin.

c. Go to the materials station and choose a shell to press halfway into the clay to make a mold in the clay. Remove the shell. This represents an organism that died, was buried in sediments, disintegrated or decomposed, leaving a space.

NOTE: Be sure students don't push the object too far into the clay. They need to be able to lift it out now and when they pull the two "rocks" apart to see the cast.

d. When everyone has prepared a mold, the teacher will mix up a batch of plaster of paris and distribute a short pour into each cup to cover the clay about 1 cm deep. The plaster represents another layer of sediment settling over the first layer.

e. When the sediments have set for "a million years," we'll tear the cup away and look for the casts in the hardened rock.

8. Make model fossils

Have Getters go to the materials station and get paper cups and cubes of clay for their group. Students should write their names on their cups, work the clay in their hands until it is soft and press it into the paper cup. Then they can go to the materials station and choose a shell to make a mold.

CROSSCUTTING CONCEPTS
Cause and effect
Stability and change

Materials for Steps 8–9
- *Paper cups*
- *Clay pieces*
- *Imprint objects*
- *Plaster of paris*
- *Water*
- *Container for mixing*
- *Stick for mixing*

Soils, Rocks, and Landforms Module—FOSS Next Generation

INVESTIGATION 2 – Landforms

9. Add the plaster of paris sediment layer
When all of the students have made their imprints, mix the plaster of paris as described in Getting Ready, Step 5. You will need to work quickly once the plaster is mixed. Have students line up and pour enough plaster into each cup to cover the mold with about a centimeter of plaster.

10. Clean up
Have students place their fossils in progress on a window sill or other area out of the way to cure overnight. Return any other materials to the materials station. Discard any remaining plaster of paris. (Caution: be sure you don't wash it down the sink.)

BREAKPOINT

11. Uncover the casts
The next day, when the plaster of paris has set, have students retrieve their model fossils. Have them tear off the paper cup surrounding the "rock" and break the two layers of rock apart to reveal their 3-dimensional "fossils."

12. Have a sense-making discussion
Have students discuss the benefits and limitations of models.

➤ *What causes fossils to form?* [Most fossils form when a dead organism ends up on the bottom of a water environment and gets buried in mud and sediments. Over long periods of time, the pressure of the layers of sediment may cause remains (bones, teeth, wood) to become rock (petrification) or the sediments in the space left by a shell may become rock (cast).]

➤ *How are the model fossils we made similar to fossils in nature?* [The clay and plaster of paris represented different layers of sediments and the shells represent organisms that might have been trapped in the sediments.]

➤ *How are the model fossils we made different from those that occur in nature?* [Fossils and sedimentary rocks are created over millions of years, not just a day; you wouldn't necessarily see two different kinds of sediment (we had to do that to be able to take them apart later); organisms die and get buried at random, not deliberately like we made them.]

➤ *What other ways might we model the process of sedimentary rock and fossil formation?*

SCIENCE AND ENGINEERING PRACTICES
Developing and using models
Constructing explanations

CROSSCUTTING CONCEPTS
Cause and effect
Stability and change

Part 4: Fossil Evidence

13. Review vocabulary
Review the vocabulary words introduced in this part. Make sure that these words are on the word wall. Add any other words students might need.

14. Answer the focus question
Have students answer the focus question in their notebooks.

➤ How do fossils get in rocks and what can they tell us about the past?

15. Assess progress: notebook entry
Have students hand in their notebooks open to the page on which they answered the focus question. After class, review students' entries to see if students understand how fossils are formed and how they provide evidence of changes in landscapes over time.

What to Look For

- Students can describe the process by which fossils are formed in sediments that become rock based on the model they created (e.g., they are buried in sediment, a mold forms as the organism decomposes, other minerals fill in the mold as the rock hardens).

- Students explain that fossils are evidence about the past. For example, if you find a fish fossil in a desert, that is evidence that the land was once covered by water, even though now the landscape is dry.

---

basin
cast
fossil
imprint
mold
petrification
preserved remains
sedimentary rock
superposition

EL NOTE
For students who need scaffolding, provide a sentence frame such as Fossils get in rocks when _____. Fossils can tell us _____.

INVESTIGATION 2 – Landforms

Fossils Tell a Story

Earth scientists called **geologists** and **paleontologists** ask questions about the history of Earth. Geologists focus their studies on the structures of rocks and how they formed. Paleontologists ask questions about what life was like millions of years ago. These scientists can't travel back in time, so they look for answers to their questions in rocks. From the **evidence** they find, paleontologists build models to explain how they think plants and animals lived and what their world was like.

Digging through the Layers

Over billions of years, Earth's surface has changed dramatically. Areas once covered by water are now dry. Continents have moved around on the globe. Large areas of Earth's surface were once covered by thick sheets of ice, which have now melted.

Through all these changes, layers of new rock have formed on Earth. Some layers were formed by molten **lava** that erupted from **volcanoes**, covered Earth's surface, cooled, and turned into solid rock. Other layers of rock formed from tiny particles of rock called sediments that settled in Earth's ocean, lakes, and swamps. After millions of years these layers of sediments changed into **sedimentary rock**. By digging through layers of sedimentary rock, earth scientists have determined the age of Earth and discovered the **organisms** that lived here at the time the sediments were deposited.

Scientists dig through layers of soil and rock.

23

ELA CONNECTION

These suggested reading strategies address the Common Core State Standards for ELA.

RI 1: Refer to details and examples in a text when explaining what the text says explicitly and when drawing inferences from text.

RI 2: Determine the main idea of a text and explain how it is supported by key details; summarize the text.

RI 3: Explain procedures, ideas, or concepts in a scientific text.

RI 6: Compare and contrast a firsthand and secondhand account of the same topic.

RI 7: Interpret information presented visually, and explain how the information contributes to an understanding of the text.

SL 1: Engage in collaborative discussions.

SL 5: Add visual displays to presentations.

READING in Science Resources

16. Read "Fossils Tell a Story"

Read "Fossils Tell a Story" during a reading period after completing the active investigation. Before reading, tell students that this article will review the ideas they have been exploring about fossils and what they tell us about changes in the land.

Introduce the title of the article and ask students to discuss what they think it means. Then, have them preview the text by looking at and discussing the photographs and reading the captions. Ask students to write down a few questions they have about fossils that might be answered in the text. They can also look for more information to add to their focus question response.

17. Use a reading comprehension strategy

Read the text aloud or have students read independently. For students who need support, conduct small guided reading groups. Then have students reread the text a second time. Pause after each section or paragraph and have students turn and talk to their reading partner about the main idea and any questions they have. Encourage students to visualize the processes the text is explaining and to refer to the photographs for clarification.

18. Discuss the reading

Give students a few minutes to review their questions. Then, write these questions on the board and have students discuss them in small groups. Give each student a role in the discussion. A Starter poses the question; a Facilitator makes sure everyone is participating; a Recorder takes notes; a Reporter prepares to share during the whole group discussion. Encourage students to use evidence from the text to support their ideas. Provide chart paper or whiteboards for students to make their thinking visible.

▶ *How are geologists and paleontologists alike and how are they different?* [Both study Earth history. One focuses on rocks and the other on life and environments.]

▶ *How are earth scientists able to determine the age of organisms that lived long ago?* [Relative age of layers above; being younger than layers below; and the use of index fossils.]

▶ *Review the text and photographs about mold and cast fossils. Compare what you did in class to the information presented in the text. How do they compare?*

▶ *Sometimes a scientist finds fossils of ocean animals high up on a mountain. What do you think she might conclude about the history of that area?* [The mountain was once lower and covered by ocean.]

Part 4: Fossil Evidence

Call on Reporters to share their group's answer and allow other group Reporters to respond by either agreeing or disagreeing, or offering alternative ideas. To support collaborative discussions, provide sentence frames such as:

We think _____ because _____ .

We agree with _____ because _____ .

We disagree with _____ because _____ .

I would like to add that _____ .

Why do you think _____ ?

Do you have evidence that _____ ?

19. Revisit the focus question
Give students a chance to review their answer to the focus question after the reading. They can add to their original response with any new information or ideas they have gathered.

BREAKPOINT

20. Read "Pieces of a Dinosaur Puzzle"
Read "Pieces of a Dinosaur Puzzle" during a reading period after completing the active investigation.

Introduce "Pieces of a Dinosaur Puzzle" and ask students to discuss why they think the author gave the article this title. Have students do a quick write in their notebook including three things they know about dinosaurs and one thing they are wondering about dinosaurs. Give students a few minutes to share their ideas with their reading partners.

Have students preview the text by looking at and discussing the photographs and diagrams. They should also read the captions and the words in bold. Focus students' attention on the drawing on page 29 and ask what the purpose of the penny is in this illustration. Discuss how illustrators use an object the reader is familiar with to show the relative size of what is being presented.

Soils, Rocks, and Landforms Module—FOSS Next Generation

INVESTIGATION 2 – Landforms

ELA CONNECTION

These suggested reading strategies address the Common Core State Standards for ELA.

RI 3: Explain procedures, ideas, or concepts in a scientific text.

RI 4: Determine the meaning of general academic domain-specific words or phrases.

RI 7: Interpret information presented visually, and explain how the information contributes to an understanding of the text.

RI 10: Read and comprehend science text

SCIENCE AND ENGINEERING PRACTICES

Obtaining, evaluating, and communicating information

21. **Use a reading comprehension strategy**
    Read the article aloud or have students read independently or in guided reading groups. If students are reading independently, give them self-stick notes to jot down unknown words or phrases and questions they have. Encourage students to pause as they read to visualize what the author is describing. Students can also annotate the text with the self-stick notes using symbols and phrases about the text as they read. Here are some examples:
    - ★ interesting
    - ? question
    - L learning something new
    - W wondering
    - S surprising

22. **Discuss the reading**
    After the reading, review any unknown words or phrases and discuss the ways to determine their meaning, e.g., context, glossary, root words, etc. Have students share out one of their self-stick notes in their small groups. If time permits, ask a few students to share their questions and conduct a science talk if the question would be useful for the whole class in developing their understanding of how fossils contribute to our understanding of organisms from long ago. Or, collect the questions and add them to an inquiry chart for later discussion or investigation topics.

BREAKPOINT

Part 4: Fossil Evidence

WRAP-UP

23. Review Investigation 2

Distribute one or two self-stick notes to each student, and have them cut each note into three pieces, making sure each part has a sticky end.

Ask students to take 3 minutes to look back through their notebook entries to find the most important things they learned in Investigation 2. Students should include at least one science and engineering practice, one disciplinary core idea, and one crosscutting concept. They should tag those pages with self-stick notes.

When students have completed their notebook review and placed their tags, lead a short class discussion to create a list of three-dimensional statements that summarize what students have learned in this investigation. Here are the big ideas that should come forward in review discussions.

- We used stream tables to model erosion and deposition; these processes produce new and altered landforms, and contribute to soil formation. (Developing and using models; stability and change.)
- The data we collected showed that the size of earth material particles determines which particles settle out of water first (large particles), and which travel farther (small pieces). (Analyzing and interpreting data; patterns.)
- Explaining soil formation required us to think about weathering (Investigation 1) as well as erosion and deposition—pieces break off, then they are carried away and deposited elsewhere, often becoming part of the soil. (Constructing explanations; systems and system models.)
- Controlling the slope and the amount of water that ran through the stream table provided evidence that both variables affect the amount of erosion and deposition. (Planning and carrying out investigations; cause and effect.)
- We obtained information and created models to show that sediments can turn back into rocks; organisms trapped in sediments can become fossils. Fossils provide evidence of organisms that lived long ago as well as clues to changes in the landscape and environment. (Obtaining, evaluating, and communicating information; scale, proportion, and quantity.)

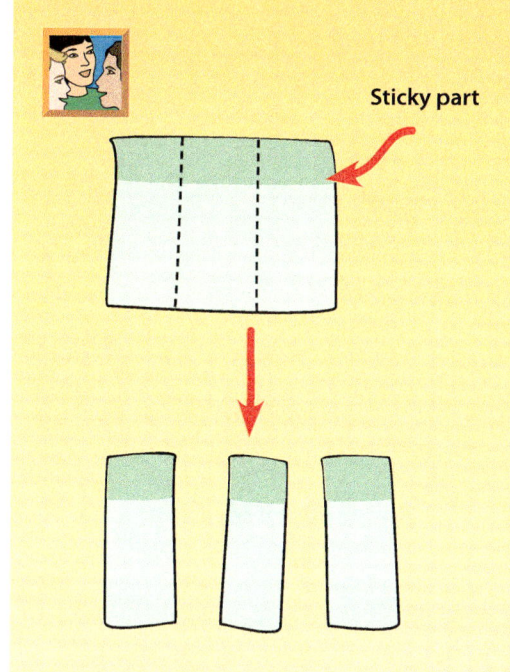

Sticky part

DISCIPLINARY CORE IDEAS

ESS1.C: The history of planet Earth

ESS2.A: Earth materials and systems

ESS2.B: Plate tectonics and large-scale system interactions

INVESTIGATION 2 – Landforms

Students should discuss the guiding questions about the investigative phenomena with a partner before responding to them in their notebooks.

➤ *How do erosion and deposition impact landforms?*

➤ *What do the locations of fossils in rock layers tell us about past life on Earth?*

BREAKPOINT

24. Assess progress: I-Check

Give the I-Check assessment at least one day after the wrap-up review. Distribute a copy of *Investigation 2 I-Check* to each student. You can read the items aloud, but students should respond to the items independently in writing. Alternatively, you can schedule the I-Check on FOSSmap for students to take the test and then you can access data through helpful reports to look carefully at students' progress.

If students take the test on paper, collect the I-Checks. Code the items using the coding guides you downloaded from FOSSweb, and plan for a self-assessment session, identifying disciplinary core ideas, science and engineering practices, and crosscutting concepts students may need additional help with. (If you are using FOSSmap, you will only need to code open-response items and confirm codes for short-answer items as other items are automatically coded.) Refer to the Assessment chapter for next-step/self-assessment strategies.

TEACHING NOTE

During or after these next-steps with the I-Check, you might ask students to make choices for possible derivative products based on their notebooks for inclusion in a summative portfolio. See the Assessment chapter for more information about creating and evaluating portfolios.

INTERDISCIPLINARY EXTENSIONS

Language Extension

- **Write an investigation report**
  Have students write up an investigation report after they have run their stream-table experiments. They can use the plans in their notebooks to help them write the report. Remind students to include appropriate landform vocabulary, and to use all the knowledge they have acquired from investigation discussions and readings to interpret their plans.

Math Extensions

- **Problem of the week**
  A class investigated how the slope of the stream table affected deposition. Each group tested the same four slopes by elevating the end of the tray 2, 3, 4, and 5 centimeters (cm). They measured the length of the alluvial fan after each test. The results are in the data table.
  - Calculate the average length of the alluvial fans formed. Round your answers to the nearest 0.1 cm.
  - Prepare a graph of the average alluvial fan lengths.
  - Predict the length of the alluvial fan if the tray were elevated 1 cm.

Slope height	Group 1	Group 2	Group 3	Group 4
2 cm high	4.4 cm	4.8 cm	4.2 cm	4.6 cm
3 cm high	5.2 cm	5.6 cm	5.4 cm	5.2 cm
4 cm high	6.3 cm	6.4 cm	6.1 cm	6.1 cm
5 cm high	7.4 cm	7.6 cm	7.3 cm	7.0 cm

> **TEACHING NOTE**
> Refer to the teacher resources on FOSSweb for a list of appropriate trade books that relate to this module.

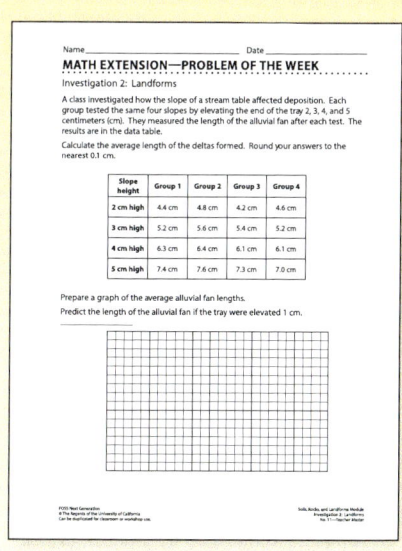

No. 11—Teacher Master

INVESTIGATION 2 – Landforms

Notes on the problem. The average lengths of the alluvial fan (rounded off to the nearest tenth) are 4.5 cm (2 cm slope), 5.4 (3 cm slope), 6.2 (4 cm slope), and 7.3 (5 cm slope). The final graph should look like the one below. From the graph, students should be able to predict that when the tray is elevated 1 cm, the length of the alluvial fan would be about 3.5 cm.

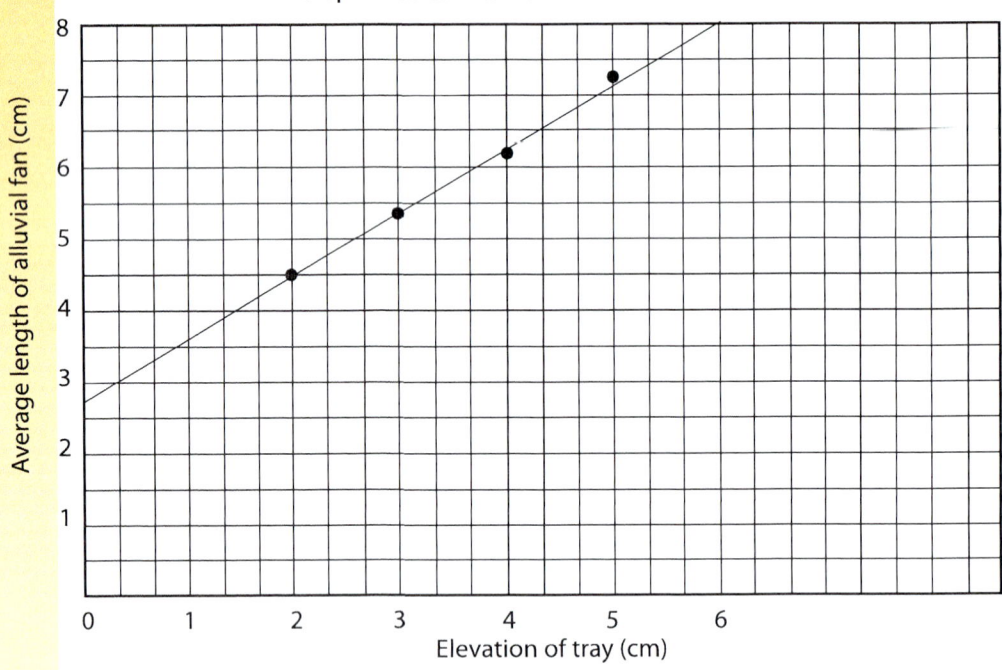

Slope of stream table and alluvial fan formation

- ### How much is a million?
  Geologists suggest that it took 5–6 million years for the Grand Canyon, a well-known landform, to form. Help students visualize big numbers, such as 1,000,000. Have them figure out how many years it takes to live 1,000,000 seconds. Challenge them to come up with some comparisons and estimates of their own, like how many grains of sand are in a 1 L container.

Interdisciplinary Extensions

Science and Engineering Extensions

- **Take stream-table photos**
  Have students use a camera to take video or still images of the stream tables in action. They can use the sequential images in a report to share with the class.

- **Find a local erosion-control expert**
  Find out which agencies are responsible for erosion control in your area. Check the Internet for your county flood-control and water district. Invite a guest speaker to describe the work the agency performs and the specific problems the group is working on. Students should develop questions for the speaker that focus on how their local area has changed over the years due to erosion and deposition.

- **Go on a 15-minute field trip**
  Take your class on a schoolyard field trip right after a hard rain. What evidence of erosion and deposition can students find? Look for places where sand and soil have been deposited. How are these places similar to what they saw in their stream tables?

- **Engage with the "Tutorial: Soil Formation"**
  For students who need more time exploring the concepts, have them review how soil is formed and how that relates to erosion and deposition by using this tutorial on FOSSweb. This is best used after Part 3.

- **Engage with the "Tutorial: Fossils"**
  For students who need more time exploring the concepts in Part 4, have them work with this tutorial on FOSSweb about different kinds of fossils and how they form.

TEACHING NOTE
Review the online activities for students on FOSSweb for module-specific science extensions.

TEACHING NOTE
Encourage students to use the Science and Engineering Careers Database on FOSSweb.

Soils, Rocks, and Landforms Module—FOSS Next Generation

INVESTIGATION 2 – *Landforms*

Environmental Literacy Extensions

- **Plan a field trip**

  Look into field trips in your local area. Possibilities might include visiting a nature center located on a river, an Army Corps of Engineers project such as a dam/reservoir site or stream model, or the water utilities offices. Encourage the expert at the facility to share some of the positive and negative effects of any human influences in the area.

- **Research floods**

  Conduct research about the negative and positive effects of floods caused by beavers, swollen rivers, or major weather events. Think about the advantages and disadvantages of living in or near flood plains. If students experienced the **FOSS Water and Climate Module** in grade 3, they can begin with those ideas and then expand their research to new areas.

Home/School Connection

Students explore media and the Internet for news about landforms and the processes that create and change them. They prepare a report that includes the details of the event, where it happened, what types of landforms were described, and how the landforms were created or changed by natural or human-caused processes.

Print or make copies of teacher master 12, *Home/School Connection* for Investigation 2, and send it home with students at the end of Part 4.

> **TEACHING NOTE**
>
> Families can get more information on Home/School Connections from FOSSweb.

No. 12—Teacher Master

206 Full Option Science System

INVESTIGATION 3 – Mapping Earth's Surface

Part 1
Making a Topographic Map .. 218

Part 2
Drawing a Profile 229

Part 3
Mount St. Helens Case Study 241

Part 4
Rapid Changes 250

Guiding questions for phenomena: *How do maps help us observe Earth's surface features? What might reduce the impact of catastrophic Earth surface events?*

Science and Engineering Practices
- Developing and using models
- Planning and carrying out investigations
- Analyzing and interpreting data
- Using mathematics and computational thinking
- Constructing explanations
- Engaging in argument from evidence
- Obtaining, evaluating, and communicating information

Disciplinary Core Ideas
ESS1: What is the universe and what is Earth's place in it?
ESS1.C: The history of planet Earth
ESS2: How and why is Earth constantly changing?
ESS2.B: Plate tectonics and large-scale system interactions
ESS3: How do Earth's surface processes and human activities affect each other?
ESS3.B: Natural hazards
ETS1: How do engineers solve problems?
ETS1.B: Developing possible solutions

Crosscutting Concepts
- Patterns
- Cause and effect
- Scale, proportion, and quantity
- Stability and change

PURPOSE

Students investigate the phenomena of Earth's mountains by representing landforms in three dimensions using topographic maps, aerial photos, and physical models. They study catastrophic phenomena to Earth's surface—earthquakes, volcanic eruptions, floods—and consider ways to mitigate their impacts.

Content

- A topographic map uses contour lines to show the shape and elevation of the land.
- A profile is a side view or cross-section of a landform, drawn from information given on a topographic map.
- The surface of Earth is constantly changing—some changes occur over a long period of time.
- Catastrophic events have the potential to change Earth's surface quickly. Scientists and engineers can do things to reduce the impacts of natural Earth processes on humans.

Practices

- Represent landforms in ways to gather new information.
- Make observations and interpret them to develop explanations in the way that scientists do.

FOSS Full Option Science System

INVESTIGATION 3 – Mapping Earth's Surface

	Investigation Summary	Time	Focus Question for Phenomenon, Practices
PART 1	**Making a Topographic Map** Students build a model mountain of Mount Shasta by stacking and orienting six foam layers. They trace outlines of the six pieces onto paper, creating a topographic map of the mountain.	Active Inv. 1 Session * Reading 1 Session	**How can we represent the different elevations of landforms?** **Practices** Developing and using models Obtaining, evaluating, and communicating information
PART 2	**Drawing a Profile** Students use their topographic maps to produce two-dimensional profiles, or cross-sections, of their foam mountain. The profile reveals a side view of Mount Shasta, a dormant volcano. Students gather information about volcanoes from a video.	Active Inv. 1 Session Reading 2 Sessions	**How can we draw the profile of a mountain from a topographic map?** **Practices** Developing and using models Analyzing and interpreting data Using mathematics and computational thinking Obtaining, evaluating, and communicating information
PART 3	**Mount St. Helens Case Study** Students compare two topographic maps and have a short debate about whether or not they show the same mountain. After learning that the two topo maps are the same mountain, they draw profiles of Mount St. Helens before and after its devastating eruption in 1980. Students watch a USGS video that explains how scientists were involved in predicting the eruption.	Active Inv. 1 Session	**How can scientists and engineers help reduce the impacts that events like volcanic eruptions might have on people?** **Practices** Developing and using models Planning and carrying out investigations Analyzing and interpreting data Using mathematics and computational thinking Constructing explanations and designing solutions Engaging in argument from evidence Obtaining, evaluating, and communicating information
PART 4	**Rapid Changes** Students think about processes that cause rapid changes to Earth's surface: landslides, earthquakes, floods, and volcanoes.	Active Inv. 1 Session Reading 1 Session Assessment 2 Sessions	**What events can change Earth's surface quickly?** **Practices** Analyzing and interpreting data Constructing explanations

A class session is 45–50 minutes.

At a Glance

Content Related to DCIs	Writing/Reading	Assessment
• A topographic map uses contour lines to show the shape and elevation of the land. • The change in elevation between two adjacent contour lines is always uniform. • The closer the contour lines, the steeper the slope and vice versa.	**Science Notebook Entry** Mountain Map **Science Resources Book** "Topographic Maps"	**Embedded Assessment** Science notebook entry
• A profile is a side view or cross-section of a landform. • A profile can be drawn from information given on a topographic map.	**Science Notebook Entry** Profile Contours and Intervals **Science Resources Book** "The Story of Mount Shasta" **Video** Volcanoes **Online Activity** "Topographer"	**Embedded Assessment** Response sheet
• A profile can be drawn from information given on a topographic map. • The surface of Earth is constantly changing; sometimes those changes take a long time to occur and sometimes they happen rapidly. • Catastrophic events have the potential to change Earth's surface quickly.	**Science Notebook Entry** Profile **Video** Mount St. Helens Impact	**Embedded Assessment** Performance assessment
• Catastrophic events have the potential to change Earth's surface quickly. • Scientists and engineers can do things to reduce the impacts of natural Earth processes on humans.	**Science Notebook Entry** Answer the focus question **Science Resources Book** "It Happened So Fast!" **Video** All about Earthquakes	**Embedded Assessment** Science notebook entry **Benchmark Assessment** Investigation 3 I-Check **NGSS Performance Expectations addressed in this investigation** 4-ESS2-2; 4-ESS3-2 3-5-ETS-1-2

Soils, Rocks, and Landforms Module—FOSS Next Generation

INVESTIGATION 3 – *Mapping Earth's Surface*

BACKGROUND *for the Teacher*

There's something about the landform we call a mountain that compels people to climb to the top of it. Mount Everest, Mount Whitney, the Matterhorn—their names and locations are familiar to many, and their peaks are the goal of outdoor adventurers. A volcanic mountain is the major phenomenon that students investigate as they work with models to represent and describe this surface feature on Earth and understand how they were formed and how they change over time.

Mount Shasta and Mount St. Helens are also mountains of great height. Mount Shasta (14,180 feet; 4,322 meters) rises impressively above most of the landforms in northern California. Mount St. Helens (8,336 feet; 2,550 meters) is in the state of Washington. Both mountains are **volcanoes** in the Cascade range. Mount Shasta is inactive, but Mount St. Helens has been quite active during the last 45 years.

How Can We Represent the Different Elevations of Landforms?

The view that we most often get of a mountain like Mount Shasta or Mount St. Helens is the **profile**. Another useful view of a mountain is from above. **Topographic maps** are two-dimensional representations of three-dimensional surfaces as seen from directly overhead. Topography is the shape of the surface of the land, its size, relief, and **elevation**. Mountains, plains, canyons, and cliffs are a few of the topographic features that contribute to the diversity of the earth's surface.

Contour lines are characteristic of topographic maps. Any one contour line represents a specific elevation, the vertical distance above **sea level**. Sea level is zero elevation; it is defined as the average distance between the highest high tide and the lowest low tide around the world.

When you trace the 9,000-foot contour line on a topographic map, everything that lies on that line is exactly 9,000 feet above sea level. Contour lines show the contour, or shape, of the land at one specific elevation.

When you look at a topographic map, you will notice that some of the contour lines are relatively close together, while others are quite far apart. Lines close together indicate regions with steep slopes. As you move from one line to another, the change in elevation remains constant, whether you move up or down. The vertical distance between contour lines is the **contour interval**.

Background for the Teacher

The US Geological Survey (USGS) has been producing a standard series of topographic maps since 1882 as part of its National Mapping Program. A long-standing goal of the USGS has been to provide complete, large-scale topographic map coverage of the United States. The original topographic mapping program (1945-1992) was created from primary data sources collected through field work. The result was a series of more than 54,000 maps that cover in detail the entire area of the 48 contiguous states and Hawaii.

In the early 2000s the USGS mapping program turned its attention to digital data and digital maps using Geographic Information Systems (GIS). GIS are at the forefront of the mapping revolution. A GIS can combine layers of digital data from different sources and manipulate and analyze how the different layers relate to each other.

In 2009, USGS began the release of a new generation of digital topographic maps (US Topo) and is now complementing them with the release of high-resolution scans of more than 178,000 historical topographic maps of the United States.

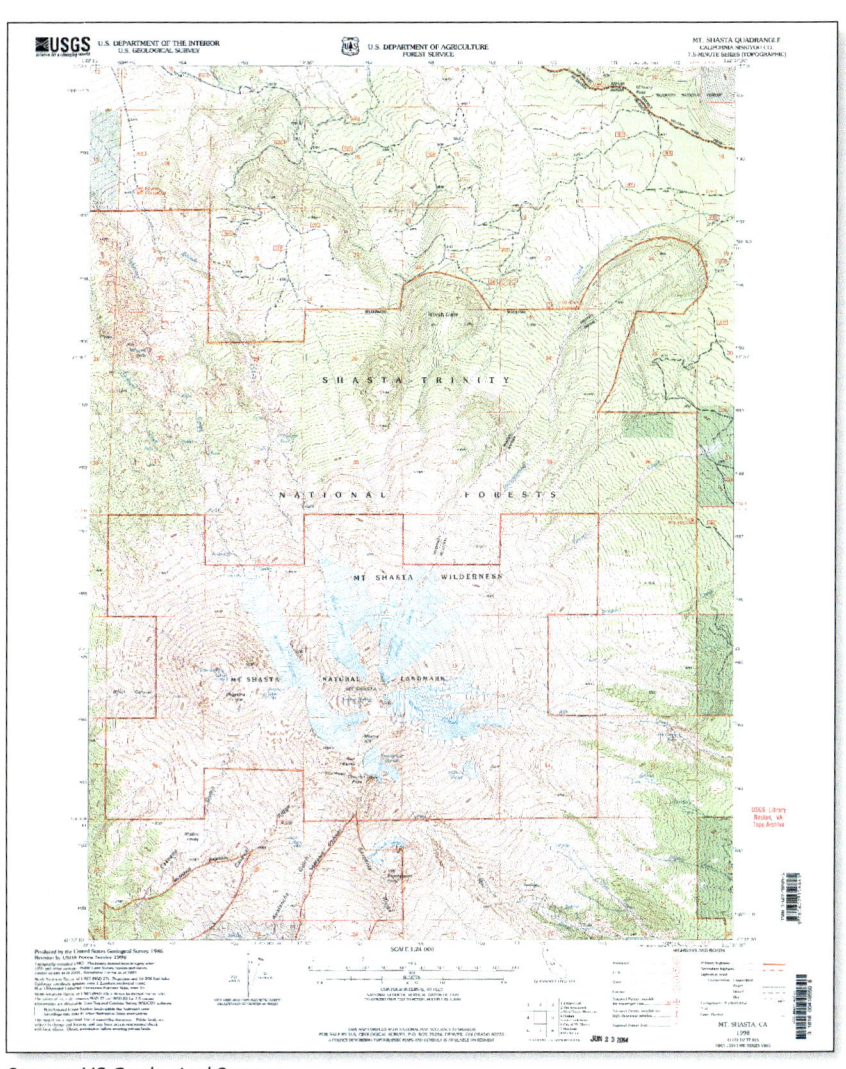

Source: US Geological Survey

Digital maps are mass-produced using automated scanners with GIS databases based on the original topographic maps providing the information. These maps can be produced much faster and a lower cost than the original hand-crafted topo maps but the trade off is a lower visual quality.

USGS maps, which students will use in this investigation, are measured in feet and meters. The USGS produces most of the topographic maps in this country in feet and is beginning to produce some maps in the metric system.

Soils, Rocks, and Landforms Module—FOSS Next Generation

INVESTIGATION 3 – Mapping Earth's Surface

How Can We Draw the Profile of a Mountain from a Topographic Map?

A topographic profile is a cross-sectional view of a mountain or other landform that is made from a topographic map. Think of a mountain as a white cake covered with chocolate frosting. You cut the cake in half, pull one half of it away, and look at the pattern made by the frosting outlining the cut-away view—that's your profile. To make a topographic profile, you "cut the cake" by drawing a line on the topographic map indicating the cross section that you want to map. Then you mark the locations on the topographic map at equal intervals along a vertical grid, and you are able to see the ups and downs of the cross section from point A to point B.

Profiles are useful to geologists as well as hikers and bicycle riders. By making a profile map, you can see where the trail gets steep and where it flattens out. You can look at multiple routes to determine which best fits your needs and abilities to meet the rigors of the hike. Today there are many websites available in which you can electronically make a profile of a hike or bike route with the click of a mouse. Google Earth™ includes this feature (called elevation profiles) if you choose to take a look at a local park or mountain for possible field trips.

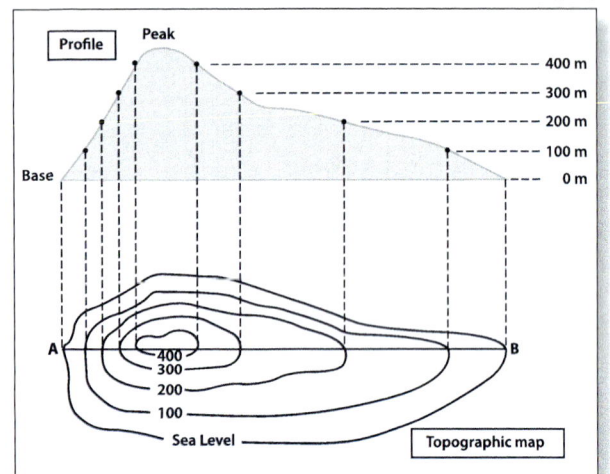

The distance between contour lines on a map is not always the same. But the difference in elevation between the lines, or contour interval, is always the same. The contour interval for this map is 100 feet.

Contour lines that are spaced closer together represent a steeper slope. You travel the same vertical distance while traveling less horizontal distance.

How Can Scientists and Engineers Help Reduce the Impact that Volcanic Eruptions Might Have on People?

Along the west coast of the United States—in California, Oregon, and Washington—volcanoes pose a threat of rapid change to Earth's surface. Indeed, Mount St. Helens changed the shape of 230 square miles in a matter of 10 minutes on May 18, 1980. Volcanoes are the result of a tremendous buildup of heat and pressure in the **mantle** below Earth's **crust**. Eventually the pressure forces the melted rock called **magma** upward through a crack or weak spot in the crust. When it does, the magma pours out as **lava** or pyroclastic flows (mixtures of extremely hot gases and dry rock fragments). This material and volcanic ash spill out onto Earth's surface, either flowing down the side of the volcano or shooting up into the air to be carried by wind currents before being deposited some distance from the eruption. A **satellite cone** can also form on the flank of a volcano where a fracture allows accumulation of volcanic material not part of the central vent. Volcanoes destroy the landforms at the site of the eruption, but they create new landforms through deposition of new rock material on Earth's surface. Eruptions

Background for the Teacher

are truly constructive events because the earth material they leave behind is not simply relocated surface material, but completely new material. The result of a volcanic eruption is creation of brand new landforms in a very short period of time and over the long run volcanic deposits also benefit soil, creating some of the richest soil on Earth for growing food crops.

What Events Can Change Earth's Surface Quickly?

Source: US Geological Survey

On occasion, Earth's surface can change rapidly, fast enough to observe in human time. Such events are sometimes lumped together under the rubric of catastrophic events. They are usually considered destructive because they result in loss and destruction of human life and property. From the point of view of the geologist, however, catastrophic events are both destructive, because they alter and dismantle existing landforms, and constructive, because they produce new landforms in their wake.

Intense rains and rapid snowmelt can produce massive floods. During these powerful natural events, human structures take a beating. Houses, washed from their foundations, end up in rivers. Railroads and highways that traverse a river's floodplain wash away. Reservoirs fill with sediments washed in by the river. Bridges, dams, and other structures are at risk in the face of the natural processes that work to change the shape of the land. Downstream from the destructive impact of the extreme water flow, the eroded material is ultimately deposited. The deposition constructs new landforms.

Landslides produce relatively fast changes to the landscape. In a matter of minutes, a mountainside can yield to gravity and tumble or slide down to the valley below. The erosion from the mountainside and the resulting deposition of masses of rock and soil below can cause tremendous damage to structures in the path of the landslide and create new deposits of rock and soil in the valley.

Earthquakes vary in intensity, but can result in extremely rapid changes to the integrity of Earth's crust. Breaks in the bedrock may result in one section of crust being lifted upward and another section dropped downward. Such events can create new landforms such as cliffs and deep crevasses.

Source: US Geological Survey

Volcanoes also change the Earth's surface rapidly as discussed in the previous section.

Soils, Rocks, and Landforms Module—FOSS Next Generation

INVESTIGATION 3 – Mapping Earth's Surface

TEACHING CHILDREN about *Topographic Maps and Rapid Change Events*

Developing Disciplinary Core Ideas (DCI)

Up to this point, students have studied how water cuts downward through earth materials. This investigation focuses on landforms that rise above their surrounding surface.

The mountain selected for this investigation is Mount Shasta in California. It was chosen because of the range of features it provides: a double peak, steep and shallow slopes, glacial valleys, river canyons, isolation from other mountains. These features provide for a range of interpretations from topographic and aerial maps.

When experts read a topographic map, they visualize the real terrain. As they turn a map, their image turns. They can mentally travel over the map as if they were flying over the terrain in an airplane. To help students visualize terrain somewhat like an expert, this module moves them through a series of experiences that transforms models to horizontal and vertical two-dimensional points of view. Each view is studied in relationship to the model. Questions invite comparisons (e.g., "Where is the steepest slope on the model?", "Where on the map?", "How can you tell a steep slope on a map?"). In a sense, these experiences encourage mental gymnastics—transformations of images from one form to another and back again in various ways.

By the end of this investigation, students will be able to develop simple topographic maps, use the cartographer's method of determining vertical images (cross-section profiles), and interpret a topographic map of a different terrain.

Volcanoes and earthquakes can change landscapes rapidly. These processes are complex and difficult to model acceptably for elementary students. They will learn about these important agents of change through images, diagrams, videos, and readings. A wealth of historical and present-day data and images bring these processes into the larger picture of a dynamic landscape.

These experiences will contribute to students' understanding of the core ideas **ESS1.C: The history of planet Earth**; **ESS2.B: Plate tectonics and large-scale system interactions**; **ESS3.B: Natural hazards**; and **ETS1.B: Developing possible solutions**.

NGSS Foundation Box for DCI

ESS1.C: The history of planet Earth
- Local, regional, and global patterns of rock formations reveal changes over time due to Earth's forces such as earthquakes. The presence and location of certain fossil types indicate the order in which rock layers were formed. (4-ESS1-1)

ESS2.B: Plate tectonics and large-scale system interactions
- The locations of mountain ranges, deep ocean trenches, ocean floor structures, earthquakes, and volcanoes occur in patterns. Most earthquakes and volcanoes occur in bands that are often along the boundaries between continents and oceans. Major mountain chains form inside continents or near their edges. Maps can help locate the different land and water features of Earth. (4-ESS2-2)

ESS3.B: Natural hazards
- A variety of hazards result from natural processes. Humans cannot eliminate the hazards but can take steps to reduce their impact. (4-ESS3-2)

ETS1.B: Developing possible solutions
- At whatever stage, communicating with peers about proposed solutions is an important part of the design process, and shared ideas can lead to improved designs. (3-5-ETS1-2)

Teaching Children about Maps and Change Events

Engaging in Science and Engineering Practices (SEP)

In this investigation, students engage in these practices.

- **Developing and using models** to visually display topographic features of landforms from different perspectives and describe the limitations of the model.
- **Planning and carrying out investigations** to determine the impact of the Mount St. Helens volcanic eruption.
- **Analyzing and interpreting data** to draw conclusions about the shape of landforms before and after catastrophic events such as volcanic eruptions.
- **Using mathematics and computational thinking** to determine contour intervals and changes in elevations in feet (Mount Shasta) and meters (Mount St. Helens).
- **Constructing explanations** using evidence to describe cause-and-effect relationships in changing landforms due to rapid changes caused by natural earth processes.
- **Engaging in argument from evidence** about the cause-and-effect relationships of earth processes and resulting landforms.
- **Obtaining, evaluating, and communicating information** from books and media and integrating that with students' firsthand experiences to construct explanations about changes to landforms due to earthquakes, landslides, floods, and volcanic eruptions. Communicate information in visual displays using topographic maps and landform profiles.

NGSS Foundation Box for SEP

- **Identify limitations of models.**
- **Develop models** to describe phenomena.
- **Make observations and/or measurements to produce data** to serve as the basis for evidence for an explanation of a phenomenon or test a design solution.
- **Analyze and interpret data** to make sense of phenomena, using logical reasoning.
- **Organize simple data** to reveal patterns that suggest relationships.
- **Construct an explanation** of observed relationships.
- **Use evidence** (e.g., observations, patterns) to support an explanation.
- **Identify the evidence** that supports particular points in an explanation.
- **Construct and/or support an argument** with evidence, data and/or models.
- **Use data to evaluate claims** about cause and effect.
- **Read and comprehend grade-appropriate complex texts** and/or other reliable media to summarize and obtain scientific and technical ideas and describe how they are supported by evidence.
- **Obtain and combine information** from books and/or other reliable media to explain phenomena.
- **Communicate scientific and/or technical information** orally and/or in written formats, including various forms of media and may include tables, diagrams, and charts.

Soils, Rocks, and Landforms Module—FOSS Next Generation

INVESTIGATION 3 – Mapping Earth's Surface

NGSS Foundation Box for CC

- **Patterns:** Similarities and differences in patterns can be used to sort and classify natural phenomena. Patterns of change can be used to make predictions.
- **Cause and effect:** Events that occur together with regularity might or might not be a cause-and-effect relationship. Cause-and-effect relationships are routinely tested, and used to explain changes.
- **Scale, proportion, and quantity:** Standard units are used to measure and describe physical quantities such as weight, time, temperature, and volume.
- **Stability and change:** Change is measured in terms of differences over time and may occur at different rates. Some systems appear stable, but over long periods of time will eventually change.

Exposing Crosscutting Concepts (CC)

In this investigation, the focus is on these crosscutting concepts.

- **Patterns.** Patterns observed in landforms can be used to sort and classify types of change due to natural processes.
- **Cause and effect.** Cause-and-effect relationships between the shape of landforms and natural earth events are useful to explain change.
- **Scale, proportion, and quantity.** Standard units are used to determine contour intervals used to display the form and shape of large mountains into a manageable visual display.
- **Stability and change.** Some natural Earth events happen over long periods of time, while others, such as floods, earthquakes, volcano eruptions, and landslides can occur rapidly, and cause major changes to landforms.

Connections to the Nature of Science

- **Scientific investigations use a variety of methods.** Scientific methods are determined by questions. Scientific investigations use a variety of methods, tools, and techniques.
- **Scientific knowledge assumes an order and consistency in natural systems.** Science assumes consistent patterns in natural systems.

Connections to Engineering, Technology, and Applications of Science

- **Interdependence of science, engineering, and technology.** Science and technology support each other. Knowledge of relevant scientific concepts and research findings is important in engineering. Tools and instruments are used to answer scientific questions, while scientific discoveries lead to the development of new technologies.

New Word
Say it · See it · Hear it · Write it

Contour interval
Contour line
Crust
Earthquake
Elevation
Landslide
Lava
Magma
Mantle
Profile
Satellite cone
Sea level
Topographic map
Volcano

Teaching Children about Maps and Change Events

Conceptual Flow

A volcanic mountain is the major phenomenon that students investigate as they work with models to represent and describe this surface feature on Earth and understand how volcanic mountains were formed and how they change over time. The guiding questions for the investigation are how do maps help us observe Earth's surface features? and what might reduce the impact of catastrophic Earth surface events?

The **conceptual flow** for this investigation starts as students are introduced to the study of topography by building a model of a landform—a mountain. They use a foam model of Mount Shasta to **create a topographic map**. They are introduced to **elevation above sea level** and the conventions of counter lines and contour intervals. Students develop and use this model to gather information about the Earth's surface.

In Part 2, students transform the bird's-eye view of Mount Shasta into **a representation of a side-view—a profile**. The perspective they take is one of many profiles that could be generated from the information provided by the topographic map. Students read about this history of Mount Shasta, a dormant volcano, and begin a general study of **volcanoes**.

In Part 3, students study an active volcano, Mount St. Helens, and learn about the changes that took place in recent times as a result of the volcanic eruption. Students use the tools developed in earlier parts to compare the profiles of this mountain before and after the 1980 eruption. The **impact of catastrophic natural events on humans** becomes a focus of study.

In Part 4, students consider forces and **events that can change Earth's surface very quickly**. Erosion and deposition produce landforms over millions of years. Other events, such as **earthquakes**, **landslides**, **floods**, and **volcanoes**, can produce changes to Earth's surface very quickly. Scientists and engineers study these events to determine ways that the impact on humans can be reduced.

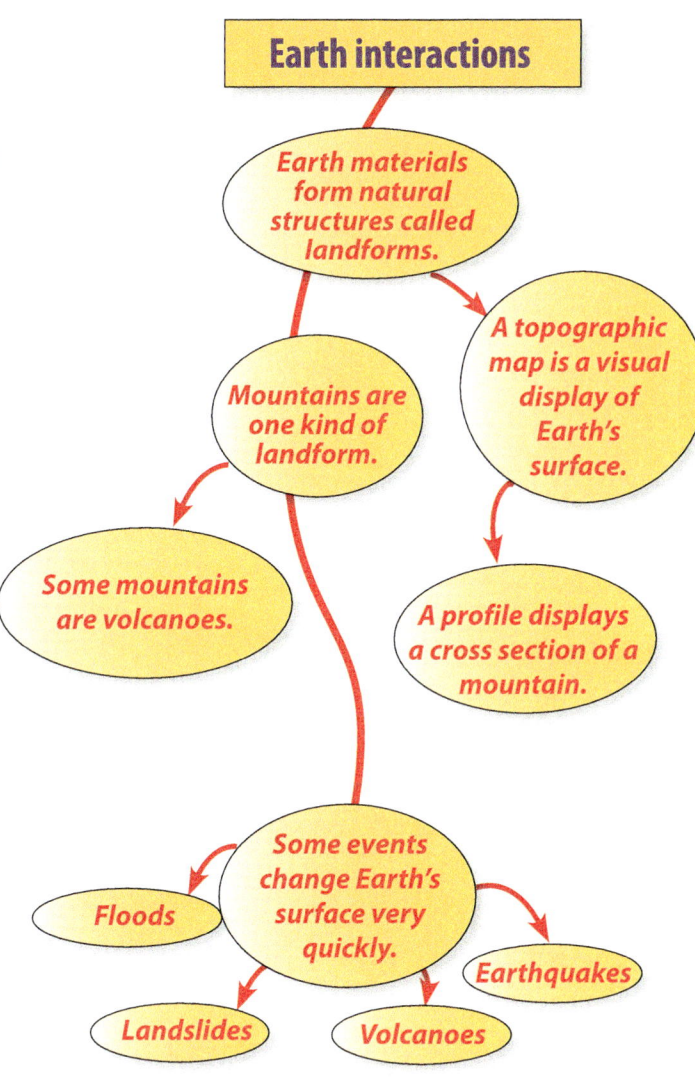

Soils, Rocks, and Landforms Module—FOSS Next Generation

INVESTIGATION 3 – Mapping Earth's Surface

MATERIALS for
Part 1: Making a Topographic Map

For each group
- 1 Foam mountain set, 6 pieces, and dowel in large zip bag
- 4 Pencils ★
- 1 Topographic map of Mount Shasta, CA (7.5-minute quadrangle)
- 1 *FOSS Science Resources: Soils, Rocks, and Landforms*
  - "Topographic Maps"

For the class
- 1 Poster, Mount Shasta
- Aluminum foil (optional) ★
- 8 *Foam-Mountain Topographic Map* transparencies (same as teacher master 13)
- 8 Teacher master 13, *Foam-Mountain Topographic Map*
- Teacher master 14, *Mountain Map*

For embedded assessment
- ❏ • *Embedded Assessment Notes*

★ Supplied by the teacher. ❏ Use the duplication master to make copies.

No. 13—Teacher Master

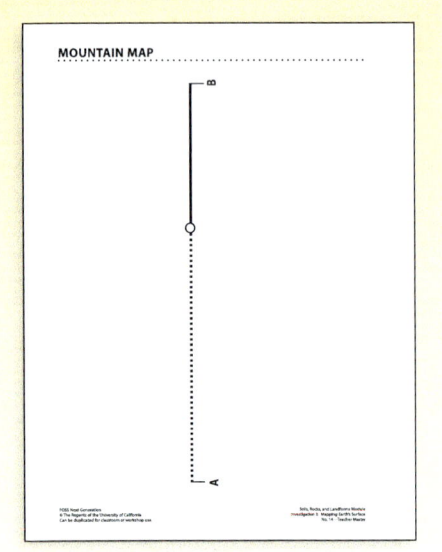

No. 14—Teacher Master

218 Full Option Science System

Part 1: Making a Topographic Map

GETTING READY for
Part 1: Making a Topographic Map

1. **Schedule the investigation**
   This part takes one active investigation session and one session for the reading.

2. **Preview Part 1**
   Students build a model mountain of Mount Shasta by stacking and orienting six foam layers. They trace outlines of the six pieces onto paper, creating a topographic map of the mountain. The focus question for this part is **How can we represent the different elevations of landforms?**

3. **Prepare foam mountains**
   Each collaborative group will build a model mountain by stacking six foam layers on a dowel. Each layer has a center hole for the dowel and a notch for alignment.

   The foam pieces are labeled in a new kit. It is very important to make sure each piece is correctly oriented and right side up. Refer to teacher sheet 13, *Foam-Mountain Topographic Map*.

   When the mountain is assembled, the elevations can be read.

▶ **NOTE**
To prepare for this investigation, view the teacher preparation video on FOSSweb.

▶ **NOTE**
The topographic maps in this part show elevation in feet rather than meters. That is because the USGS maps for Mount Shasta are in feet.

Soils, Rocks, and Landforms Module—FOSS Next Generation

INVESTIGATION 3 — *Mapping Earth's Surface*

> **NOTE**
> Save your practice *Mountain Map* for the next part.

4. **Practice drawing a mountain map**
   Print or make copies of teacher master 14, *Mountain Map*. You will need enough for each student to have one, plus a few extras for those who may need to start over.

   Practice making your own topographic map by following Step 9 in Guiding the Investigation.

   Check your *Mountain Map* by laying a *Foam-Mountain Topographic Map* transparency on top of the map you drew. If you lined up the notches and traced carefully around the foam pieces the two maps should match. (Make sure you have the transparency right side up.)

5. **Plan to read *Science Resources*: "Topographic Maps"**
   Plan to read the article "Topographic Maps" during a reading period after completing the active investigation for this part.

6. **Plan assessment: notebook entry**
   In Step 19, students answer the focus question. Check their understanding of how topographic maps represent the three-dimensional landforms. Make notes on an *Embedded Assessment Notes* sheet.

Part 1: Making a Topographic Map

GUIDING *the Investigation*
Part 1: *Making a Topographic Map*

1. **Review stream-table maps**
   Ask students to recall the maps they made when they recorded their observations of the stream-table models. Have them turn to one of the stream-table observation maps in their notebooks. Have them discuss in their groups the kinds of information those maps provided.

   ➤ *What landforms were you able to map on your stream-table observation sheet?*

   Call for attention and have a few students share the different landforms they recorded on their maps (canyons, valleys, rivers, and deltas, for example).

2. **Discuss map limitations**
   Weave into the discussion the idea that the stream-table map shows the shape of the different landforms, but it does not provide any information about three-dimensional aspects. For example, there is no indication of how deep a canyon might be.

   Tell students when we talk about vertical distance (going up or down in a canyon, mountain, or other landform) we are talking about **elevation**. Elevation is the vertical distance or height above (or below) **sea level**.

   Tell students that today they are going to make a model and map of another type of landform.

3. **Focus question: How can we represent the different elevations of landforms?**
   Ask the focus question as you project or write it on the board, and have students copy it into their notebooks.

   ➤ *How can we represent the different elevations of landforms?*

4. **Demonstrate how to build a model mountain**
   Show students a set of six foam pieces. Point out that the pieces are all different sizes and shapes, but each is the same thickness. Point out the numbers on the layers; the numbers should remain side up.

   Demonstrate how to punch the dowel up through the hole near the center of each foam piece. Tell them to start with the piece labeled with the lowest number (11,000), then the next lowest (11,500), continuing until all six are in place. Show them how to align the notches.

   Have Getters pick up the materials and start building models.

Soils, Rocks, and Landforms Module—FOSS Next Generation

FOCUS QUESTION
How can we represent the different elevations of landforms?

Materials for Step 4
- *Foam-mountain sets*

INVESTIGATION 3 – Mapping Earth's Surface

> **NOTE**
> Encourage students to push the dowel through the holes from the bottom of the foam pieces.
>
> You can mold a piece of aluminum foil around a foam mountain to give an impression of what the mountain would look like if it had a smooth slope.

SCIENCE AND ENGINEERING PRACTICES
Developing and using models

CROSSCUTTING CONCEPTS
Scale, proportion, and quantity

Say it • See it • Hear it • Write it — **New Word**

Materials for Step 5
- *Mount Shasta Poster*

> **NOTE**
> Keep the poster of Mount Shasta displayed on the wall throughout this investigation.

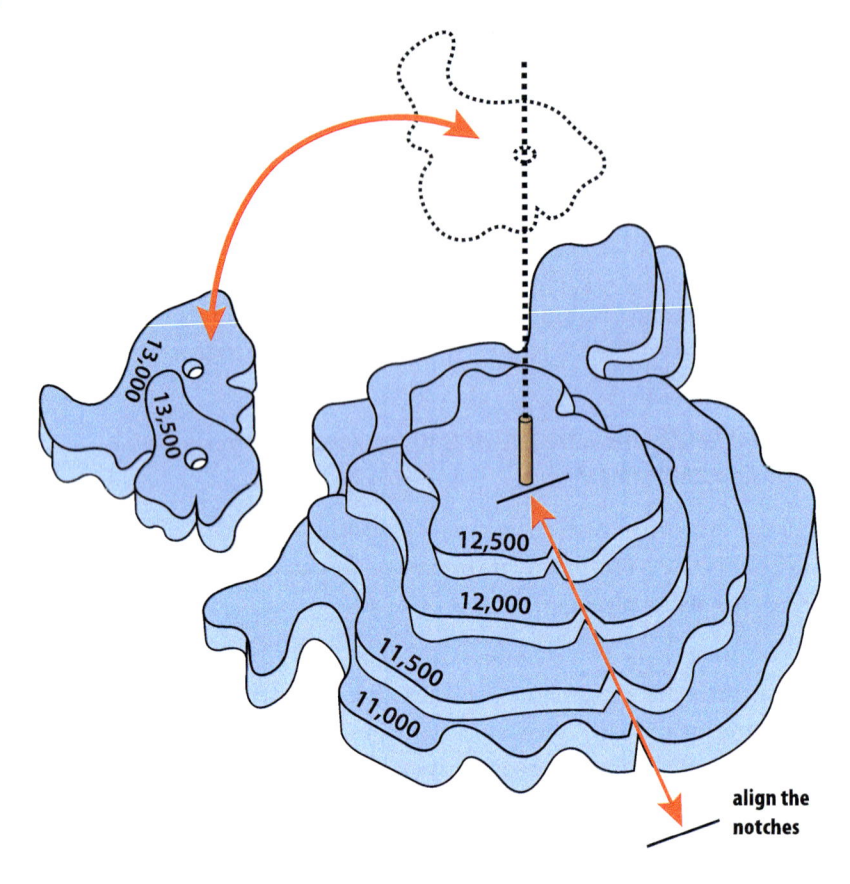

align the notches

5. Describe the model landform

Ask students what landform they have modeled. Confirm that they have built a model of a mountain. Some students might suggest that it looks like a **volcano**. Show them the poster and project the photo of Mount Shasta on page 34 of the *FOSS Science Resources* book ("The Story of Mount Shasta") and tell them the name of the mountain. Mount Shasta is a volcano in the Cascade mountain range in the western United States. Have students compare the foam mountain to the real mountain on the poster.

Ask,

▶ *How does our model mountain compare to a real mountain?* [Most mountains aren't built like stair steps, but have smoother slopes. Real mountains are much bigger; they aren't blue, and so forth.]

Point out the two pieces of the mountain that extend way beyond the other pieces at the 11,000- and 11,500-foot levels. These layers represent the **satellite cone** of Mount Shasta called Shastina. Encourage students to find Shastina on the poster. Shastina is 12,329 feet high. Were it considered a separate mountain, it would be the third highest in the Cascade range! If we wanted to make the foam-mountain model more accurate, we could add another foam piece to show Shastina at the 12,000-foot level.

Full Option Science System

Part 1: Making a Topographic Map

6. Circle the mountain
Draw students' attention to the numbers printed on the top edges of the layers. Tell them that each layer is labeled with an elevation above sea level. For example, the top of the first layer represents 11,000 feet. Trace around the top edge of the first layer as you explain.

If you travel all the way around the mountain on top of this layer, you will always be 11,000 feet above sea level.

Have students take finger walks around different layers of the foam mountains and identify their elevations.

7. Discuss equal intervals
Ask students to determine the difference in the elevation between the top of the first layer (11,000) and the top of the second layer (11,500).

➤ *Is the vertical distance from the top of any one layer of foam to the top of the next layer always the same?*

Confirm that, because the layers are the same thickness, the vertical distance from the top surface of any layer to the top of the next layer is always the same, in this case 500 feet.

8. Introduce *topographic map*
Tell students that you would like to make a map that would show both the shape of the land and the elevation at the same time. Have them discuss ideas in their groups for a few minutes. Share their ideas and guide the discussion to topographic maps.

*There is a kind of map that shows both the elevation and the shape of the land. It is called a **topographic map**. You can use the foam pieces to draw a topographic map of your model mountain.*

9. Demonstrate the mapmaking procedure
Explain as you demonstrate the procedure.
   a. Take the foam mountain apart.
   b. Poke a pencil point from the top through the hole of one of the layers; make sure the elevation label is showing (on the top side).
   c. Set the point of the pencil on the center circle of the *Mountain Map*.
   d. Line up the notch with the solid dark line that runs from the circle to point B.
   e. Hold the foam piece securely, and carefully trace around its outside edge.

Materials for Step 9
- *Mountain Map* sheets

Soils, Rocks, and Landforms Module—FOSS Next Generation

INVESTIGATION 3 – Mapping Earth's Surface

f. Repeat this procedure for all six pieces of foam. The order in which the pieces are traced doesn't matter, as long as you always line up the hole in the foam mountain with the circle and the notch with the solid black line.

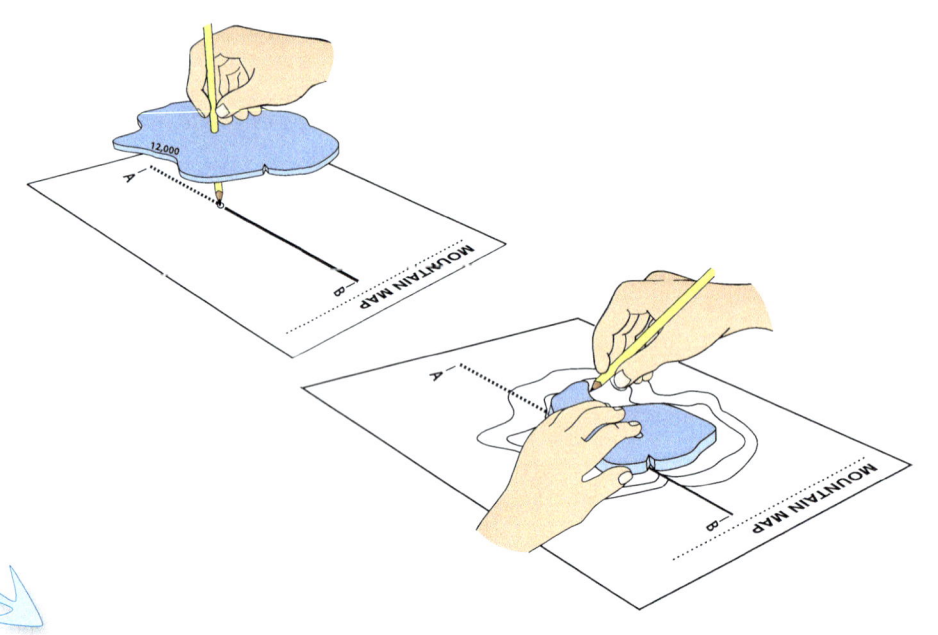

▶ **NOTE**
Students should use pencils, not pens, to trace the foam pieces.

TEACHING NOTE

Monitor students' progress as they work, making sure they have the proper alignment and side face up for the foam pieces. You might want to stop work briefly after about 5 minutes to do a quick check.

10. Draw the topographic map

Have Getters pick up four *Mountain Map* sheets. Remind students to use pencils for tracing. Visit students as they begin tracing to make sure they have the foam pieces right side up and are aligning the pieces correctly. When they finish, ask the groups to reassemble the model.

As students begin to finish their maps, distribute a *Foam-Mountain Topographic Map* transparency to each group. Students can use this transparency to check their work. If they lay the transparency over their map (make sure you can read the transparency title right side up), the lines students drew should closely match the lines on the transparency. Point out to students that the peak of Shastina is indicated on the transparency. They can mark the approximate location of Shastina on their topo maps.

Materials for Step 10
- *Mountain Map* sheets
- *Foam-Mountain Topographic Map* transparencies

11. Compare maps and mountains

Ask students to compare their topographic map with the view of the foam mountain from directly overhead looking down on the top of the mountain. Have them locate the highest and lowest areas on their topographic maps.

SCIENCE AND ENGINEERING PRACTICES
Developing and using models

224   Full Option Science System

Part 1: Making a Topographic Map

12. Introduce contour lines
Explain that the lines that they have traced on the paper are **contour lines**. Tell them that contour lines are imaginary lines that follow the surface of the mountain at a precise elevation. For example, wherever the 11,000-foot contour line is, they know the elevation of the land is 11,000 feet.

13. Introduce contour interval
Remind students that the thickness, or height, of each piece of foam is the same. On topographic maps, the difference in elevation from one contour line to the next is always the same. This distance is the **contour interval**. Ask,

➤ *What is the contour interval on your topographic map?* [500 feet.]

Set foam mountains off to one side, on a countertop or other location close by.

14. Compare topographic maps
Distribute one USGS topographic map of Mount Shasta to each group. Challenge the groups to position the *Topographic Map Transparency* to match the USGS map. If students have difficulty doing this, have them locate the peaks of Mount Shasta and Shastina on the topographic map (each is marked with a + sign, the name of the peak, and its elevation). Once students have located those two peaks, they should be able to orient the transparency on the map and see how the contour lines match.

➤ *What is the contour interval of the USGS map?*

➤ *How were you able to match the topographic map transparency to the USGS map?*

➤ *Can you match the foam mountain to the USGS map?*

➤ *Can you find glaciers and lava flows on the map?*

Have students label the elevations of the peaks of Mount Shasta (14,179 ft/4,322 m) and Shastina (12,329 ft/3,758 m) on the topographic maps they drew. Ask,

➤ *How do topographic maps help people "see" the shape of the land?*

15. Notice the slope
Ask students to look at the topographic maps they drew and ask,

➤ *Which is the steeper side of the foam mountain?* [The side with the notches, side B is the steeper.]

➤ *How can you tell just by looking at a topographic map where the mountain is steeper?* [The lines are closer together.]

TEACHING NOTE

Emphasize that contour lines are not always the same distance apart on the map, but they always represent equal steps in elevation.

Say it → See it → Hear it → Write it → New Word

▶ **NOTE**
Foam mountains will be used again in Step 15, so don't take them apart yet.

Materials for Step 14
- USGS Topographic Maps of Mount Shasta

▶ **NOTE**
You may find slightly different elevations for the mountain peaks in different references.

▶ **NOTE**
Have students write the peak elevators in feet and meters. Students will work with meters rather than feet in Part 3.

CROSSCUTTING CONCEPTS
Scale, proportion, and quantity

Soils, Rocks, and Landforms Module—FOSS Next Generation

INVESTIGATION 3 – Mapping Earth's Surface

SCIENCE AND ENGINEERING PRACTICES
Developing and using models

contour interval
contour lines
elevation
satellite cone
sea level
topographic map
volcano

SCIENCE AND ENGINEERING PRACTICES
Developing and using models

CROSSCUTTING CONCEPTS
Scale, proportion, and quantity

Have students retrieve the foam mountains and confirm that B is the steeper side. Have students look at the foam mountain from directly overhead. The contour lines suggested by the edges of the foam pieces will be closer together on the steep side of the mountain.

Remind students that the contour interval is the same whether the contour lines are close together or far apart. Contour lines that are close together represent a steep slope. On the more gently sloped areas of the mountain, the lines are farther apart.

16. Practice reading the slope
In groups, have two students at a time each point to a place on the USGS topographic map. The two students not pointing should call out which place they think is steeper. Pointers agree or disagree, and discussion continues until all have come to consensus. Then they switch roles. Continue the activity until you feel confident that students can determine relative steepness based on the contour line spacing.

17. Clean up
Have students return all materials to the materials station. Foam mountains should be taken apart and bagged.

18. Review vocabulary
Review key vocabulary introduced in this part. If these words are not on the word wall, add them now.

19. Answer the focus question
Have students answer the focus question in their notebooks.

➤ *How can we represent the different elevations of landforms?*

20. Assess progress: notebook entry
Have students hand in their notebooks open to the page on which they answered the focus question. Review students' notebooks after class and check their understanding of topographic maps.

What to Look For
- *Topographic maps represent the shape of the land.*
- *Each line represents one elevation, for example 11,000 feet.*
- *The interval between contour lines is always the same.*
- *You can tell the steepness of a slope by how close together the lines are; the closer the lines, the steeper the slope.*

Part 1: Making a Topographic Map

READING *in Science Resources*

21. Read "Topographic Maps"
Tell students that this article will help them learn more about how topographic maps are made and the standardized symbols used to communicate information. Before reading, have students do a quick write in their notebooks, jotting down three things they know so far about topographic maps and a question they have about how they are made or how to interpret the symbols.

Give students a few minutes to preview the text by looking at and discussing the diagrams with a reading partner.

Ask them to compare the topographic maps they made from the Mount Shasta model to the one in the article. Have them make predictions about what they think the different colors on the map represent.

22. Use a focused note-making strategy
Have students read the story independently or in guided reading groups. Tell students to think like a cartographer (a person who makes maps) as they are reading and to jot down in their notebook or on self-stick notes important information they would need to make a map.

23. Discuss the reading
Have students compare their notes to a reading partners' notes. Based on those notes, have them work together to determine what they think the most important things a cartographer would need to know to make a topographic map. Have students share the features of a topographic map that they learned from the article and make a class list on chart paper. The list should include:

- Contour lines
- Index contours
- Scale
- Elevation
- Symbols and colors for landforms and structures

Discuss the role of the USGS and ask students what they think topographic maps are used for. Visit the USGS website and let students view topographic maps from all over the world, including your local area.

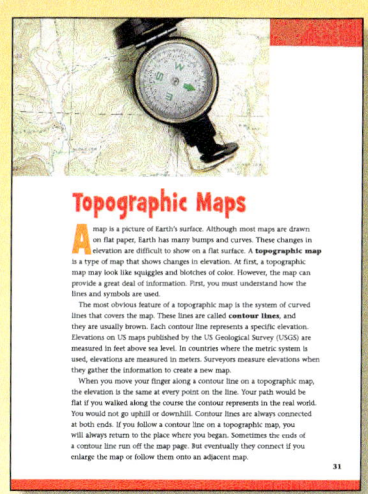

ELA CONNECTION

These suggested reading strategies address the Common Core State Standards for ELA.

RI 1: Refer to details and examples in a text when explaining what the text says explicitly and when drawing inferences from text.

RI 2: Determine the main idea of a text and explain how it is supported by key details; summarize the text.

RI 3: Explain procedures, ideas, or concepts in a scientific text.

RI 7: Interpret information presented visually, and explain how the information contributes to an understanding of the text.

W 8: Gather relevant information from experiences and print, and categorize the information.

SCIENCE AND ENGINEERING PRACTICES

Obtaining, evaluating, and communicating information

Soils, Rocks, and Landforms Module—FOSS Next Generation

INVESTIGATION 3 – Mapping Earth's Surface

WRAP-UP/WARM-UP

24. Share notebook entries

Conclude Part 1 or start Part 2 by having students discuss notebook entries. Ask students to open their science notebooks to their focus question answer. Read the focus question together.

➤ *How can we represent the different elevations of landforms?*

Ask students to pair up with a partner to discuss these questions.

➤ *What is the difference between contour lines and contour intervals?*

➤ *If you were going on a hike and took a topographic map with you, how would you use it to determine the route you take on the hike?*

Part 2: Drawing a Profile

MATERIALS for
Part 2: *Drawing a Profile*

For each group

- 1 Foam mountain set
- 1 *Mount Shasta* topographic map
- 4 *Mountain Maps,* completed (from Part 1)
- 4 Transparencies, *Mount Shasta*
- ❏ 4 Notebook sheet 15, *Contours and Intervals*
- 4 *FOSS Science Resources: Soils, Rocks, and Landforms*
  - "The Story of Mount Shasta"

For the class

- 1 Poster, Mount Shasta
- 1 Computer with Internet access ★
- 1 Projection system (optional) ★
- ❏ 1 Teacher master 15, *Mountain Maps with Elevations*
- ❏ 1 Teacher master 16, *Profile*
- ❏ 1 Teacher master 17, *Video Discussion Guide: Volcanoes*

For embedded assessment

- ❏ • Notebook sheet 16, *Response Sheet A—Investigation 3*
- ❏ • Notebook sheet 17, *Response Sheet B—Investigation 3*
- ❏ • *Embedded Assessment Notes*

★ Supplied by the teacher.   ❏ Use the duplication master to make copies.

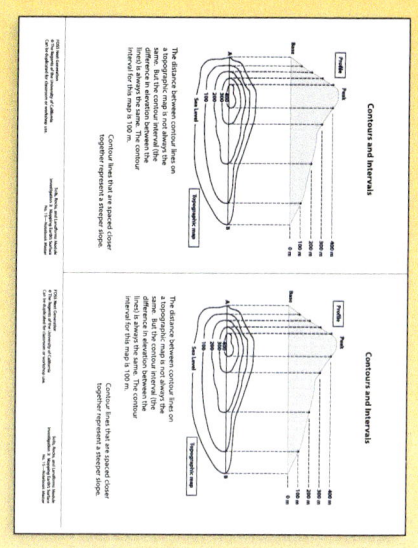

No. 15—Notebook Master

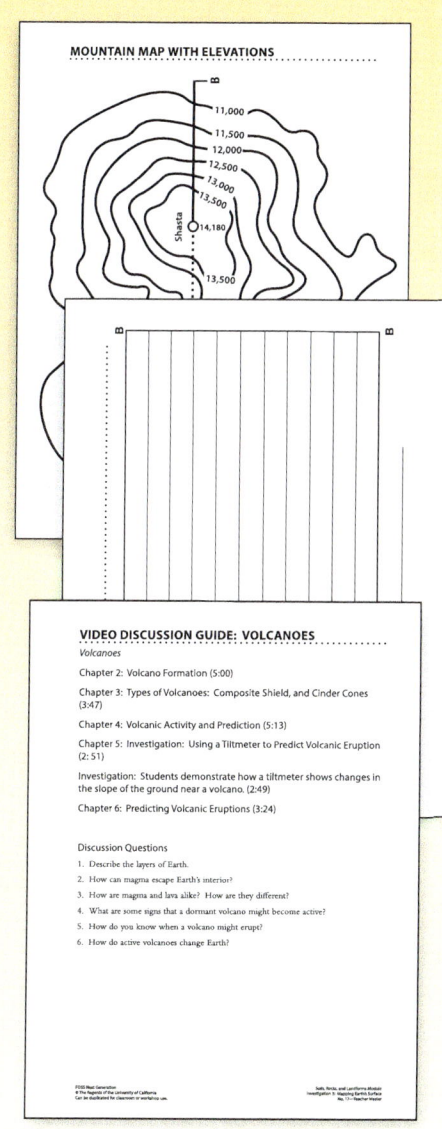

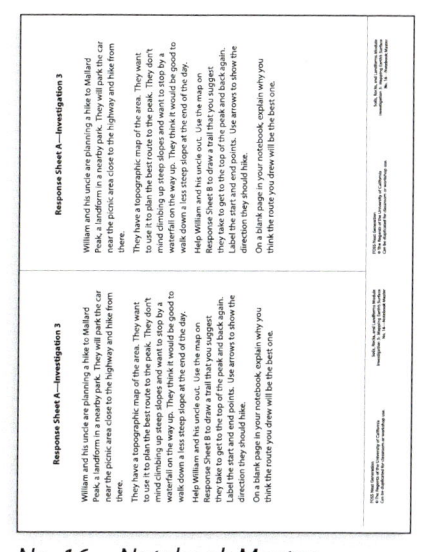

No. 16—Notebook Master

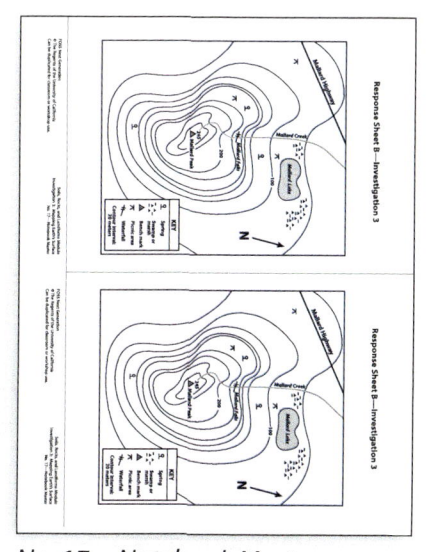

No. 17—Notebook Master

Nos. 15–17—Teacher Masters

Soils, Rocks, and Landforms Module—FOSS Next Generation

INVESTIGATION 3 – *Mapping Earth's Surface*

GETTING READY for
Part 2: *Drawing a Profile*

1. **Schedule the investigation**
   This part will take one active investigation session, one for the reading, and for one for the video on volcanoes.

2. **Preview Part 2**
   Students use their topographic maps to produce two-dimensional profiles, or cross-sections, of their foam mountain. The profile reveals a side view of Mount Shasta, a dormant volcano. Students gather information about volcanoes from a video. The focus question for this part is **How can we draw the profile of a mountain from a topographic map?**

3. **Practice drawing maps and profiles**
   Use the practice *Mountain Map* you drew in Part 1 to create a profile of the mountain. (See Step 3 of Guiding the Investigation.)

4. **Print or make copies**
   Students will write the elevations on their *Mountain Maps*, but you may want to print or make several copies of teacher master 15, *Mountain Maps with Elevations*. You can give this sheet to students who had difficulty drawing the topographic maps or you anticipate will have difficulty writing the elevations on the *Mountain Map*. The new sheet includes accurate elevation labels on the contour lines.

   Print or make a copy of teacher master 16, *Profile,* for each student, and a few extra copies for early finishers who might enjoy the challenge of drawing another profile map that includes Shastina.

5. **Plan to read *Science Resources*: "The Story of Mount Shasta"**
   Plan to read "The Story of Mount Shasta" during a reading period after completing the active investigation for this part.

6. **Preview the video**
   Preview the *Volcanoes* video (duration 24 minutes). In *Volcanoes*, students get an up-close look at these fiery formations, discovering how they form and what causes them to erupt. Students learn the difference between magma and lava, and discover the characteristics that enable volcanologists to classify a volcano as cinder cone, shield, or composite. Students see a model of a tiltmeter that is used to predict when a volcano might erupt by charting changes in the slope of the surrounding ground, and virtually visit Horseshoe Lake at Mammoth Mountain, California, to see the impact of an invisible sign of volcanic activity—poisonous carbon dioxide fumes.

Part 2: Drawing a Profile

The video will introduce students to ways to monitor volcano activity to reduce the impact of an eruption on humans.

A link to the video is found on FOSSweb in the Resources by Investigation section. Have students view the video after they complete the Mount Shasta reading. Plan to use teacher master 17, *Video Discussion Guide: Volcanoes*, to lead a class discussion after watching the video with the class.

7. **Preview the online activity**
   Preview the online activity "Topographer" by going to the Resources by Investigation section of FOSSweb. This activity will give students practice making profiles from topographic maps.

8. **Plan assessment: response sheet**
   Use notebook sheets 16 and 17, *Response Sheet A and B—Investigation 3* for a closer look at students' ability to interpret and use topographic maps. Plan to have students spend time reflecting on their responses after you have reviewed them. For more information about next-step and self-assessment strategies, see the Assessment chapter.

Soils, Rocks, and Landforms Module—FOSS Next Generation

INVESTIGATION 3 – Mapping Earth's Surface

FOCUS QUESTION

How can we draw the profile of a mountain from a topographic map?

Say it · See it · Hear it · Write it
New Word

Materials for Step 3
- *Mountain Maps* from Part 1
- *Profile* sheets

▶ **NOTE**
If students didn't label the peak in Part 1, have them do that now. The peak of Mount Shasta is 14,179 feet (4,322 m).

GUIDING *the Investigation*
Part 2: *Drawing a Profile*

1. **Introduce "profile"**
   Point to the Mount Shasta poster and tell students,

   *This picture was taken by someone looking toward one side of Mount Shasta. This one-sided view is called a **profile**. The topographic map you made of the foam mountain is a top-down, or "bird's-eye" view of Mount Shasta. Both views of a mountain can give us information about the mountain that is useful. For example, hikers use topographic maps to plot their routes from one place to another. Bicycle riders like to use profile maps to understand how big the hills are that they are going to have to ride over. We can make a profile of Mount Shasta from the topographic maps we made.*

2. **Focus question: How can we draw the profile of a mountain from a topographic map?**
   Ask the focus question as you project or write it on the board, and have students copy it into their notebooks.

   ▶ *How can we draw a profile of a mountain from a topographic map?*

3. **Label the contour lines**
   Have students fold their *Mountain Maps* along line *AB* so that the bottom half of the sheet is visible (the *A* and *B* are right side up). Show them how to write the elevations on the contour lines at both places where they cross line *AB*.

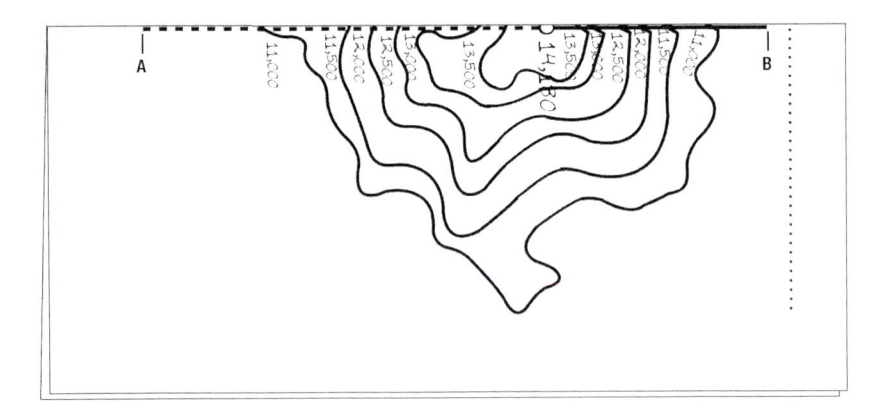

   Students should start with 11,000 on the outside lines and work their way to the center (13,500). Remind students that the interval between each line is 500 feet. The labels should be close to the point where the contour line crosses the line *AB*.

232   Full Option Science System

Part 2: Drawing a Profile

4. Demonstrate drawing a profile

Using a document camera or other means of projection, demonstrate drawing a profile using a completed *Mountain Map* (or *Mountain Map with Elevations*).

a. Write the elevations up the left side of the *Profile* sheet. Use the *Mountain Map* to determine what elevations to write. Start with 11,000 and go up to 14,000 (to more easily include the peak) using 500-foot intervals.

b. Point out the A and B lines on the *Profile* sheet.

c. Fold the *Mountain Map* along line *AB* on that sheet. If folded correctly, you should be able to see the *A* and *B* at either end of the fold line. Make sure the contour lines on the *Mountain Map* are labeled with the elevations on both sides of the peak.

d. Line up the *A* and *B* on the *Mountain Map* with the bottom of the *A* and *B* lines on the *Profile* sheet. The fold should be situated along the 11,000 line on the *Profile* sheet.

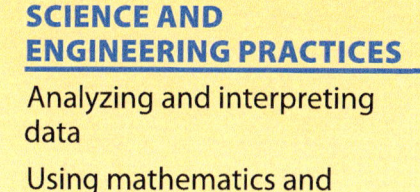

SCIENCE AND ENGINEERING PRACTICES

Analyzing and interpreting data

Using mathematics and computational thinking

CROSSCUTTING CONCEPTS

Scale, proportion, and quantity

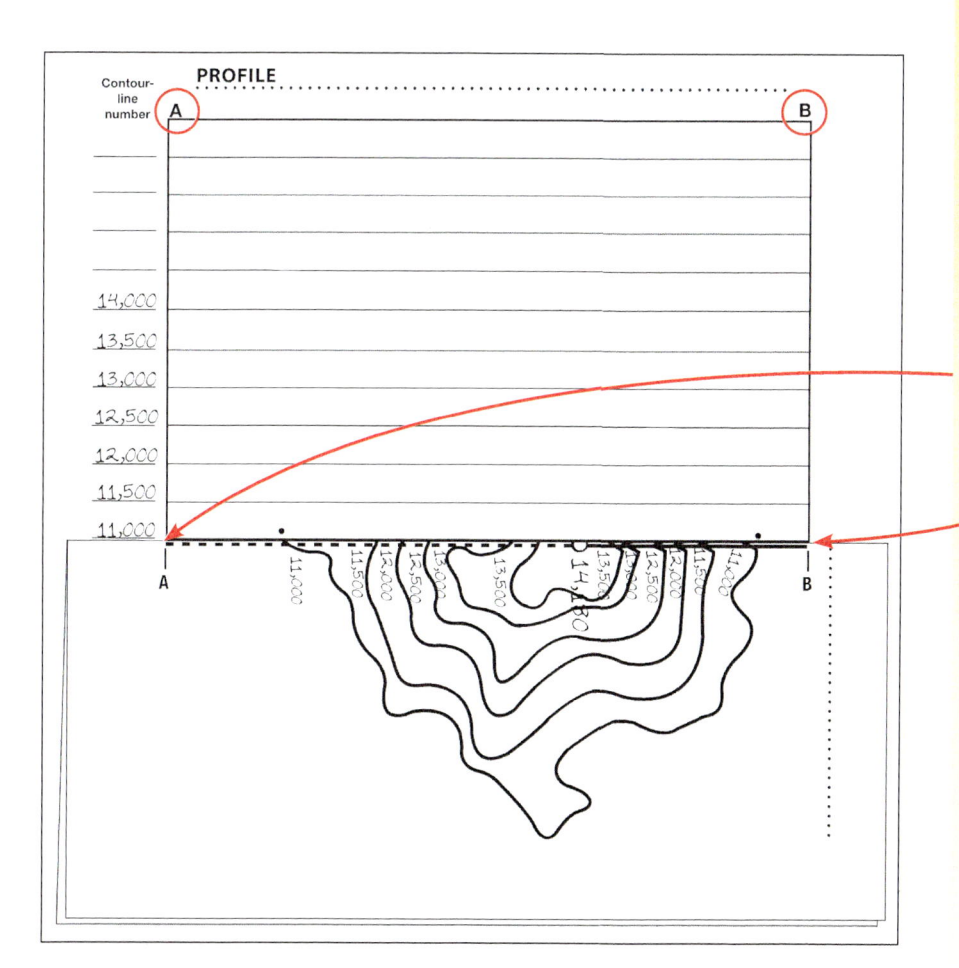

Make sure that the *A* and *B* on the *Mountain Map* are aligned with the *A* and *B* lines on the *Profile* sheet at all times.

Soils, Rocks, and Landforms Module—FOSS Next Generation

INVESTIGATION 3 — Mapping Earth's Surface

e. Find the two places where the 11,000 contour line crosses the horizontal 11,000 line on the *Profile*. Use a pencil to mark the point at both intersections on the profile line.

f. Slide the folded *Mountain Map* to the line labeled 11,500 and mark where the 11,500 contour line intersects its corresponding line in both places.

g. Continue moving the *Mountain Map* up the *Profile* sheet marking both points on the intersections of the corresponding contour lines.

h. Mark one point for the peak. Students may need a few minutes to ponder where that point should go. [Not on, but just above the 14,000 horizontal line.]

i. Connect the intersection points as smoothly as possible. This is the profile.

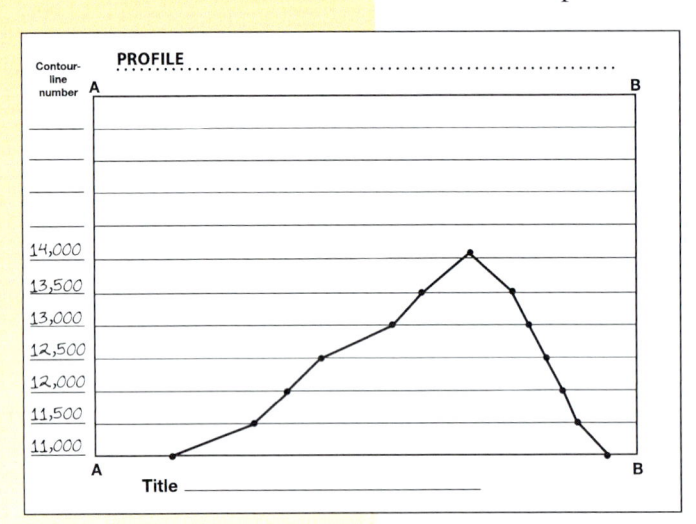

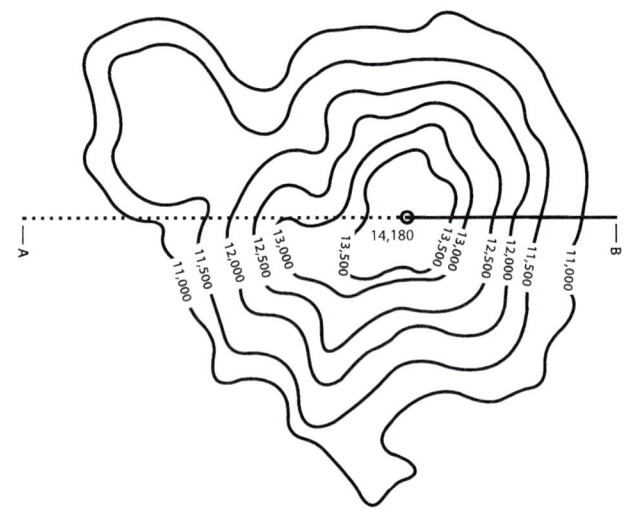

5. Start drawing

Make sure students have their *Mountain Maps* available. Provide a copy of teacher master 15, *Mountain Map with Elevations,* to students who may have unusable *Mountain Maps*. Distribute a *Profile* sheet to each student. Have them use pencils to start making their profiles.

Part 2: Drawing a Profile

6. Offer a second challenge (optional)
Some students will finish this task quickly and others will need significantly more time. As the first few students finish up, offer a second challenge. Ask them if they can draw another profile that includes Shastina. Discuss the fact that they must make their new *AB* line the same length as it was originally and when they draw the new line, they need to make sure that they include the peaks of both mountains. They won't be able to mark points for the 12,000-foot elevation or the peak (12,329 ft) for Shastina since that piece was left off the model, so they will have to make an educated guess about those. Send them off to take up the challenge.

SCIENCE AND ENGINEERING PRACTICES
Developing and using models

▶ **NOTE**
Shastina's peak is 12,329 feet (3,758 m).

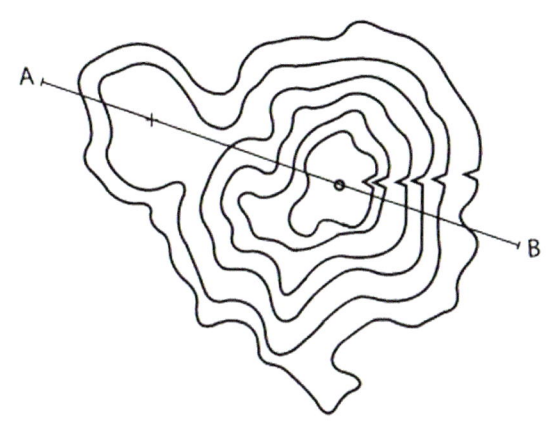

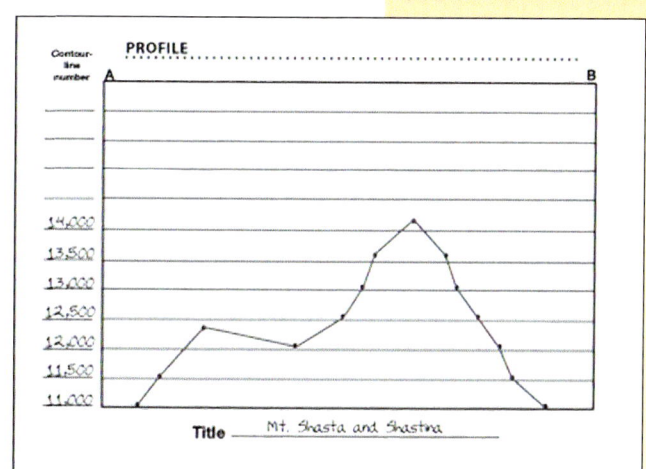

7. Compare profiles and mountains
When the groups have finished drawing profiles, distribute the foam mountains and have the groups assemble them. Ask them to compare the profiles with side views of the foam mountains and with the poster. Ask,

➤ *What new information does the profile provide about Mount Shasta?* [A profile gives you a picture of what the slopes of the mountain look like.]

Distribute a USGS *Mount Shasta* topographic map to each group. Have students compare the steepness of the slopes on the profile with the spacing of the lines on the topographic map. They should once again note that the steeper side has contour lines that are spaced closer together.

Materials for Step 7
- *Mount Shasta* topographic maps

CROSSCUTTING CONCEPTS
Scale, proportion, and quantity

▶ **NOTE**
Go to FOSSweb for *Teacher Resources* and look for the Crosscutting Concepts—Grade 4 chapter for details on how to engage students with the concept of scale, proportion, and quantity.

Soils, Rocks, and Landforms Module—FOSS Next Generation

INVESTIGATION 3 – Mapping Earth's Surface

TEACHING NOTE

Refer to the Sense-Making Discussions for Three-Dimensional Learning chapter in Teacher Resources on FOSSweb for more information about how to facilitate this with students.

SCIENCE AND ENGINEERING PRACTICES

Developing and using models

Analyzing and interpreting data

CROSSCUTTING CONCEPTS

Scale, proportion, and quantity

profile

8. **Have a sense-making discussion**
   Ask,

   ➤ *What is the contour interval for the topographic map you drew?* [500 feet.]

   ➤ *Is it the same for the USGS topographic map?* [No, the USGS map interval is 40 feet.]

   ➤ *Which side of the mountain shown on the Profile sheet is steeper?* [The B or right side.]

   ➤ *How could you tell this is the steeper side by just looking at either of the topographic maps?* [The contour lines are more closely spaced.]

   ➤ *What do the contour lines on the A (or left) side of the map tell you about the slope of the mountain?* [It is less steep.]

   ➤ *Which side of the mountain would you want to walk up and why?*

   ➤ *Why might people involved in outdoor activities use topographic maps and mountain profiles?*

   ➤ *Why might scientists use topographic maps and mountain profiles?*

   Have students who did the second profile that included Shastina show and compare their work. Have them summarize the process needed in order to draw the profile including both peaks. Have them debate any differences in technique. Finally discuss how topographic maps and profiles are helpful to people in their work and outdoor recreation.

   Ask,

   ➤ *Our mountain profiles are models. What are the strengths and limitations of these models?* [The models give an overall view of the steepness of the slope (strength) but from only one perspective (limitation) and not the fine detail of the terrain (limitation).]

   Distribute a copy of notebook sheet 15, *Contours and Intervals* to each student. Project a copy of the sheet and review it together. Have students glue this sheet into their notebooks as a reference sheet.

9. **Review vocabulary**
   Review key vocabulary introduced in this part. If the word is not on the word wall, add it now.

10. **Answer the focus question**
    Have students answer the focus question in their notebooks.

    ➤ *How can we draw the profile of a mountain from a topographic map?*

236   Full Option Science System

Part 2: Drawing a Profile

11. Clean up
Have students take the foam mountains apart and put them and the dowel in the original bag. Have the Getters return the foam mountains and the USGS topographic maps to the materials station.

12. Assess progress: response sheet
Distribute a copy of *Response Sheet A—Investigation 3* and *Response Sheet B—Investigation 3* to each student. This assessment should be completed during class. Collect the response sheets after class and check students' ability to interpret a topographic map and draw a trail for hikers given the interests of the hikers.

What to Look For

- *Students start the trail at the picnic area near the highway.*
- *A steep path up the mountain is indicated, passing the waterfall on the way.*
- *A less steep path is indicated on the way back down to the picnic area.*
- *Students write an explanation about why they drew the path the way they did, including something about the closer together the contour lines are, the steeper the slope indicated.*

SCIENCE AND ENGINEERING PRACTICES

Analyzing and interpreting data

Soils, Rocks, and Landforms Module—FOSS Next Generation

INVESTIGATION 3 – Mapping Earth's Surface

READING *in Science Resources*

13. Read "The Story of Mount Shasta"

This article provides a description of Mount Shasta from John Muir's (1838–1914) point of view. Muir was a naturalist and explorer who traveled throughout California, describing its natural beauty through published articles based on his journal notes.

Refer students to the poster of Mount Shasta and tell them that more than 100 years ago John Muir explored many areas of California including Mount Shasta. Have students close their eyes and imagine what Muir might have observed as he climbed the mountain. Give students a few minutes to share their thoughts and their own experiences. Read the title and the first paragraph to set the stage. Emphasize the time period this article was written (more than 130 years ago!) Point out that it is written in the first person and includes a photograph of Muir and his illustrations of Mount Shasta.

Read the story aloud or have students read independently or in guided reading groups. If students are reading independently, give them self-stick notes to jot down unknown words or phrases and questions they have. Encourage students to pause as they read to visualize what the author is describing.

14. Use a "summary" and "personal response" strategy

This is a good opportunity to have students practice summarizing and responding to text. Have students set up their notebook in two-column notes, titled "Summary" and "Personal Response." Have them start with the second column and write their own personal response to the text. Ask,

➤ *What did the article make you think about?*

Give students a few minutes to share with a partner what they wrote. Ask students to share a few descriptions in Muir's writing that gave them a feeling for what it might have been like to be on the mountain. Encourage students to cite details from the text. Next, have them write a short summary of the article. Remind students that when summarizing they should pick out what they think are the main ideas; keep it brief; and say it in their own words.

15. Discuss the reading

Have students discuss in their groups what they learned about how Mount Shasta was formed. Ask,

➤ *What descriptions in the article identify Mount Shasta as a volcanic mountain?* [There were fumaroles belching steam; red lava beds are described.]

Summary	Personal Response

ELA CONNECTION

These suggested reading strategies address the Common Core State Standards for ELA.

RI 1: Refer to details and examples in a text when explaining what the text says explicitly and when drawing inferences from text.

RI 2: Determine the main idea of a text and explain how it is supported by key details; summarize the text.

RI 4: Determine the meaning of general academic domain-specific words or phrases.

Part 2: Drawing a Profile

▶ *John Muir and his companion Fay were surprised by a violent storm. Describe the storm and how the two explorers survived.* [Temperature drop, lightning, gale winds; they huddled near the hot springs for protection to keep warm.]

Students can also discuss what they think the author's purpose for this article was and the audience he was targeting. They might explore the topic of national parks by reading another text. See FOSSweb for a list of recommended books. One suggestion is *America's National Parks: The Spectacular Forces that Shaped Our Treasured Lands* by Paul Schullery.

16. View video about volcanoes

Tell students that volcanoes play an important role in the formation of landforms around the world. Most of the volcanic activity that formed the major mountains on the west coast of the United States happened a long time ago, but some are still volcanically active. The Hawaiian Islands continue to build through volcanic action, even today.

Tell them you have a video with information about volcanoes. Before showing the video, use a pictorial of the Earth's layers in cross section (see page 67 in the *FOSS Science Resources*) to introduce the new vocabulary words.

Ask students to watch the video to find answers to two questions.

▶ *What is a volcano?*

▶ *How do active volcanoes change Earth?*

The video can be accessed on FOSSweb, Resources by Investigation for this module. After the video, use teacher master 17, *Video Discussion Guide: Volcanoes*, to guide the discussion. Here is a summary of the video chapters.

- Chapter 1: Introduction
- Chapter 2: Volcano Formation
- Chapter 3: Types of Volcanoes
- Chapter 4: Volcanic Activity and Prediction
- Chapter 5: Using a Tiltmeter to Predict Volcanic Eruption
- Investigation: Students demonstrate how a tiltmeter shows changes in the slope of the ground near a volcano.
- Chapter 6: Predicting Volcanic Eruptions
- Chapter 7: Conclusion

Soils, Rocks, and Landforms Module—FOSS Next Generation

ELA CONNECTION

These suggested reading strategies address the Common Core State Standards for ELA.

RI 5: Describe the overall structure of information in a text.

RI 9: Integrate information from two texts on the same topic.

SCIENCE AND ENGINEERING PRACTICES

Obtaining, evaluating, and communicating information

INVESTIGATION 3 – *Mapping Earth's Surface*

crust
lava
magma
mantle

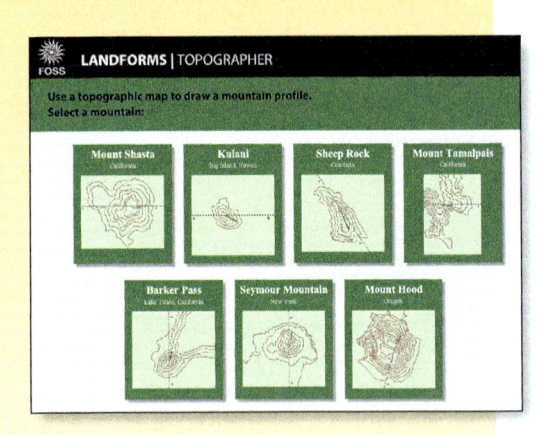

17. Review vocabulary

After the video, review the two questions asked in Step 16 and additional questions found on teacher master 17, *Video Discussion Guide: Volcanoes*. This is a good way to review the vocabulary. You can provide each group with a copy of the sheet.

➤ *Describe the layers of Earth.* [Inner core, outer core, **mantle**, and **crust**.]

➤ *How can magma escape Earth's interior?* [Pressure and heat can cause the **magma** to move up to the surface through weak spots in Earth's crust. Over time, the **lava** layers can build up and form volcanoes. Sometimes volcanoes are plugged up, and the pressure keeps building until it erupts explosively.]

➤ *How are magma and lava alike? How are they different?* [Both magma and lava are melted (molten) rock. Magma is found underground; when it escapes Earth's interior, it is called lava.]

➤ *What are some signs that a dormant volcano might become active?* [Hot soil, bubbling mud pots, hot springs.]

➤ *How can you know when a volcano might erupt?* [Tiltmeters measure ground swelling. Seismographs record earthquakes in the ground just before a volcano erupts.]

➤ *How do active volcanoes change Earth or affect humans?* [Active volcanoes build new land, create fertile soils, provide ash and lava to pave roads, and can be used as sources of energy. They can also destroy homes and human strutures.]

➤ *Why do some people choose to live near the slopes of volcanoes?* [Spectacular views; fertile soil for crops; to study the volcano.]

18. View online activity

Have students engage with the online activity "Topographer" to gain more experience making profile maps.

WRAP-UP/WARM-UP

19. Share notebook entries

Conclude Part 2 or start Part 3 by having students share notebook entries. Ask students to open their science notebooks to the most recent entry. Read the focus question together.

➤ *How can we draw the profile of a mountain from a topographic map?*

Ask students to pair up with a partner to discuss

- the process of making profiles from topographic maps;
- what they learned about volcanoes that was particularly interesting or that they didn't know before.

240   **Full Option Science System**

Part 3: Mount St. Helens Case Study

MATERIALS for
Part 3: Mount St. Helens Case Study

For each group
- 4 Copies of *Profile*
- 4 Copies of *Topographic Map A*
- 4 Copies of *Topographic Map B*

For the class
- 1 Poster, *Mount St. Helens*
- Chart paper ★
- Index cards or half sentence strips ★
- 1 Computer with Internet access ★
- 1 Projection system ★
- ❏ 1 Teacher master 16, *Profile*
- ❏ 1 Teacher master 18, *Topographic Map A*
- ❏ 1 Teacher master 19, *Topographic Map B*

For embedded assessment
- ❏ • Performance Assessment Checklist

❏ Use the duplication master to make copies.

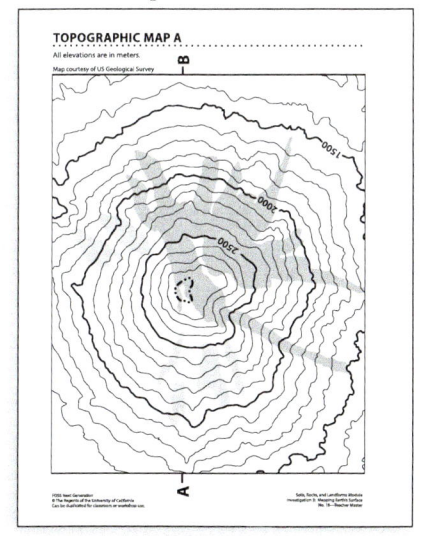

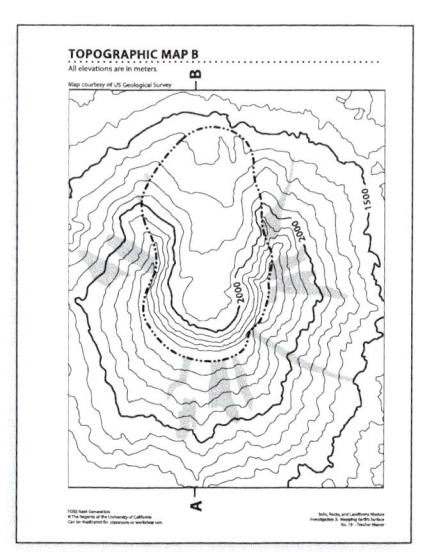

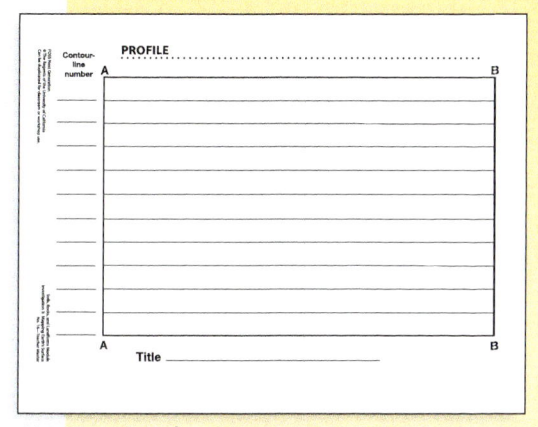

No. 16—Teacher Master

No. 18—Teacher Master

No. 19—Teacher Master

▶ **NOTE**
These topographic maps are scaled in meters. For comparison, Mt. Shasta is 4,332 m high.

Soils, Rocks, and Landforms Module—FOSS Next Generation

INVESTIGATION 3 – Mapping Earth's Surface

GETTING READY for
Part 3: Mount St. Helens Case Study

1. **Schedule the investigation**
   This part will take one session for the active investigation.

2. **Preview Part 3**
   Students compare two topographic maps and have a short debate about whether or not they show the same mountain. After learning that the two topographic maps are the same mountain, they draw profiles of Mount St. Helens before and after its devastating eruption in 1980. Students watch a USGS video that explains how scientists were involved in predicting the eruption and discuss the focus question: **How can scientists and engineers help reduce the impacts that events like volcanic eruptions might have on people?**

3. **Print or make copies of teacher masters**
   Print or make copies of the *Profile* sheet as well as *Topographic Map A* and *Topographic Map B* for each student. In Step 1, you will hand out one copy of *Topographic Maps A* and *B* to each group, then in Step 3 you will distribute additional copies so each student has one of each as well as a profile sheet.

4. **Prepare an engineering solutions poster**
   In Step 10, students will share their ideas about what scientists and engineers can do to help reduce the impacts of events like volcanoes. You will need to post a large piece of chart paper on the wall and provide index cards or half sentence strips to pairs of students so that they can announce their ideas and post them on the chart paper. This poster will be used again in Part 4.

5. **Preview the USGS video:** *Mount St. Helens Impact*
   Preview the USGS video describing the eruption of Mount St. Helens on May 18, 1980 (duration 7:30 minutes). Students will view this video twice. The first time in Step 6 is to help students understand the significance of this event from the point of view of scientists who were involved in monitoring the mountain.

   The second time, in Step 12, the purpose is to have students focus on what scientists were doing at the time and how that had an impact on human life (it could have been much worse) and what scientists continue to do to further the study of volcanoes and their possible impact on people.

Part 3: Mount St. Helens Case Study

6. Plan assessment: performance assessment

Students use topographic maps to look at the changes that occurred in the 1980 eruption of Mount St. Helens. This will provide an opportunity for you to observe students' science practices.

In Step 1 of Guiding the Investigation, students look for patterns, comparing two topographic maps to argue whether they are representations of the same mountain. This is the first opportunity to conduct the performance assessment (see Step 2). In Steps 4 and 9 of Guiding the Investigation, students analyze and interpret data to draw new profile maps. These steps present two other opportunities to conduct the performance assessment. (See Step 5.)

Carry the *Performance Assessment Checklist* with you as you visit the groups while they are working. For more information about what to look for during observation, see the Assessment chapter.

Soils, Rocks, and Landforms Module—FOSS Next Generation

INVESTIGATION 3 – Mapping Earth's Surface

FOCUS QUESTION

How can scientists and engineers help reduce the impacts that events like volcanic eruptions might have on people?

Materials for Step 1
- *Topographic Map A* sheets
- *Topographic Map B* sheets

SCIENCE AND ENGINEERING PRACTICES

Analyzing and interpreting data

Engaging in argument from evidence

CROSSCUTTING CONCEPTS

Patterns

GUIDING *the Investigation*
Part 3: *Mount St. Helens Case Study*

1. **Introduce new topographic maps**
   Distribute one copy of teacher masters 18 and 19, *Topographic Map A* and *Topographic Map B* to each group. Ask students to study the two maps and then engage in argument in their groups to determine if the two maps are the same mountain or two different mountains.

2. **Assess progress: performance assessment**
   This is the first opportunity to conduct this performance assessment. The other opportunities are in Steps 5, 8, and 9 of Guiding the Investigation.

   Circulate from group to group. Listen to the group discussions and observe how students argue whether the maps describe the same mountain. Write notes on the *Performance Assessment Checklist* while students work.

   What to Look For

   - Students argue for or against the supposition that the maps are describing the same mountain; they use evidence from their analysis of the topographic maps to support their claims. (Analyzing and interpreting data; engaging in argument from evidence.)

3. **Discuss the maps**
   Have groups report their conclusions to the class, including their reasons for those conclusions. If students don't come to the conclusion that these could be the same mountain, point out some of the prominent patterns on the south, east, and west sides of the mountains that are almost the same on both maps. It would be very unusual for two mountains to have features that similar. Then acknowledge that the north side of the mountain on map A is very different from the north side of the mountain on map B.

   Inform students that, in fact, both of these maps are of Mount St. Helens, a volcano that is part of the same mountain range as Mount Shasta. They are both part of the Cascade mountain range on the west coast of the United States. On May 18, 1980, Mount St. Helens erupted, providing an unprecedented opportunity for scientists and engineers to learn about volcanoes and how to deal with impact.

Part 3: Mount St. Helens Case Study

4. **Draw the first profile**
   Tell students,

   Before May 1980, the last time Mount St. Helens had erupted was 1857. It had been quiet for more than 120 years. Generations of people had lived in its shadow, enjoying its snow-topped peaks, but never imagining it might erupt again. How could we show what Mount St. Helens looked like from the surface of the ground before May 1980?

   Students should say they could use *Topographic Map A* to make a profile using the A-B line. Let students know that the north side is on the right (the B side). So the profile would show what the mountain looked like if you were standing in the east, facing west.

   Distribute additional copies of *Topographic Map A* so that every student has one. Distribute a copy of *Profile* to each student as well. Give students time to make a profile of Mount St. Helens. Remind students that they should follow the same procedure that they did when they drew the profile of Mount Shasta.

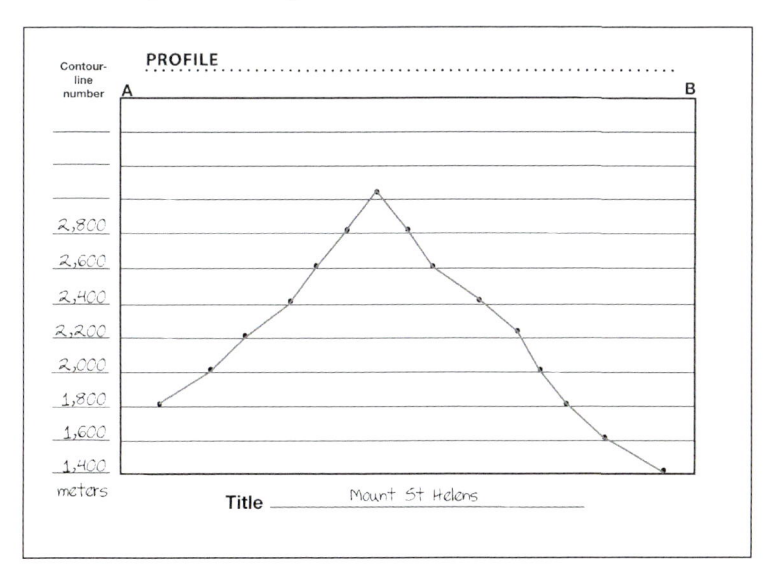

> **TEACHING NOTE**
>
> The AB line runs North to South. As the students draw the profile, the north side will be on the right (the B side).

Materials for Step 4
- *Profile* sheets
- Additional *Topographic Map A* sheets

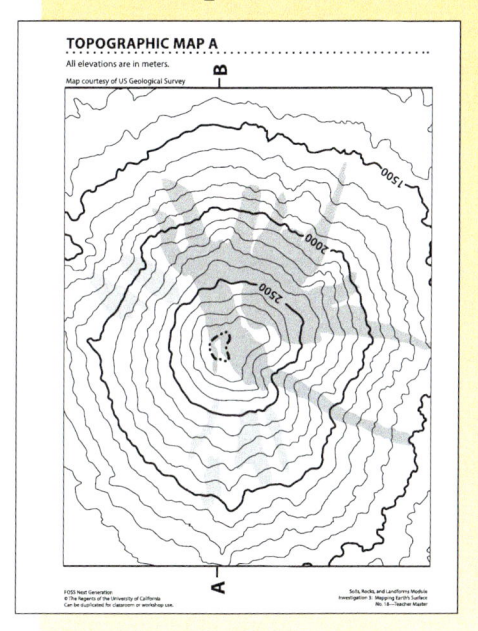

5. **Assess progress: performance assessment**
   Use this opportunity and those in Steps 8 and 9 to assess students' ability to draw profile maps and explain how topographic maps and profiles can be used to represent features on Earth's surface.

   What to Look For
   - Students can draw profile maps from the topographic maps (Steps 4 and 8). (Planning and carrying out investigations.)
   - Students can interpret the maps to tell a story about what happened to Mount St. Helens and how much of the mountain was blown off during the eruption in Steps 8 and 9. (Analyzing and interpreting data; ESS1.C: The history of planet Earth; scale, proportion, and quantity.)

SCIENCE AND ENGINEERING PRACTICES

Planning and carrying out investigations

Analyzing and interpreting data

DISCIPLINARY CORE IDEAS

ESS1.C: The history of planet Earth

CROSSCUTTING CONCEPTS

Scale, proportion, and quantity

Soils, Rocks, and Landforms Module—FOSS Next Generation

INVESTIGATION 3 – Mapping Earth's Surface

▶ **NOTE**
Keep this video at the ready to show again in Step 12.

TEACHING NOTE

Pyroclastic flows consist of extremely hot gases and rock that flow downhill, usually hugging the ground. But in the case of Mount St. Helens it included ash catapulted into the atmosphere that traveled for miles before settling back to Earth. The word "pyroclast" comes from two Greek words meaning "fire" and "broken pieces."

SCIENCE AND ENGINEERING PRACTICES
Obtaining, evaluating, and communicating information

CROSSCUTTING CONCEPTS
Stability and change

6. View the USGS video of the eruption
Tell students,

On March 20, 1980, a moderate (4.2) earthquake shook Mount St. Helens. Scientists took note, but became even more curious about the mountain when they noticed steam vents a week later. By the end of April, the mountain began to bulge. Scientists were using a new, supposedly very precise instrument to measure changes in the mountain, but they just couldn't believe that the bulge was growing five to six feet a day. These measurements were so out of the ordinary, that they sent the instrument back to the factory to get fixed. It turned out the instrument was working perfectly. Magma was building up inside the mountain at a very rapid rate. The eruption of Mount St. Helens was an event that scientists anticipated, but completely underestimated. Let's watch a video in which USGS scientists describe what happened on May 18, 1980.

Show students the "Mount St. Helens Impact" video. It can be accessed on FOSSweb in the Resources by Investigation or the multimedia section for this module.

7. Discuss the video
Use these questions as needed to guide the discussion. Included are a few additional facts that you can use to emphasize how unusual and awesome this event was.

➤ *What did you find most interesting about the video?*

➤ *When most people think about volcanoes, they think about liquid lava splashing out of the top of a mountain with red and black lava flowing slowly down its sides. How was the eruption of Mount St. Helens different?* [The magma in Mount St. Helens exploded in a pyroclastic flow that flattened everything in its path, (some 230 square miles) in less than 10 minutes; huge trees were knocked over as if they were toothpicks.]

➤ *Several people who lived near Mount St. Helens at the time refused to leave when scientists warned them to go. Why do you think they decided to stay?* [Mount St. Helens had been inactive for so long they didn't believe what the scientists were telling them. They didn't think that it would really erupt.]

➤ *Why were scientists so eager to get out on the mountain after each eruption after the big one in May 1980?* [It provided the rare opportunity to observe a volcano when it was active; it provided the opportunity to learn how to reduce the impacts that volcanoes might have on people; at least be able to warn them of future eruptions so they can get out of the way.]

Full Option Science System

Part 3: Mount St. Helens Case Study

8. **Draw the new profile**
   Distribute additional copies of *Topographic Map B* so that each student has one. Have them redraw the profile right on top of the first one on the same *Profile* sheet.

   As students begin this work, they should see that most of the points they mark on the south side of the mountain will be the same, while marks on the north side have dramatically changed.

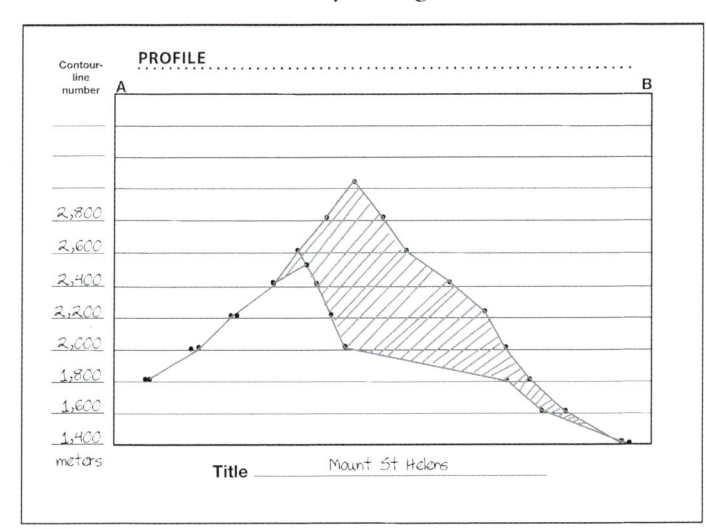

SCIENCE AND ENGINEERING PRACTICES
Developing and using models

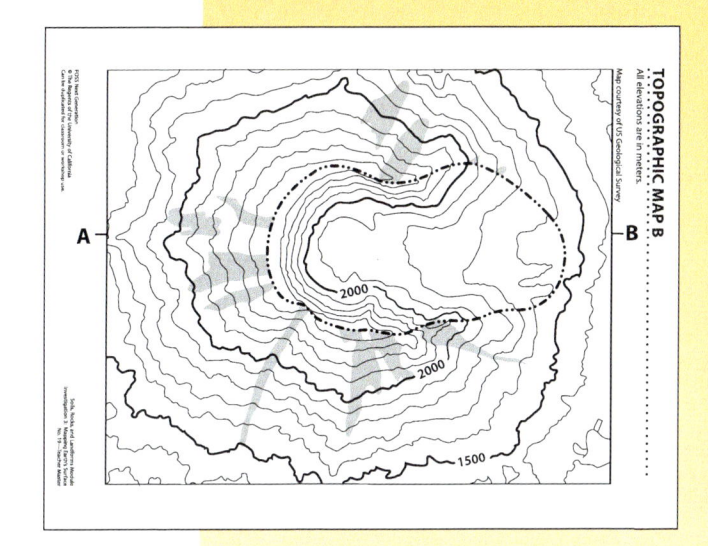

9. **Discuss the new profiles**
   Have students share their dual profiles with other students in their group. Have them discuss how the mountain changed. Have a few students share their groups' observations with the class.

   Tell students,

   Volcanic eruptions are generally thought of as a mountain-building process. For example, you remember from the video we watched, the Hawaiian Islands continue to be built by volcanic activity. One of the things that scientists did not expect when Mount St. Helens erupted was blowing off its top, reducing its height by about 1,300 feet, nor did they expect the collapse of the north side, which caused the mudflows and a great deal of damage to trees and human structures such as bridges, highways, and homes.

Soils, Rocks, and Landforms Module—FOSS Next Generation

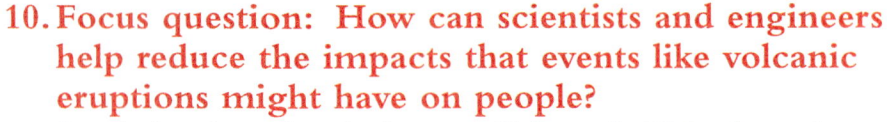

INVESTIGATION 3 – *Mapping Earth's Surface*

10. **Focus question: How can scientists and engineers help reduce the impacts that events like volcanic eruptions might have on people?**

    Remind students that the impact of Mount St. Helens' eruption devastated life around the mountain as well as had a large economic impact (lumber and structures lost). Every living thing within that 230 square mile area on the north side including 57 people, lumber camps, and more than 200 homes were destroyed. So an important question for us to ask is,

    ➤ *How can scientists and engineers help reduce the impacts that events like volcanic eruptions might have on people?*

    Have students write this focus question in their notebooks.

11. **Watch the USGS video one more time**

    Have students keep their notebooks open to take notes as you have the class watch the USGS video one more time. This time, have students look for things that the scientists were doing then and have done since.

12. **Have a sense-making discussion**

    Have students talk with a partner and compare the notes they took when they watched the video a second time. Ask them to discuss why the eruption of Mount St. Helens was such an important event in the world of geology.

    While students are having their partner discussions, post a large sheet of chart paper on a wall. Give each pair an index card or half a sentence strip to write down one way scientists can help reduce impacts. Have each pair go to the chart paper, state their idea for the class and post it on the chart paper. Save this poster to add to it in the next part.

13. **Answer the focus question**

    Give students a few minutes to answer the focus question.

    ➤ *How can scientists and engineers help reduce the impacts that events like volcanic eruptions might have on people?*

    Students will continue to think about this question in the next part as they learn about more events that happen rapidly and can have a large impact on human life and property.

    Students should describe some way in which scientists and engineers can help reduce the impacts of events such as volcanic eruptions. For example, they can continue to build instruments that can help monitor volcanic mountains that may appear to be dormant, but have the potential to become active again.

TEACHING NOTE

Refer to the Sense-Making Discussions for Three-Dimensional Learning chapter in Teacher Resources *on FOSSweb for more information about how to facilitate this with students.*

Materials for Step 12
- *Chart paper*
- *Index cards or sentence strips*

CROSSCUTTING CONCEPTS
Stability and change

SCIENCE AND ENGINEERING PRACTICES
Constructing explanations and designing solutions

Part 3: Mount St. Helens Case Study

WRAP-UP/WARM-UP

14. Share notebook entries

Conclude Part 3 or start Part 4 by having students share notebook entries. Ask students to open their science notebooks to the most recent entry. Read the focus question together.

➤ *How can scientists and engineers help reduce the impacts that events like volcanic eruptions might have on people?*

Ask students to pair up with a partner to

- discuss why the eruption of Mount St. Helens was such an important event in the world of geology.
- discuss what scientists and engineers can do to help reduce the impact of volcanic eruptions on people.

TEACHING NOTE

If students are interested in finding out more about volcanoes, introduce them to the USGS and NOAA websites with maps and current information about volcano activity alerts. Refer to the environmental literacy section in the Extensions at the end of the investigation.

Soils, Rocks, and Landforms Module—FOSS Next Generation

INVESTIGATION 3 – *Mapping Earth's Surface*

MATERIALS for
Part 4: *Rapid Changes*

For each group
- 4 Markers ★
- 4 *FOSS Science Resources: Soils, Rocks, and Landforms*
  - "It Happened So Fast!"

For the class
- 2 Sets of rapid change cards, 16 cards/set
- • Transparent tape ★
- 4 Sheets of chart paper ★
- 1 Computer with Internet access and projector ★

For embedded assessment
- ❏ • Embedded Assessment Notes

For benchmark assessment
- ❏ • Investigation 3 I-Check
- ❏ • Assessment Record

★ Supplied by the teacher.   ❏ Use the duplication master to make copies.

Part 4: Rapid Changes

GETTING READY for
Part 4: *Rapid Changes*

1. **Schedule the investigation**
   This part will take one session for instruction, one for reading and viewing the video, one for review, and one for the I-Check.

2. **Preview Part 4**
   Students think about processes that cause rapid changes to Earth's surface: landslides, earthquakes, floods, and volcanoes. The focus question is **What events can change Earth's surface quickly?**

3. **Preview rapid change cards**
   The kit contains two sets of rapid change cards with 16 cards in each set. The cards show the results of earthquakes, floods, landslides, and volcanic eruptions. The cards are numbered so that you can easily sort them into the rapid-change categories.

   Cards 1–4: Earthquakes (green)
   1. A faultline near Flaming Gorge, Utah
   2. A 1976 earthquake in El Progresso, Guatemala, shows the 105-cm shift in alignment of rows in a cultivated field
   3. The aftermath of an earthquake in San Fernando, California, in February 1971
   4. Surface cracks from an earthquake in Loma Prieta, California, on October 17, 1989

   Cards 5–8: Floods (blue)
   5. A washed-out road caused by flooding
   6. A flash flood in California
   7. A flooded road caused by heavy rains
   8. A flash flood in Arizona

   Cards 9–12: Landslides (purple)
   9. A mountain landslide in Banff National Park, Alberta, Canada
   10. The largest landslide in Daly City, San Mateo County, California, in October 1989
   11. A landslide toppled trees along a river
   12. A landslide near Fort Funston, California, as a result of an earthquake in October 1989

   Cards 13–16: Volcanic eruptions (red)
   13. The volcanoes of Bromo National Park in Java, Indonesia
   14. Flowing lava from a volcano near Kona on the Island of Hawaii
   15. Mount St. Helens, Washington, after an eruption in March 1982
   16. A view of Crater Lake in Oregon

Soils, Rocks, and Landforms Module—FOSS Next Generation

INVESTIGATION 3 — *Mapping Earth's Surface*

4. **Prepare the observation gallery of rapid changes**

   Set up an observation gallery of rapid changes by taping the rapid change cards onto four large pieces of paper (chart paper or butcher paper). Place the four volcano pictures on one sheet, the four earthquake pictures on another, and so forth. Leave enough room between the cards for students to write their observations.

   Hang the four observation charts in different areas of the room where students will be able to view and write on them. The kit includes an extra set of cards to make additional charts if you want to have students rotate to each chart in smaller groups.

   Alternatively place rapid change cards (enough so there is at least one for each student) in the middle of each group's table. Have students take turns observing the cards and then have them pick one to write about in their notebooks, using the same rapid-change-gallery prompts.

5. **Plan to Read *Science Resources*: "It Happened So Fast!"**

   Plan to read "It Happened So Fast!" during a reading period after completing the active investigation for this part.

6. **Plan to view *All about Earthquakes* video**

   Preview the *All about Earthquakes* video (chapters 1–8, duration about 22 minutes). A link to the video is on FOSSweb in the Resources by Investigation section. Have students view the video in Step 6.

7. **Plan assessment: notebook entry**

   In Step 8, students consider a variety of solutions for reducing impact from one of the rapid change events studied in this part. Use *Embedded Assessment Notes* to record your observations of students' understanding of how scientists and engineers can reduce the impact of those events.

8. **Plan benchmark assessment: I-Check**

   When students have completed all the activities and readings for this investigation, plan a review session, and then give them *Investigation 3 I-Check*. Students should complete their responses to the items independently. After you review and code students' work, plan which items you will bring back for additional discussion and self-assessment. See the Assessment chapter for self-assessment guidance and information about online tools. The assessment coding guides are available on FOSSweb.

Full Option Science System

Part 4: Rapid Changes

GUIDING the Investigation
Part 4: Rapid Changes

1. **Discuss rapid and slow changes**
   Tell students,

   Weathering and erosion usually change Earth's surface a tiny bit at a time. It can take millions of years to create a canyon or large delta. Most of the time, you can't see it happening, only the results.

   A volcanic eruption is an event that can change Earth's surface rapidly. We talked about that in our last lesson about Mount St. Helens.

2. **Focus question: What events can change Earth's surface quickly?**
   Have students write the focus question in their notebooks as you say it and project or write it on the board.

   ➤ *What events can change Earth's surface quickly?*

   Give students a couple of minutes to think of other land-altering events they may have experienced or heard about in the news.

3. **Introduce the rapid change gallery**
   Focus students' attention on the observation-gallery charts posted around the classroom. Tell them that they will be observing images of rapid changes in Earth's surface on these posters. Write these questions on the board for students to consider as they observe the images on each poster.

   ➤ *What do you notice?*

   ➤ *What can you infer from the picture?*

   ➤ *What questions do you have?*

   Tell students that they will stay in their groups (4–5 students each) and rotate on your signal to their assigned poster. They will have only 3 minutes at a poster, and on your signal they will move on to the next poster. Each student will use a marker to write his or her observations, inferences, or questions about one of the pictures on each chart. Make sure students write on the paper, not on the cards.

4. **Activate prior knowledge**
   Assign each group one chart to begin, and give them approximately 3 minutes to observe and record. Begin the observation-gallery rotation. Monitor students' progress as they write their observations, inferences, and questions on the chart paper. Provide a signal every 3 minutes and ask the groups to move to the next poster of images.

FOCUS QUESTION
What events can change Earth's surface quickly?

Materials for Step 3
- Rapid change posters

SCIENCE AND ENGINEERING PRACTICES
Analyzing and interpreting data

Constructing explanations

CROSSCUTTING CONCEPTS
Cause and effect

Soils, Rocks, and Landforms Module—FOSS Next Generation

INVESTIGATION 3 – *Mapping Earth's Surface*

TEACHING NOTE

Students generally think in terms of destruction when they hear about earthquakes, volcanoes, floods, and landslides. It is important to help them understand that these are constructive as well as destructive processes.

CROSSCUTTING CONCEPTS

Cause and effect

CROSSCUTTING CONCEPTS

Stability and change

▶ **NOTE**
Go to FOSSweb for *Teacher Resources* and look for the Crosscutting Concepts—Grade 4 chapter for details on how to engage students with the concept of stability and change.

SCIENCE AND ENGINEERING PRACTICES

Constructing explanations

If students need a little prompting about what to write, ask questions.

➤ *What is happening or has happened?*
➤ *Were earth materials moved? In what way?*
➤ *What caused the movement of earth material?*
➤ *Were earth materials deposited?*
➤ *Is this event destructive or constructive? Is any landform being broken down and carried away or built up?*

5. **Discuss rapid changes**
   Call students to attention and talk about the rapid changes they saw depicted in the images. They should report that earthquakes, **landslides**, floods, and volcanoes can all produce rapid changes to Earth's landforms. Ask,

   ➤ *How do landslides and mudslides change landforms?* [Rock material moves from high locations to low locations during slides, eroding earth material from the top and depositing it at the bottom.]

   ➤ *How do earthquakes change landforms?* [The shaking of the ground can cause rock to rise or drop and trigger landslides if the ground is unstable. As a result, mountains and cliffs build up, and rock materials can move downward to different locations.]

   ➤ *How do volcanoes change landforms?* [Volcanoes create mountains by bringing new rock material to the surface. Volcanoes also throw out lots of rocks and ash, which fall to Earth. These earth materials form a new earth surface and enrich soils.]

   ➤ *How do floods change landforms?* [Floods move large amounts of earth materials (rock and soil) from one place to another. Floods in rivers may create new meanders, sandbars, and so forth when the eroded materials are deposited. This process can damage or enrich soils.]

6. **View video on earthquakes**
   Introduce the video *All about Earthquakes*. Ask students to listen for information that will explain how scientists study earthquake events so they can predict how often and where they might occur. Students should listen for information about how scientists and engineers try to reduce the destruction from big Earth-shaking events.

   Show students the video (chapters 1–8, duration about 22 minutes). The video can be accessed on FOSSweb in the Resources by Investigation for this module.

Part 4: Rapid Changes

The topics in the video include:
- What causes earthquakes?
- Types of earthquakes and the effects on Earth
- Energy waves from earthquakes
- Measuring earthquake waves with the Richter Scale and seismographs
- Earthquake prediction and preparation
- Investigation: testing earthquake-proof structures

After watching the video, give students time to share the information they gathered from the video. Also have them share their personal experiences with earthquakes. Review the safety procedures they should follow if they are in an earthquake.

7. **Answer the focus question**
   Have students answer the focus question in their notebooks.

   ➤ *What events can change Earth's surface quickly?*

8. **Revisit the poster from Part 3**
   Review the poster students started in Part 3. Read through the things that students suggested scientists and engineers could do to help reduce the impact of volcanic eruptions on humans.

   Have students work in their groups to think about the other rapid changes they have learned about in this part. Give each group or pair another index card or sentence strip. Have them think about these new changes and think of a way that engineers and scientists might work to reduce the impact of these Earth changes.

9. **Assess progress: notebook entry**
   When the group work is completed, have students rewrite the focus question from Part 3 in their notebooks.

   ➤ *How can scientists and engineers help reduce the impacts that events like volcanic eruptions might have on people?*

   Have students choose a rapid change event, consider all the suggested ways to reduce the impact, then describe which they think is the best suggestion and why. Have students turn in their notebooks open to the page on which they wrote about this question.

   What to Look For
   - Students write about one rapid change event, describe what they think is the best way to reduce impact, and why they think that is the best solution.

EL NOTE

For students who need scaffolds, provide a sentence frame such as: Earth's surface can be changed quickly by _____.

▶ **NOTE**
This is an opportunity for students to use the Internet to research things that scientists and engineers are doing to reduce the impact of natural disasters.

Soils, Rocks, and Landforms Module—FOSS Next Generation

INVESTIGATION 3 – Mapping Earth's Surface

READING in Science Resources

9. Read "It Happened So Fast!"

Read "It Happened So Fast!" during a reading period after completing the active investigation.

Before reading, have students preview what they know so far about rapid changes to Earth's surface. One option is to set up a class concept grid (students can also make one in their notebooks). See the example below. Start with a description of each event based on the class discussion of the rapid change images and the volcano video. Next, refer back to the Mount St. Helen's article and fill in how the volcanic eruption changed the land around it and how it impacted humans and other life forms. Tell students that this article will describe the other ways the land can be transformed rapidly—floods, earthquakes, and landslides.

Rapid changes	Descriptions	Effects on landforms	Effects on humans (and other organisms)
Volcanoes			
Landslides			
Earthquakes			
Floods			

10. Use a group or jigsaw reading strategy

Give students a few minutes to preview the text by looking at the pictures and reading the captions. Have them discuss what they may already know and what questions they have about the images with a partner.

Read the article aloud or have students read independently or in guided reading groups. Tell students to be looking for the ways the rapid changes are both constructive and destructive processes.

Another option is to jigsaw the article. Assign each group a section to read and discuss. Then, have each group develop a presentation as if they were a news station reporting on the event.

11. Discuss the reading

Tell students that catastrophic events are usually remembered for the destruction they cause in a short period of time. Often the destruction is measured in human losses—loss of life, buildings, and infrastructure. There are significant changes to the land as well. Some of the changes involve construction of new landforms. Thus catastrophic events often involve a combination of destructive and constructive outcomes.

ELA CONNECTION

These suggested reading strategies address the Common Core State Standards for ELA.

RI 2: Determine the main idea of a text and explain how it is supported by key details; summarize the text.

RI 3: Explain procedures, ideas, or concepts in a scientific text.

RI 7: Interpret information presented visually, and explain how the information contributes to an understanding of the text.

W 8: Gather relevant information from experiences and print, and categorize the information.

SL 4: Report on a text in an organized manner, using appropriate facts and relevant, descriptive details to support main ideas or themes; speak clearly at an understandable pace.

Part 4: Rapid Changes

Review how each type of rapid change affects landforms and record the information on the concept grid. Include both destructive and constructive impacts. Encourage students to cite details and examples from the text to support their ideas. If students jigsawed the article, have each group focus on one of the events and then share out with the whole class.

12. **Explore the USGS earthquake website**
    If students are interested in finding out more about earthquake activity and statistics, introduce them to the USGS website and show them how to use the interactive maps. Have students generate questions they would like to answer using the maps. For example, a student living in California might ask, how many earthquakes occurred in California today? For more information on the USGS website, go to the Environmental Literacy section of the Extensions at the end of this investigation.

> **TEACHING NOTE**
>
> See the **Home/School Connection** for Investigation 3 at the end of the Interdisciplinary Extensions section. This is a good time to send it home with students.
>
> In this activity, students use provided data on longitude and latitude to locate Earth's twenty largest recorded earthquakes and volcanoes on a world map to look for patterns.

INVESTIGATION 3 – Mapping Earth's Surface

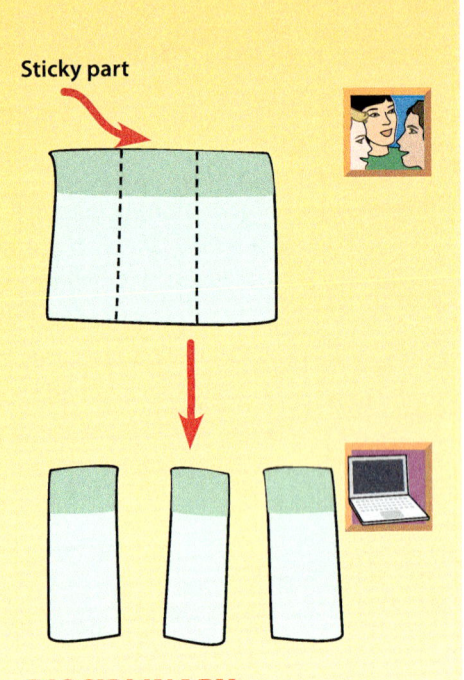

Sticky part

DISCIPLINARY CORE IDEAS

ESS1.C: The history of planet Earth

ESS2.A: Earth materials and systems

ESS2.B: Plate tectonics and large-scale system interactions

ESS3.B: Natural hazards

ETS1.B: Developing possible solutions

WRAP-UP

13. Review Investigation 3

Distribute one or two self-stick notes to each student. Ask students to cut each note into three pieces. Have them take 3 minutes to look back through their notebook entries to find the three most important things they learned in Investigation 3. They should tag those pages using the three self-stick notes.

Ask students to take a few minutes to look back through their notebook entries to find the most important things they learned in Investigation 3. Students should include at least one science and engineering practice, one disciplinary core idea, and one crosscutting concept. They should tag those pages with self-stick notes.

When students have completed their notebook review and placed their tags, lead a short class discussion to create a list of three-dimensional statements that summarize what students have learned in this investigation.

- We used topographic maps to model the shape of the land; each contour line shows the shape of the land at a certain elevation; the closer the lines, the steeper the slope. (Developing and using models; patterns.)

- We created a different model by making profiles (side views) from information given on topographic maps. (Developing and using models; analyzing and interpreting data; scale, proportion, and quantity.)

- We obtained and communicated information that the surface of Earth is constantly changing; sometimes those changes take a long time to occur and sometimes they happen rapidly. (Obtaining, evaluating, and communicating information; cause and effect.)

- We argued that scientists and engineers look for multiple solutions to reduce the impacts of natural Earth processes on humans. (Constructing explanations and designing solutions; stability and change.)

Students should discuss the guiding questions about the investigative phenomena with a partner before responding to them in their notebooks.

➤ *How do maps help us observe Earth's surface features?*

➤ *What might reduce the impact of catastrophic Earth surface events?*

BREAKPOINT

Part 4: Rapid Changes

14. Assess progress: I-Check

Give the I-Check assessment at least one day after the wrap-up review. Distribute a copy of *Investigation 3 I-Check* to each student. You can read the items aloud, but students should respond to the items independently in writing. Alternatively, you can schedule the I-Check on FOSSmap for students to take the test and then you can access data through helpful reports to look carefully at students' progress.

If students take the test on paper, collect the I-Checks. Code the items using the coding guides you downloaded from FOSSweb, and plan for a self-assessment session, identifying disciplinary core ideas, science and engineering practices, and crosscutting concepts students may need additional help with. (If you are using FOSSmap, you will only need to code open-response items and confirm codes for short-answer items as other items are automatically coded.) Refer to the Assessment chapter for next-step/self-assessment strategies.

> **TEACHING NOTE**
>
> *During or after these next-steps with the I-Check, you might ask students to make choices for possible derivative products based on their notebooks for inclusion in a summative portfolio. See the Assessment chapter for more information about creating and evaluating portfolios.*

INVESTIGATION 3 – Mapping Earth's Surface

TEACHING NOTE

Refer to the teacher resources on FOSSweb for a list of appropriate trade books that relate to this module.

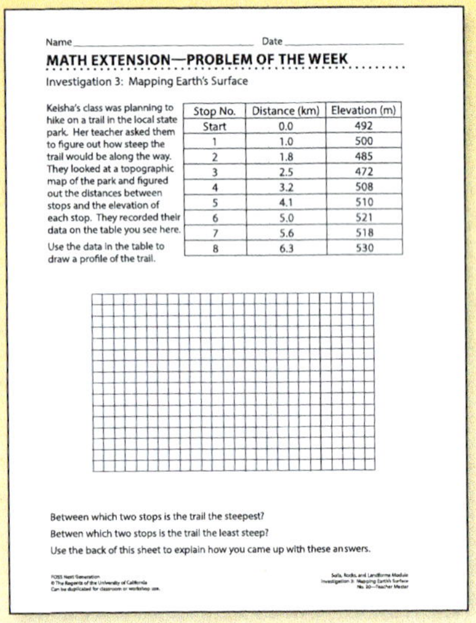

INTERDISCIPLINARY EXTENSIONS

Math Extension

- **Problem of the week**

    Keisha's class was planning to hike on a trail in the local state park. Her teacher asked them to figure out how steep the trail would be along the way. They looked at a topographic map of the park and figured out the distances between stops and the elevation of each stop. They recorded their data on the table you see here.

    - Use the data in the table to draw a profile of the trail.
    - Between which two stops is the trail the steepest?
    - Between which two stops is the trail the least steep?
    - Use the back of this sheet to explain how you came up with these answers.

    Notes on the problem. By creating a graph using the cumulative distance on the trail from stop to stop and the elevation of the stop, students create an exaggerated (since it's not to scale) profile of the trail. The graph might look like the one shown below.

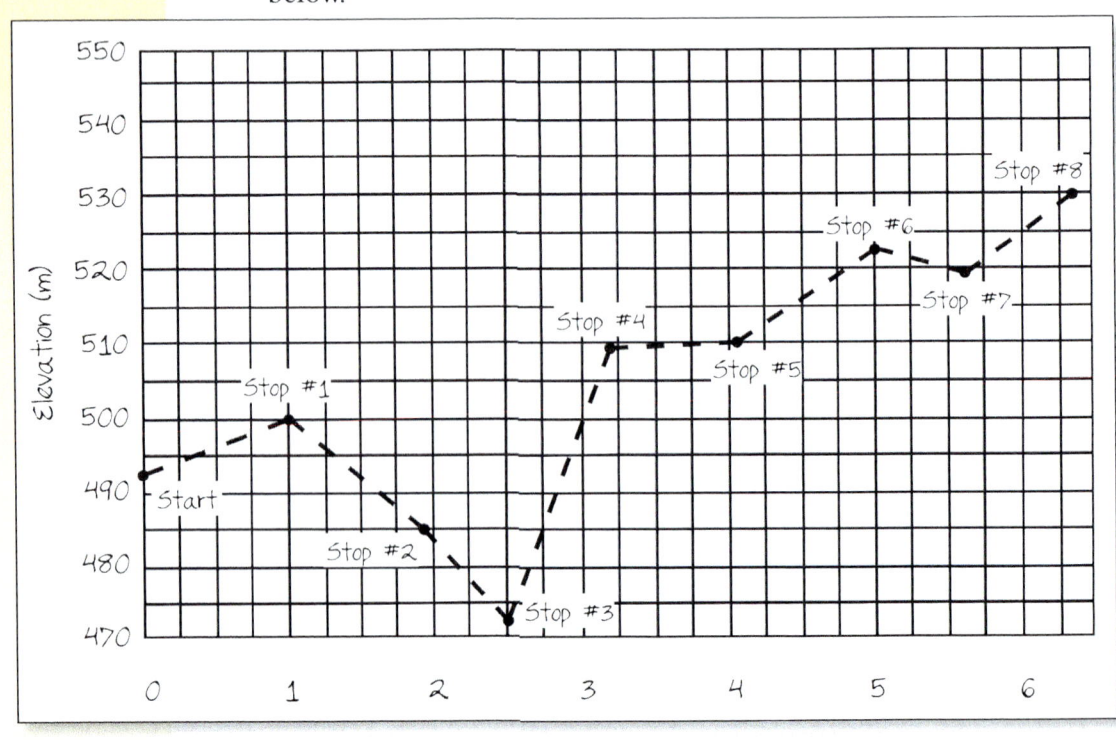

No. 20—Teacher Master

Stop No.	Distance (km)	Elevation (m)
Start	0.0	492
1	1.0	500
2	1.8	485
3	2.5	472
4	3.2	508
5	4.1	510
6	5.0	521
7	5.6	518
8	6.3	530

260   Full Option Science System

Interdisciplinary Extensions

Language Extension

- **Create maps of fictional places**
  Have students create their own topographic maps of fictional places. They can develop a story to go along with the map and a set of questions about the area for other students to answer. Profiles can also be generated from these maps.

Science Extensions

- **Find out what surveyors do**
  Surveyors are important members of a mapping team. Have students find out what they do and what education they needed to get their jobs. Invite a surveyor to your classroom to demonstrate the tools of the trade.

- **Construct other profiles**
  Challenge students to draw other profile lines across their *Mountain Maps* or *Topographic Maps A* and *B* to construct the profiles using this new side-view perspective. How do the resulting profiles compare to the original profile?

- **View online activities**
  As extensions, students can study the information presented in the Investigation 3 online activities on: rock database, sand types, volcanoes, faulting and folding.

Environmental Literacy Extensions

- **Hike or bike a mountain or trail**
  Find some hiking savvy adults or an environmental or outdoor education center to lead your class on a nearby hike. Have students use topographic maps of the area to learn about the mountain before the hike to anticipate the strenuous parts of the trails. Students could make a profile of various sides of the mountain to plan the hike.

- **Research active volcanoes**
  Have students interact with the following websites to find out more about volcanoes. Students can find these links on FOSSweb in recommended websites.
    - United States Geological Survey (USGS), Volcanoes: The main page shows a map of US volcanoes and current activity alerts. Students can interact with the map to zoom in to certain geographical areas to see volcano locations and find out the activity status of those volcanoes.

> **TEACHING NOTE**
>
> Encourage students to use the Science and Engineering Careers Database on FOSSweb.

Soils, Rocks, and Landforms Module—FOSS Next Generation

INVESTIGATION 3 – *Mapping Earth's Surface*

By using the navigation buttons at the top of the screen, students can find out more about "Volcano Updates," "Monitoring," "Hazards," "Education," and "Multimedia."

- Global Volcanism Program (Smithsonian/USGS Partnership): This site looks at recent global volcanic activity for a given week. The most current week (updated every Wednesday) is what shows when the site is accessed. Recent activity is detailed on the page.

  Students can use the navigation buttons at the top of the screen to find out more about world volcanoes, past and present.

- National Oceanic and Atmospheric Administration (NOAA): This interactive map will give students a look at global volcanic activity, as well as tsunami events and earthquake activity.

- **Research earthquakes**

  If your students are interested in finding out more about earthquakes, introduce them to the United States Geological Survey (USGS) website.

- **Find out about ShakeAlert**

  The USGS, working with a group of state and university partners, is developing and testing an Earthquake Early Warning (EEW) system called ShakeAlert for the west coast of the United States. The project has some pilot projects and hopes to become a public warning system in the years to come. Get an update on the system from their website and watch a video describing how ShakeAlert works (English or Spanish).

  The primary project partners include: United States Geological Survey; California Governor's Office of Emergency Services (CalOES); California Geological Survey; California Institute of Technology; University of California Berkeley; University of Washington; University of Oregon; Gordon and Betty Moore Foundation.

Interdisciplinary Extensions

Home/School Connection

Students use provided longitude and latitude data to plot the locations of 20 active volcanoes in the world and the location of the 20 strongest earthquakes in Earth's recorded history. Note that the latitude and longitude data has been rounded to the nearest 5° to make it easier for students to plot. After locating these events and features, students look for patterns.

Print or make copies of *Home/School Connection* for Investigation 3, teacher masters 21, 22, and 23, and send it home with students after Part 4. (You can print or copy masters 21 and 22 back to back, but make sure that the world map is a separate sheet.) You may need to get students started by first reviewing the information on the world map and then demonstrating the process for plotting a few locations using the data provided. Note that the map the students are working on is Pacific Ocean-centric in order to help students see the pattern known as the Ring of Fire. You may need to help orient them by using a globe or traditional map that puts the Prime Meridian at the center of the map. Students will need to be careful that they check for East, West, North, and South when they plot latitude and longitude.

See the other Environmental Literacy Extensions to introduce students to websites they can use at home with their families to find out more about earthquakes and volcanoes.

> **TEACHING NOTE**
>
> Families can get more information on Home/School Connections from FOSSweb.

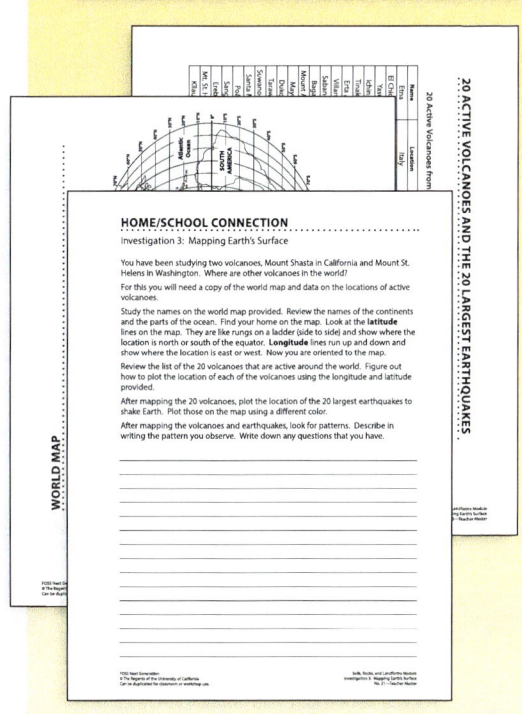

Nos. 21–23—Teacher Masters

Soils, Rocks, and Landforms Module—FOSS Next Generation

263

INVESTIGATION 3 – Mapping Earth's Surface

INVESTIGATION 4 – Natural Resources

Part 1
Introduction to Natural
Resources............................ 274

Part 2
Making Concrete................. 284

Part 3
Earth Materials in Use 292

Guiding question for phenomenon:
What makes rock a natural resource and how is this resource used by people?

PURPOSE

Students find that many earth materials can be classified as natural results when they are put to use to perform tasks and solve human problems. Students experience this phenomenon by engaging in one of the most basic earth materials technologies—making concrete.

Content

- Natural resources are natural materials taken from the environment and used by humans. Rocks and minerals are natural resources important for shelter and transportation.
- Concrete is an important building material made from earth materials (limestone, sand, gravel and water).
- There are re renewable and nonrenewable natural resources. Scientists and engineers work together to improve how people use natural resources.
- Alternative sources of energy include solar, wind, and geothermal energy.

Practices

- Observe how earth materials are used in the community around the school.
- Consider the ways people impact natural resources and how humans can conserve them.

Science and Engineering Practices
- Planning and carrying out investigations
- Constructing explanations and designing solutions
- Engaging in argument from evidence
- Obtaining, evaluating, and communicating information

Disciplinary Core Ideas
ESS3: How do Earth's surface processes and human activities affect each other?
ESS3.A: Natural resources
ETS1: How do engineers solve problems?
ETS1.A: Defining and delimiting engineering problems

Crosscutting Concepts
- Scale, proportion, and quantity
- Structure and function

FOSS *Full Option Science System*

265

INVESTIGATION 4 – *Natural Resources*

	Investigation Summary	Time	Focus Question for Phenomenon, Practices
PART 1	**Introduction to Natural Resources** Students review what they have learned in the module about soils, rocks, and landforms. They write a story or draw a concept map to bring the ideas together. They focus on earth materials as renewable and nonrenewable natural resources by viewing and discussing a video.	**Active Inv.** 2 Sessions * **Reading** 1 Session	**What are natural resources and what is important to know about them?** **Practices** Constructing explanations Engaging in argument from evidence Obtaining, evaluating, and communicating information
PART 2	**Making Concrete** Students focus on the earth resources that make up a very important material used for walkways, buildings, and bridges—concrete. The class uses local natural resources to make one concrete stepping stone.	**Active Inv.** 1 Session **Reading** 1 Session	**How are natural resources used to make concrete?** **Practices** Planning and carrying out investigations Constructing explanations and designing solutions Obtaining, evaluating, and communicating information
PART 3	**Earth Materials in Use** Students go on a walk around the school and schoolyard, searching for earth materials in use. They search for various objects and structures and consider what natural resources were used to construct them.	**Active Inv.** 1 Session **Reading** 1 Session **Assessment** 2 Sessions	**How do people use natural resources to make or build things?** **Practices** Planning and carrying out investigations Constructing explanations and designing solutions Obtaining, evaluating, and communicating information

* A class session is 45–50 minutes.

At a Glance

Content Related to DCIs	Writing/Reading	Assessment
• Natural resources are natural materials taken from the environment and used by humans. • Some natural resources are renewable (sunlight, air and wind, water, soil, plants, and animals) and some are nonrenewable (minerals and fossil fuels). • Geoscientists study earth materials in part to help humans use those resources wisely. • Alternative sources of energy include solar, wind, and geothermal energy.	**Science Notebook Entry** Concept map or story **Science Resources Book** "Monumental Rocks" "Geoscientists at Work" **Video** *Natural Resources* **Online Activity** "Resource ID"	**Embedded Assessment** Response sheet
• Concrete is an important building material made from earth materials (limestone to make cement, sand and gravel for aggregates, and water for mixing).	**Science Notebook Entry** Answer the focus question **Science Resources Book** "Making Concrete"	**Embedded Assessment** Science notebook entry
• Rocks and minerals are natural resources important for shelter and transportation. • Earth materials are resources for artists. • Scientists and engineers work together to improve how people use natural resources.	**Science Notebook Entry** Answer the focus question **Science Resources Book** "Earth Materials in Art" "Where Do Rocks Come From?" (optional) **Online Activity** "Virtual Investigation: Natural Resources"	**Embedded Assessment** Performance assessment **Benchmark Assessment** *Posttest* **NGSS Performance Expectations addressed in this investigation** 4-ESS3-1 3–5-ETS1-1

Soils, Rocks, and Landforms Module—FOSS Next Generation

INVESTIGATION 4 – Natural Resources

BACKGROUND *for the Teacher*

The Sun, air, land, and water provide organisms with the resources needed to survive and thrive on our planet Earth. These are the **natural resources** that come from the physical environment. But when we use the term *natural resources*, we are usually referring to the resources derived from the environment that are used by humans not only for survival but also to satisfy the demands of a complex technological society. So the human factor and the economic factor are important considerations when discussing natural resources. We will define natural resources as the natural living and nonliving materials harvested or extracted from Earth by humans.

What Are Natural Resources and What Is Important to Know about Them?

Natural resources can be classified in a number of ways. One way is to consider the source as either living (biotic) or nonliving (abiotic). Biotic resources include the plants (forests for wood products, plant crops for food, fiber, and biofuels; herbs and other plants for medicines) and animals for food, fabric, transportation, and power. Abiotic resources include sunshine, water, air (wind), rocks, and minerals, including ores for valuable metals such as gold, iron, aluminum, copper, and silver.

Fossil fuels (coal, petroleum, and natural gas) are harder to put into a living or nonliving category because they originated as plants and animals that became buried under sediments millions of years ago, then slowly transformed into deposits of carbon-rich substances. Soil is also derived from both nonliving (clay, silt, sand, gravel) and living (organic humus) sources.

Another way of classifying natural resources is to consider their renewability. If the resource is naturally replenished continuously and quickly, then it is considered a **renewable resource**. Sunlight is the most plentiful and reliable renewable resource on Earth and is the major source of energy for all living things. Solar energy drives the water cycle and produces wind, ocean waves, and flowing water in rivers and streams, which are used to produce electric energy for human societies. Water and air are also renewable resources. Plants and animals on land and in the ocean are renewable resources. But the quality and quantity of water, air, plants, and animals can be greatly impacted by human activities. So understanding the source of the resource and the processes by which it is renewed is essential to the wise management of Earth's renewable resources. Knowing the conditions that enhance or retard the renewal process is critical to making informed decisions to achieve sustainability.

Background for the Teacher

If the resource is not replenished because it takes extended geological periods (millions of years) to form, then in human terms, it is a **nonrenewable resource**. Rocks, minerals, and fossil fuels are in this category because once they are extracted from Earth, they are used up or depleted. Fossil fuels are burned to produce energy, and ores are smelted to extract the valuable elements. At any rate, they are no longer available for renewal. It is possible to slow down the extraction of these nonrenewable resources by using less of them so they last longer. Finding alternative energy sources (**solar energy**, **wind power**, **geothermal power**) for human society is one of the ways to manage the depletion of fossil fuels. It is also possible to reclaim rock and mineral resources, so while they cannot be renewed, they can be recycled and reused. For example, concrete debris from the demolition of concrete structures is recycled into aggregate to be used to make new concrete. It is estimated that more than 100 million tons of concrete and worn-out asphalt pavement are recycled as aggregates each year in the United States to provide about 5 percent of the total aggregate needed for road construction and building.

And as individuals, we can change our behavior to use less energy in our homes and to purchase products that take less energy to manufacture or operate. We can reuse and recycle products made from minerals, such as metals and glass, and wood and paper products. We can use good conservation practices by supporting policies that promote the wise and sustainable use of land, air, water, and other earth materials.

How Are Natural Resources Used to Make Concrete?

Concrete is a rocklike material used in huge quantities worldwide to construct highways and roads, bridges and overpasses, dams and stadiums, and buildings, particularly foundations. We often refer to the large truck with the rotating drum that delivers the material to the construction site as a cement truck. But it is more accurate to call it a concrete truck. Inside that drum is a mixture of **cement** and **aggregates**, pieces of rock of different sizes. Small aggregates include sand and gravel. Larger aggregates include pebbles of several sizes.

Portland cement (sometimes referred to as OPC for ordinary Portland cement) is the most commonly used cement in the world. The fine gray cement powder is made primarily from limestone. Limestone is dug out of a quarry and then heated, with small amounts of some other substances, to extremely high temperatures in a furnace. The result is a chemical reaction that produces new compounds. When the resulting

INVESTIGATION 4 – *Natural Resources*

Portland cement is mixed with water, the cement forms strong bonds, turning the mixture into a rock-hard material that can withstand pressure, water immersion, and the effects of other substances.

The mixture of cement, water, and aggregates is a thick fluid that can be poured into forms and finished to make a firm and solid smooth surface. The drum on small mixers rotates to mix the ingredients for immediate use. The big container on the back of the cement truck is always turning, moving the concrete around so it won't harden into one solid mass inside the truck.

The aggregate used to make concrete is inexpensive if it doesn't have to be transported very far from the cement plant. There are regional sand and gravel quarries that provide the aggregate for each local area's concrete. Consequently, concrete varies from region to region because different aggregates have been used to make it.

And where did the name come from? Portland cement was developed from natural rock material found in Britain in the early part of the 19th century. The rock resembled Portland stone, a type of building stone that was quarried on the Isle of Portland in Dorset, England. So it was not named for the city Portland in Oregon or in Maine.

How Do People Use Natural Resources to Make or Build Things?

The following list includes some of the common earth materials used for construction, manufacturing, and landscaping.

Aggregate A mass of rock particles, such as pebbles, gravel, and sand.

Asphalt or **blacktop** A dark-brown to black hydrocarbon (oil)-based material with a consistency varying from thick liquid to solid. Sand and gravel are added to asphalt to change its texture and strength.

Glass A mixture of quartz sand, lime, and sodium carbonate, melted together and formed into containers, windows, and other objects.

Mortar Lime or cement, or a combination of both, mixed with sand and water and used as a bonding agent between bricks or rocks.

Portland cement A mixture of ground limestone and shale that has been heated until almost fused and then finely ground. Cement sets and hardens after being mixed with water.

Potter's clay An iron-free, shapable clay especially suitable for making pottery, modeling, or throwing on a potter's wheel. Sand is often added to the clay to help the potter get a grip on the clay as it turns on the wheel. When clay is heated to a high temperature (fired), the particles fuse and no longer disintegrate in water.

New Word — Say it, Write it, See it, Hear it

Aggregate
Cement
Concrete
Fossil fuel
Geothermal power
Natural resource
Nonrenewable resource
Renewable resource
Solar energy
Wind power

Teaching Children about Natural Resources

TEACHING CHILDREN *about Natural Resources*

Developing Disciplinary Core Ideas (DCI)

To students, things are what they are. Glass is glass, brick is brick, concrete is concrete, and on and on. And why not? Most of us are isolated from the processes that produce these materials.

One of the roles of science education is to help students take their world apart and see what it is made of, and to help them understand how human actions impact the resources that go into making the things of our society. Materials like concrete become comprehensible when they are analyzed to discover their composition, and even greater knowledge of the material world is achieved when students reconstruct objects using everyday processes. Students can think about the ways that different cultures over thousands of years have used concrete to build structures and marvel at how the concrete used by the Romans has resulted in the monuments that have lasted more than 2,000 years. Modern scientists and engineers are studying the ingredients in those ancient structures to help us develop a better way to make concrete today.

Students see the world in interesting ways, and given the new lens of looking for earth materials as natural resources in use, they will search with enthusiasm. It is not the accuracy with which students analyze the contents of a building material that is important, but rather the processes of observing closely, comparing, and communicating their discoveries.

Engaging in Science and Engineering Practices (SEP)

In this investigation, students engage in these practices.

- **Planning and carrying out investigations** dealing with properties of rock material to make a product for human use—concrete.
- **Constructing explanations** using evidence to describe interactions of earth materials with each other and other substances in the environment and to find ways that humans use earth materials in everyday life. **Designing solutions** by using earth materials to produce a product.
- **Engaging in argument from evidence** about the process of soil formation and the impact of weathering, erosion, deposition and volcanic eruptions on the surface of Earth. Work with others to critique relevant information and how it supports explanations.

> **NGSS Foundation Box for DCI**
>
> **ESS3.A: Natural resources**
> - Energy and fuels that humans use are derived from natural sources, and their use affects the environment in multiple ways. Some resources are renewable over time, and others are not. (4-ESS3-1, foundational)
>
> **ETS1.A: Defining and delimiting engineering problems**
> - Possible solutions to a problem are limited by available materials and resources (constraints). The success of a designed solution is determined by considering the desired features of a solution (criteria). Different proposals for solutions can be compared on the basis of how well each one meets the specified criteria for success or how well each takes the constraints into account. (3-5-ETS1-1)

> **NGSS Foundation Box for SEP**
> - **Make observations and/or measurements to produce data** to serve as the basis for evidence for an explanation of a phenomenon or test a design solution.
> - **Identify the evidence** that supports particular points in an explanation.
> - **Apply scientific ideas** to solve design problems.
> - **Compare and refine arguments** based on an evaluation of the evidence presented.
> - **Respectfully provide and receive critiques from peers** about a proposed procedure, explanation, or model by citing relevant evidence and posing specific questions.
> - **Construct and/or support an argument** with evidence, data and/or models.

Soils, Rocks, and Landforms Module—FOSS Next Generation

INVESTIGATION 4 – *Natural Resources*

NGSS Foundation Box for SEP

- **Read and comprehend grade-appropriate complex texts** and/or other reliable media to summarize and obtain scientific and technical ideas and describe how they are supported by evidence.
- **Obtain and combine information** from books and/or other reliable media to explain phenomena.
- **Communicate scientific and/or technical information** orally and/or in written formats, including various forms of media and may include tables, diagrams, and charts.

- **Obtaining, evaluating, and communicating information** from books and media and integrating that with their firsthand experiences to construct explanations about natural earth processes and the use of natural resources by humans.

Exposing Crosscutting Concepts (CC)

In this investigation, the focus is on these crosscutting concepts.

- **Scale, proportion, and quantity:** Natural earth processes can happen quickly (floods, earthquakes, landslides, volcanoes) and other processes occur over hundreds, thousands, or even millions of years (weathering, soil formation, rock formation).
- **Structure and function:** Building materials are made from earth materials; the changes in the kinds and amounts of the ingredients can change the properties and durability of the structures.

NGSS Foundation Box for CC

- **Scale, proportion, and quantity:** Observable phenomena exist from very short to very long time periods.
- **Structure and function:** Different materials have different substructures, which can sometimes be observed.

Connections to the Nature of Science

- **Scientific investigations use a variety of methods.** Scientific methods are determined by questions. Scientific investigations use a variety of methods, tools, and techniques.
- **Science is a human endeavor.** Men and women from all cultures and backgrounds choose careers as scientists and engineers. Most scientists and engineers work in teams. Science affects everyday life. Creativity and imagination are important to science.

Connections to Engineering, Technology, and Applications of Science

- **Influence of science, engineering, and technology on society and the natural world.** Engineers improve existing technologies or develop new ones. Over time, people's needs and wants change, as do their demands for new and improved technologies. When new technologies become available, they can bring about changes in the way people live and interact with one another.
- **Interdependence of science, engineering, and technology.** Science and technology support each other. Knowledge of relevant scientific concepts and research findings is important in engineering. Tools and instruments are used to answer scientific questions, while scientific discoveries lead to the development of new technologies.

Teaching Children about Natural Resources

Conceptual Flow

Students explore the phenomenon of earth materials as natural resources. The driving questions for the module are what are Earth's land surfaces made of? and why are landforms not the same everywhere? The guiding question for this investigations is what makes rock a natural resource and how is this resource used by people?

The **conceptual flow** for this fourth investigation starts with students reviewing what they have learned about earth materials in the first three investigations by reviewing their notebooks and writing a story or making a concept map. Then they are introduced to **natural resources**, and through a video, learn about two classifications—**renewable resources** (**sunlight**, air and **wind**, **water**, **geothermal** energy, **soil**, **plants**, and **animals**) and **nonrenewable resources** (**fossil fuels** such as coal, oil, and natural gas; **rocks** and **minerals**), and the human actions that impact the management of those resources.

In Part 2, students experience **concrete**, one of the most widely used construction materials in the world (**buildings**, **roads**, and **walkways**), and discover that it is made from earth materials. While engaged in a class project to make a concrete stepping stone, students learn that concrete is a mixture of **cement** (made from limestone), **aggregates** of gravel and sand, and water.

In Part 3, students go outdoors to search for earth materials (either singly or in a mixture) that are serving a useful function. They develop a list of these objects (**clay** flower pots, concrete steps, gravel paths, **brick** walls, mortar between bricks, garden landscaping rocks, **asphalt** paving, and sand on the playground). Earth materials are essential for a variety of products that are important in everyday life.

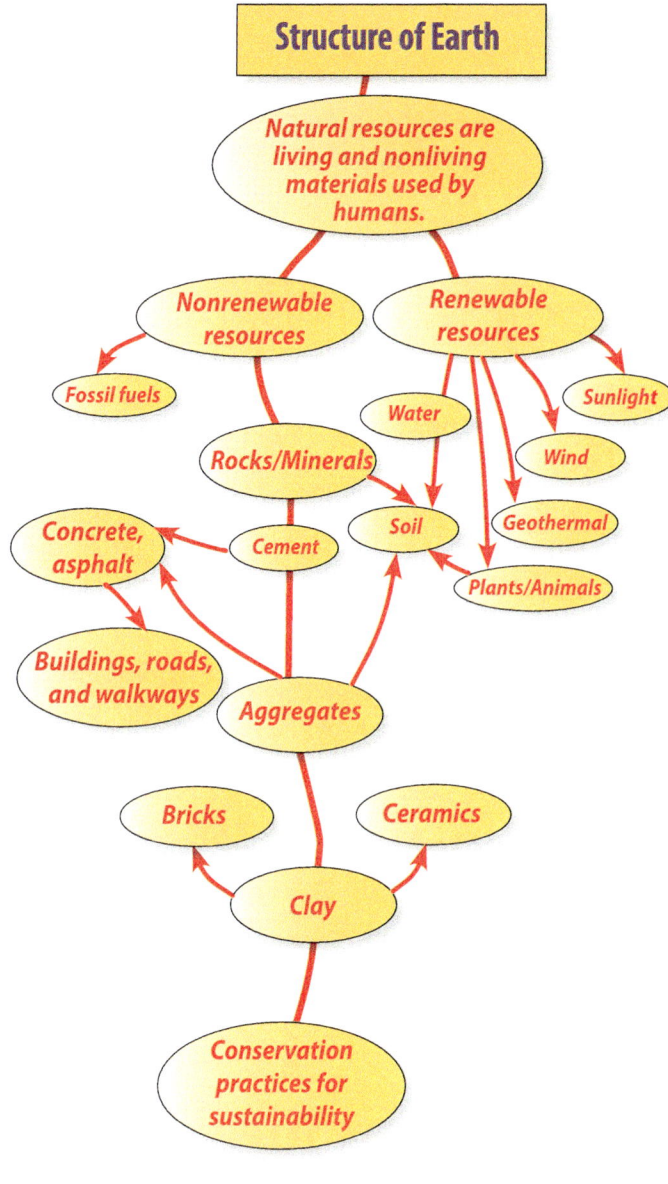

Soils, Rocks, and Landforms Module—FOSS Next Generation

INVESTIGATION 4 – *Natural Resources*

No. 24—Teacher Master

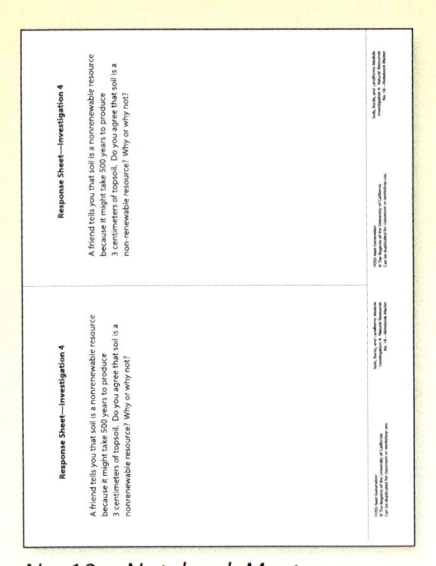

No. 18—Notebook Master

MATERIALS *for*
Part 1: *Introduction to Natural Resources*

For each student

    1   *FOSS Science Resources: Soils, Rocks, and Landforms*
- "Monumental Rocks"
- "Geoscientists at Work"

For the class

    1   Computer with Internet access ★
    ❏ 1   Teacher master 24, *Video Discussion Guide: Natural Resources*

For assessment

    ❏ • Notebook sheet 18, *Response Sheet—Investigation 4*
    ❏ • *Embedded Assessment Notes*

★ Supplied by the teacher.   ❏ Use the duplication master to make copies.

274   Full Option Science System

Part 1: Introduction to Natural Resources

GETTING READY for
Part 1: Introduction to Natural Resources

1. **Schedule the investigation**
   This part will take two active investigation sessions; one session for students to review the major concepts and write a story or make a concept map, and a second session to view and discuss the 23-minute video, *Natural Resources*. The reading requires a third session.

2. **Preview Part 1**
   Students review what they have learned in the module about soils, rocks, and landforms. They write a story or draw a concept map to bring the ideas together. They focus on earth materials as renewable and nonrenewable natural resources by viewing and discussing a video. The focus question for this part is **What are natural resources and what is important to know about them?**

3. **Plan for class discussion and writing**
   In Step 1, of Guiding the Investigation the class creates a list of the big ideas they have learned in the module, and use this list to write or draw a conceptual story or map. They can do this individually, in groups, or as a class.

4. **Preview the video**
   Preview the *Natural Resources* video (chapters 2, 3, 5, 7, and 8). Links to the video are found on FOSSweb in the Resources by Investigation section. Have students view the video after you introduce the focus question. Plan to use teacher master 24, *Video Discussion Guide: Natural Resources*, to lead a class discussion after watching the video with the class.

5. **Plan to read *Science Resources*: "Monumental Rocks" and "Geoscientists at Work"**
   Plan to read the articles "Monumental Rocks" and "Geoscientists at Work" during a reading period toward the end of this part.

6. **Preview the online activity**
   Preview the online activity "Resource ID" by going to the Resources by Investigation section on FOSSweb. Students can engage with this activity after the reading.

7. **Plan assessment: response sheet**
   In Step 9 of Guiding the Investigation, use notebook sheet 18, *Response Sheet—Investigation 4*, for a closer look at students' understanding of renewable vs. nonrenewable resources. Plan to spend time discussing the sheet with students after you have reviewed them. Record your notes on *Embedded Assessment Notes*.

> **NOTE**
> To prepare for this investigation, view the teacher preparation video on FOSSweb.

> **TEACHING NOTE**
> You will need to make this decision based on students' writing abilities and experience with concept maps.

Soils, Rocks, and Landforms Module—FOSS Next Generation

INVESTIGATION 4 – Natural Resources

FOCUS QUESTION

What are natural resources and what is important to know about them?

TEACHING NOTE

You could also have students work together in groups to create the list, then share it with the class.

GUIDING *the Investigation*
Part 1: Introduction to Natural Resources

1. **Gather for review discussion**

   Have students get their science notebooks and prepare for discussion.

   We've been studying earth materials for many weeks now. Let's review what we've found out.

   Have students go through their notebooks and help them create a class list describing how the big ideas developed in this module. Guide students to include all of the following:

   - Soils are made of earth materials of different sizes, mixed with humus.
   - Rocks break down into smaller pieces through physical and chemical weathering.
   - Rock pieces sometimes stay in place, and sometimes move, which changes and creates landforms through the processes of erosion and deposition.
   - Humans create, analyze, and interpret data from maps to describe patterns of Earth's features.
   - Humans can take action to reduce the impacts of natural Earth processes, such as building levees to minimize flooding and monitoring active volcanoes.
   - Volcanoes contribute to building landmass and enriching the soil; they add water and gases to the atmosphere.

   A sample story about how soils form

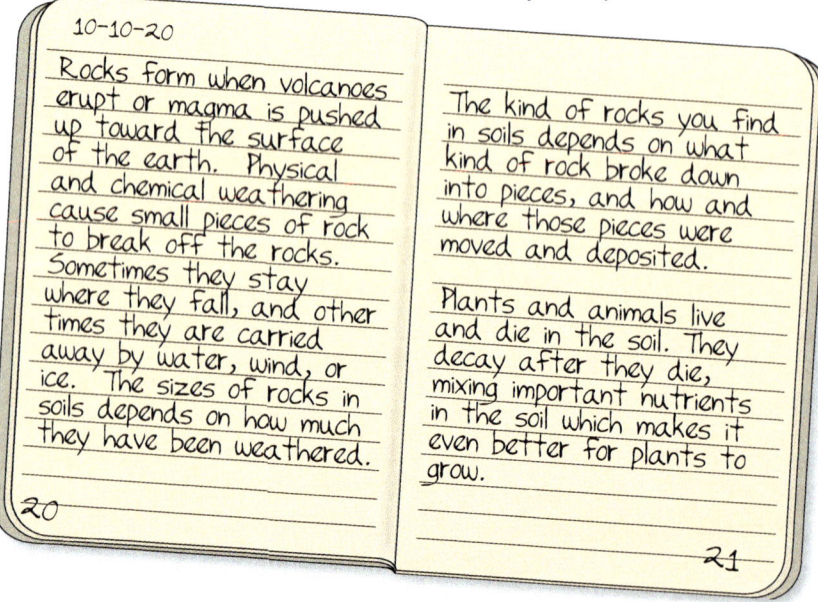

   > 10-10-20
   >
   > Rocks form when volcanoes erupt or magma is pushed up toward the surface of the earth. Physical and chemical weathering cause small pieces of rock to break off the rocks. Sometimes they stay where they fall, and other times they are carried away by water, wind, or ice. The sizes of rocks in soils depends on how much they have been weathered.
   >
   > The kind of rocks you find in soils depends on what kind of rock broke down into pieces, and how and where those pieces were moved and deposited.
   >
   > Plants and animals live and die in the soil. They decay after they die, mixing important nutrients in the soil which makes it even better for plants to grow.
   >
   > 20   21

276   Full Option Science System

Part 1: Introduction to Natural Resources

2. Build relationships among the concepts

Remind students of the soils they explored in Investigation 1 and the stream tables they used in Investigation 2. If you still have the vials with the layered soils available, distribute those to the groups.

Ask students to look at the concepts that they just listed and think about how soils form. Have individual students or groups work together for 5–10 minutes to create a story or concept map. Have half the class write a story about how soils form, and the other half write about how landforms change over time.

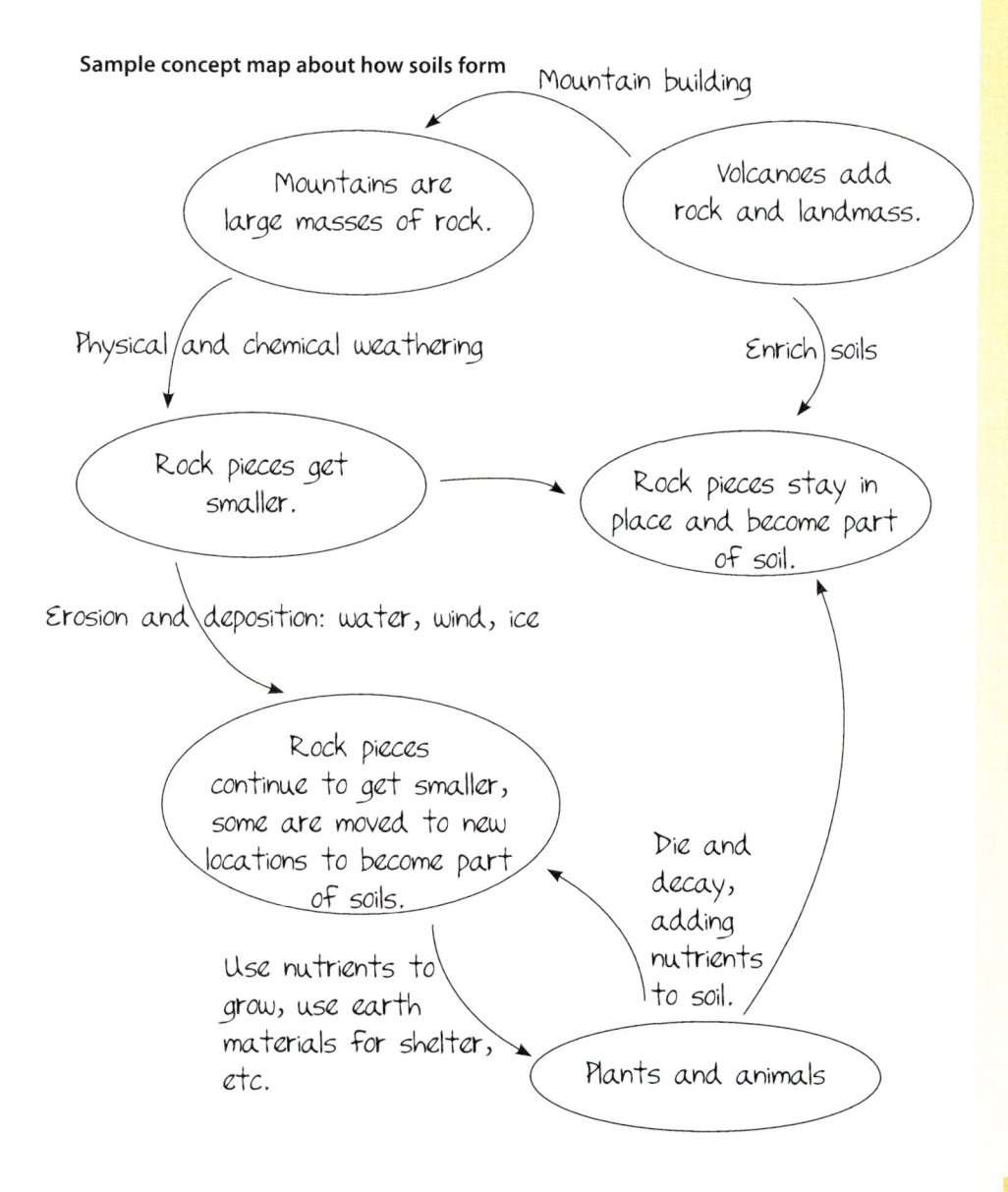

Sample concept map about how soils form

- Mountain building → Mountains are large masses of rock.
- Volcanoes add rock and landmass.
- Physical and chemical weathering → Rock pieces get smaller.
- Enrich soils → Rock pieces stay in place and become part of soil.
- Erosion and deposition: water, wind, ice → Rock pieces continue to get smaller, some are moved to new locations to become part of soils.
- Die and decay, adding nutrients to soil.
- Use nutrients to grow, use earth materials for shelter, etc. → Plants and animals

EL NOTE

Select 6–10 of the vocabulary words and write each word on a separate self-stick note. Have students work in groups of four to arrange the words by how they relate to each other. Then have them draw lines between the words and explain the relationships.

SCIENCE AND ENGINEERING PRACTICES

Constructing explanations

Engaging in argument from evidence

Obtaining, evaluating, and communicating information

POSSIBLE BREAKPOINT

Soils, Rocks, and Landforms Module—FOSS Next Generation

INVESTIGATION 4 – *Natural Resources*

3. **Revisit the anchor phenomenon**
   Use the concept map from Step 2 and the landform images from the module introduction in Investigation 1 to have students review the driving questions for the module. Ask,

   ➤ *What is Earth's surface made of?*

   ➤ *Why are landforms not the same everywhere?*

   Have a sense-making discussion about the two questions.

4. **Discuss why scientists study earth materials**
   Explain to students that there are many scientists who study earth materials. Ask,

   ➤ *What is so important about earth materials that many people would want to spend so much time studying them?*

   Allow students to discuss this question, guiding them to the idea that these are **natural resources**—materials taken from the environment, many necessary for our survival. Soils are necessary for growing much of the food people eat. We mine minerals for iron to make steel for building things, and so forth. You might also include the idea that we use natural resources for satisfying our wants. Jewelry, statues, and decorative objects are not needed for survival, but people enjoy the way they look and their durability.

5. **Focus question: What are natural resources and what is important to know about them?**
   Project or write the focus question on the board and have students copy it into their notebooks.

   ➤ *What are natural resources and what is important to know about them?*

   Tell students that they will watch several chapters of a video, have a discussion, and then write their answers to the focus question.

6. **View video about natural resources**
   Show students chapters about natural resources from the *Natural Resources* video. The video can be accessed on FOSSweb, Resources by Investigation for this module. After the video, use teacher master 22, *Video Discussion Guide: Natural Resources*, to guide the discussion.

   The topics in the video include:
   - Chapter 2: Renewable and Nonrenewable Natural Resources
   - Chapter 3: Renewable Natural Resources: Sunlight and Air
   - Chapter 5: Renewable Natural Resources: Water and Soil

278   Full Option Science System

Part 1: Introduction to Natural Resources

- Chapter 7: Nonrenewable Natural Resources: Minerals and Fossil Fuels
- Chapter 8: Alternative Sources of Energy

Ask these questions about natural resources.

➤ *What is a **renewable resource**? What are some examples?* [A renewable resource is something continually recreated and is available for use again within a relatively short time. Examples are air, sunlight, soil, water, animals, and plants.]

➤ *What is a **nonrenewable resource**? What are some examples?* [A nonrenewable resource is available only in limited amounts, and it can take a long, long time to be renewed. It can be used up. Some examples are coal, oil, natural gas, minerals, and sand. Some people include soil in this category.]

➤ *How does water become polluted?* [Water is polluted by oil spills, sewage, and runoff from land that has fertilizers, which can be poisons.]

➤ *How can soil be damaged?* [Soil can be damaged by natural processes like erosion and by poor farming techniques that allow the topsoil to erode quickly.]

➤ *What are some examples of **fossil fuels**? How are they important for people?* [Coal, oil, and natural gas. People use them as energy sources for transportation, heat, and factories.]

➤ *What kind of rock do you think coal is? Sedimentary, metamorphic, or igneous? How do you know?* [Sedimentary. It is formed in layers.]

➤ *What are some examples of alternative sources of energy?* [Nuclear power, **solar energy**, **wind power**, waves, moving water, and **geothermal power**.]

7. Review vocabulary
Review the vocabulary words introduced in the video. Make sure that new words are on the word wall. Add any other words students may need.

8. Answer the focus question
Ask students to answer the focus question in their notebooks.

➤ *What are natural resources and what is important to know about them?*

For students who need scaffolding, provide a sentence frame such as Natural resources are ____ . It is important to know ____ .

SCIENCE AND ENGINEERING PRACTICES
Obtaining, evaluating, and communicating information

CROSSCUTTING CONCEPTS
Scale, proportion, and quantity

fossil fuel
geothermal power
natural resource
nonrenewable resource
renewable resource
solar energy
wind power

Soils, Rocks, and Landforms Module—FOSS Next Generation

INVESTIGATION 4 – Natural Resources

TEACHING NOTE

This is a great opportunity for a class debate. If you check the Internet, you will see that there is an ongoing controversy about whether soils really are renewable or not. Time is the essential factor. Soils build up again, but it takes a very long time for them to be renewed.

SCIENCE AND ENGINEERING PRACTICES

Obtaining, evaluating, and communicating information

CROSSCUTTING CONCEPTS

Scale, proportion, and quantity

9. Assess progress: response sheet

Distribute notebook sheet 18, *Response Sheet—Investigation 4*. This sheet asks students to consider whether soil is a renewable or nonrenewable resource. Students should complete this sheet independently. Collect and review the sheet and plan next steps as needed.

What to Look For

- Students use information to present a written argument. They can agree or disagree, but they need to be able to justify their responses. The following bullets are the key point students should include in their arguments.

- Soil takes a long time to form from weathered and transported rock, as well as from decaying plants and animals; some scientists believe this process is long enough to consider soil as a nonrenewable resource.

- Nutrients in soils can be used; humans add fertilizers and compost to renew it.

- Soils can be washed away (eroded).

Part 1: Introduction to Natural Resources

READING *in Science Resources*

10. Read "Monumental Rocks"

Read "Monumental Rocks" during a reading period after completing the active investigation.

Before the reading, tell students that the "Monumental Rocks" article describes four monuments of worldwide significance. Have students preview the text by looking at the photographs and discussing what they know about the structures and what questions they have.

Read the introductory paragraph aloud and ask students to think about what is the same about these monuments and what is different. Continue reading aloud or have students read independently or in guided reading groups. Pause after each section to have students take turns summarizing the text.

For reading strategies to support English learners and below-grade-level readers, see the Science-Centered Language Development chapter.

11. Use a graphic organizer strategy

Make a class graphic organizer or have students work as a group or in pairs to compare and contrast the four monuments. See the example below.

Compare and Contrast Four Monuments

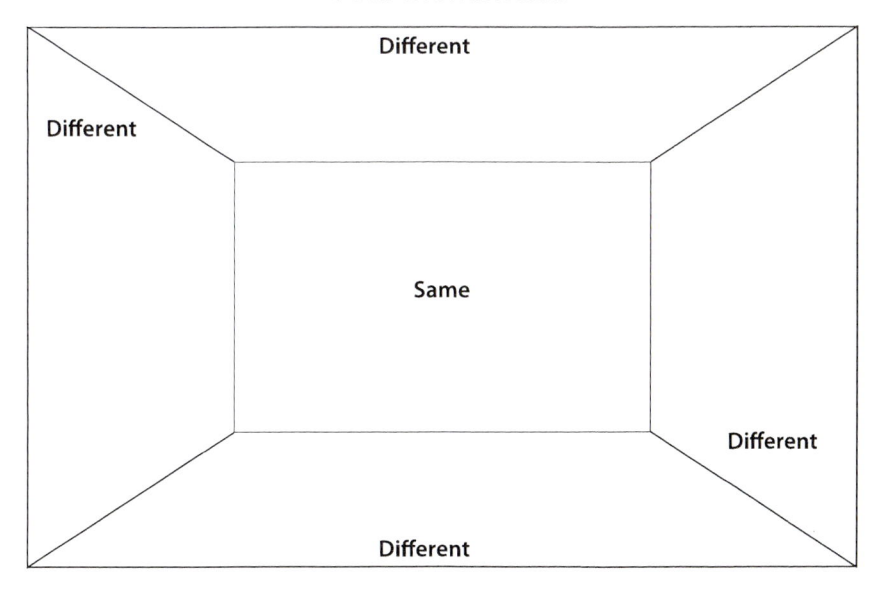

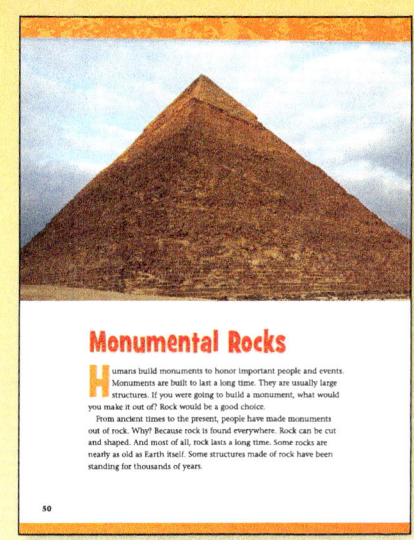

- Different (top)
- Different (left)
- Same (center)
- Different (right)
- Different (bottom)

ELA CONNECTION

These suggested reading strategies address the Common Core State Standards for ELA.

RI 2: Determine the main idea of a text and explain how it is supported by key details; summarize the text.

W 8: Gather relevant information from experiences and print, and categorize the information.

SL 4: Report on a text in an organized manner, using appropriate facts and relevant, descriptive details to support main ideas or themes; speak clearly at an understandable pace.

SCIENCE AND ENGINEERING PRACTICES

Obtaining, evaluating, and communicating information

Soils, Rocks, and Landforms Module—FOSS Next Generation

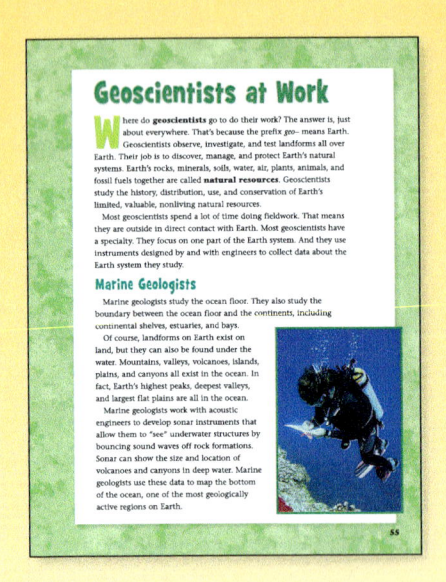

INVESTIGATION 4 – Natural Resources

ELA CONNECTION

These suggested reading strategies address the Common Core State Standards for ELA.

RI 4: Determine the meaning of general academic domain-specific words or phrases.

RI 7: Interpret information presented visually, and explain how the information contributes to an understanding of the text.

W 8: Gather relevant information from experiences and print, and categorize the information.

L 4: Determine or clarify the meaning of unknown and multiple-meaning words and phrases.

SCIENCE AND ENGINEERING PRACTICES

Obtaining, evaluating, and communicating information

12. **Read "Geoscientists at Work"**

    Explain to students that the next article, "Geoscientists at Work" describes the various careers followed by people who enjoy studying Earth and earth materials. It also describes the ways that engineers contribute to the work of geoscientists.

    Have students read independently or with a partner. Provide self-stick notes for students to jot down unknown words or phrases and questions they have about the different types of geoscientists.

    Discuss any of the unknown words and phrases. Ask students to discuss how they figured out some of the words using context clues, root words, and affixes.

13. **Engage in pair conversations**

    After reading, have students discuss the reading in pairs. Students should select the career that interests them the most, and describe it to their partners. If possible, students should include the kinds of instruments that the scientists use and their collaborations with engineers. They should speculate on what they might need to learn to be good at that career.

    The career descriptions include

    - Marine geologist
    - Atmospheric scientist
    - Seismologist
    - Structural geologist
    - Hydrologist
    - Petroleum geologist
    - Volcanologist
    - Soil scientist

14. **Discuss the readings as a class**

    Have students look again at pages 58 and 59 ("Soil Scientists" and "Soil Science at Home"). Ask them to look at the photos and the captions and describe what is shown to a partner. Then have a class discussion about soil conservation.

Part 1: Introduction to Natural Resources

15. View online activity
Have students engage with the online activity "Resource ID" and use the interactive to check their understanding of inexhaustible (perpetual), renewable, and nonrenewable natural resources. Students can work in small groups or as individuals.

WRAP-UP/WARM-UP

16. Share notebook entries
Conclude Part 1 or start Part 2 by having students share notebook entries. Ask students to open their science notebooks to the most recent entry. Read the focus question together.

➤ *What are natural resources and what is important to know about them?*

Ask students to pair up with a partner to share the answers they wrote to the focus question.

ELA CONNECTION

This suggested reading strategy addresses the Common Core State Standards for ELA.

SL 1: *Engage in collaborative discussions.*

Soils, Rocks, and Landforms Module—FOSS Next Generation

283

INVESTIGATION 4 – Natural Resources

MATERIALS for
Part 2: Making Concrete

For each student
- 1 *FOSS Science Resources: Soils, Rocks, and Landforms*
  - "Making Concrete"

For the class
- 2 Plastic cups
- 1 Trowel
- Sand ★
- Gravel ★
- Pebbles ★
- Water ★
- 1 Pitcher
- 1 Container, 1/2 L
- Concrete mix, 10 lbs dry powder ★
- 1 Bucket ★
- 1 Cardboard box, about 30 cm square ★
- 1 Plastic garbage bag ★
- 1 Pair of rubber or gardening gloves ★
- 1 Safety goggles ★

For embedded assessment
- ❑ Embedded Assessment Notes

★ Supplied by the teacher.   ❑ Use the duplication master to make copies.

Part 2: Making Concrete

GETTING READY for
Part 2: Making Concrete

1. **Schedule the investigation**
   This part will take one active investigation session and one session to observe the results of the investigation and read the article.

2. **Preview Part 2**
   Students focus on the earth resources that make up a very important material used for walkways, buildings, and bridges—concrete. The class uses local natural resources to make one concrete stepping stone. The focus question for this part is **How are natural resources used to make concrete?**

3. **Acquire concrete and other earth materials**
   For the class stepping stone, purchase a small, 10 lb bag of concrete mix at a hardware or home improvement center. It's inexpensive and contains cement and aggregate—you'll add the water. You can also add a little more local gravel and sand to the mix (one plastic cup of each), and invite students to contribute by bringing in decorative pebbles, or use pebbles from the class rock museum.

   There is also a type of concrete made especially for stepping stones that is available at craft stores. It is a bit more expensive than plain concrete.

4. **Practice safety**
   Ready-mix concrete contains Portland cement, the major ingredient in the reaction that produces concrete. Portland cement should be used with caution and handled only by an adult. When making stepping stones and working with concrete, it is essential that you use protective clothing. Concrete on bare skin can burn. If it gets on your skin, wash immediately with warm soapy water.

   The package label will suggest the use of gloves and goggles. It is good to demonstrate to students the way to safely handle cement. Some dust may enter the air when you pour the contents from the bag. Have students stand upwind as they watch.

   You might ask the school custodian to help mix the concrete and clean up the tools.

5. **Select the outdoor site**
   Find an appropriate outdoor site where you can make the concrete. You and the other adults will mix the concrete while students sit and supervise the process. Check the site the morning of the outdoor activity for unexpected, unsightly, and distracting items where students will be sitting.

> **TEACHING NOTE**
> For alternatives to making a class stepping stone, see Step 8 in this Getting Ready section.

Soils, Rocks, and Landforms Module—FOSS Next Generation

INVESTIGATION 4 – Natural Resources

Pizza box

Garbage bag

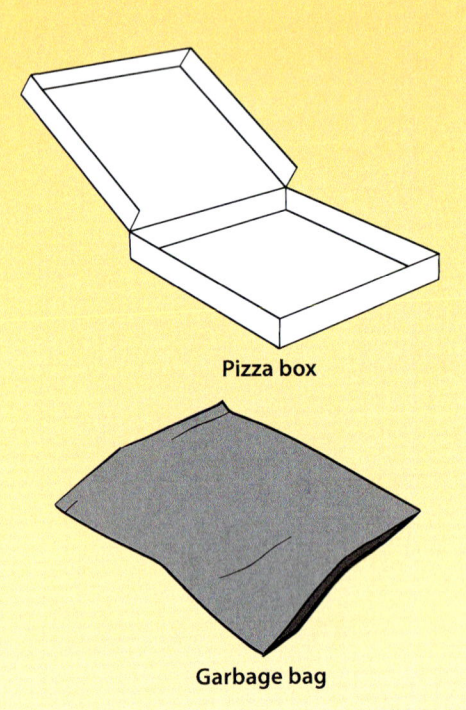

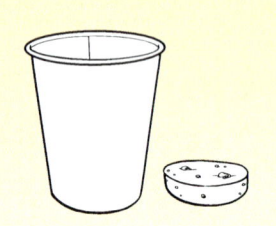

6. **Secure other tools**
   You will need a mold to hold the concrete to make the stepping stone. Stepping-stone molds can be made out of just about anything, from old cake pans to pizza boxes. The possibilities are endless. A cardboard box (about 30 cm square) lined with a garbage bag will work just fine. The box will be the mold or frame you pour the concrete in to harden. The finished stepping stone will be the size of the box, so consider how big you want to make it.

7. **Learn how to mix concrete**
   Mix the concrete in a bucket or old basin outdoors. Stir it with the trowel provided in the kit.

   Directions for correct proportions for mixing should be on the label. See Step 7 in Guiding the Investigation for general instructions. Be sure to add the water last. You want to end up with concrete the consistency of cake batter. Too much water, and the concrete will fail to cure. The easiest way to get the proportions correct is by adding water a little at a time.

   Once in the mold, allow the concrete to rest 5 minutes before having students press pebbles into the top of the mix. Allow the concrete 48 hours to cure completely before taking it out of the mold.

8. **Consider alternatives to a concrete stepping stone**
   As an alternative to making a stepping stone out of concrete, you can make small concrete-like objects using plaster of paris, sand, and gravel. See the Interdisciplinary Extensions section for information about that process.

   Another alternative is to work with the custodian to find a small concrete repair job that is needed at school. The custodian could mix the concrete and show students how it is used.

9. **Plan to read *Science Resources*: "Making Concrete"**
   Plan to read the article "Making Concrete" during a reading period after completing the active investigation for this part.

10. **Plan assessment: notebook entry**
    In Step 14 of Guiding the Investigation, students answer the focus question. Look for evidence that students can describe how concrete is made and why it is cost effective to use local aggregates.

Full Option Science System

Part 2: Making Concrete

GUIDING *the Investigation*
Part 2: *Making Concrete*

1. **Relate earth materials to natural resources**
   Ask students to think about the relationship between earth materials and natural resources. Review what a natural resource is.

   Natural resources are materials taken from the environment and put to use by people.

2. **Introduce** *concrete*
   Ask students if they have heard of the material **concrete**. Tell them that concrete is a hard, white or tan building material used to make many things. Ask,

   ➤ *What have you seen in our local area (neighborhood) that is made of concrete?*

   If students are stuck, suggest they think about roadways, sidewalks, buildings, and so forth.

3. **Focus question: How are natural resources used to make concrete?**
   Introduce the focus question and write it on the board.

   ➤ *How are natural resources used to make concrete?*

   Have students open their science notebooks to a new page and write the date and the focus question at the top of the page.

4. **Introduce ingredients in concrete**
   Tell students,

   Concrete is a very common building material. It is made of gravel and sand glued together by **cement**. *Cement is a fine gray powder made from a sedimentary rock, limestone.*

   When cement is mixed with water, it makes a sticky, mudlike mixture. Over time, the cement mixture hardens, changing into a solid, hard lump. This change is the result of a chemical reaction.

   If sand and gravel are added to the sticky mixture, the cement bonds to the pieces of rock, turning everything into an even stronger mass. This is what we call concrete. The sand and gravel used in this way are called **aggregates**.

   Today we are going to make concrete and shape it into a stepping stone.

FOCUS QUESTION
How are natural resources used to make concrete?

EL NOTE
Use a graphic to illustrate and label the components of concrete.

Soils, Rocks, and Landforms Module—FOSS Next Generation

INVESTIGATION 4 – Natural Resources

SCIENCE AND ENGINEERING PRACTICES

Planning and carrying out investigations

Materials for Steps 5–7
- Cardboard box
- Garbage bag
- Bucket
- Trowel
- Concrete mix
- Cup of sand
- Cup of gravel
- Water
- Safety goggles
- Gloves
- Pebbles

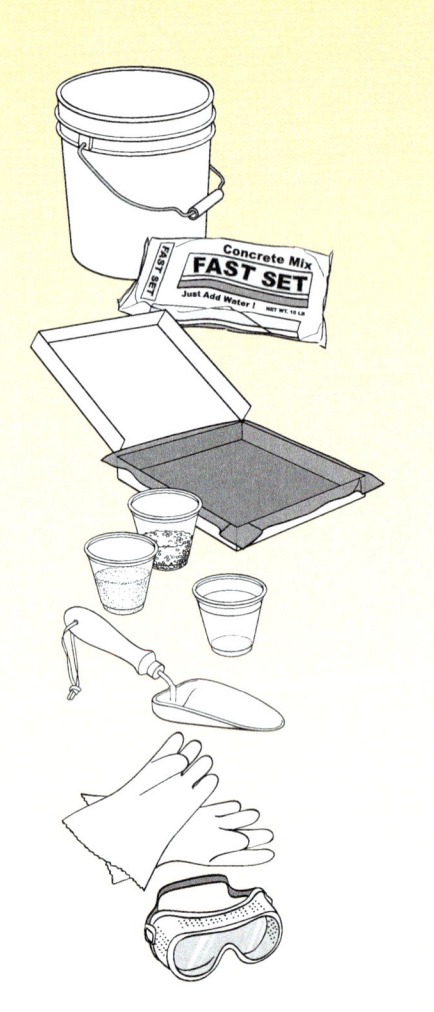

5. **Introduce the tools**

   Show students the mold (cardboard box and garbage bag), mixing tools, and concrete ingredients (bag of concrete mix, sand, gravel, and water).

   Tell students that you have read the instructions on the bag and it says to use protective eyewear and gloves to handle the cement. And it also says to avoid breathing the fine powder. Make sure they understand that cement is to be used with caution.

6. **Choose pebbles for the concrete**

   Tell students that they can put some pebbles in the concrete as decorative rocks. Each group (or pair of students) can select one of the class rocks to decorate the stepping stone. Students should carry these rocks outside where the concrete will be made.

7. **Go outdoors**

   Go outdoors with the concrete-making tools. Gather students at the sharing circle. Describe the steps as you make them.

   a. Pour one cup of gravel and one cup of sand into the bucket.

   b. Add the concrete mix.

   c. Add the measured amount of water based on the label instructions (about half a liter for 10 pounds of mix).

   d. Mix with the trowel.

   e. Add a little more water to get the mixture to be like cake batter.

   f. Get the mold ready, and use the trowel to transfer the mixture to the mold.

   g. Wait 5 minutes, then one at a time, ask each group to press their selected pebble into the wet concrete.

   h. Clean the trowel and bucket using water. Find a suitable place outdoors to recycle the water.

   i. Wash hands with soap and water.

8. **Return to class**

   Take the stepping stone back into the classroom. Students can observe it over the next 24–48 hours.

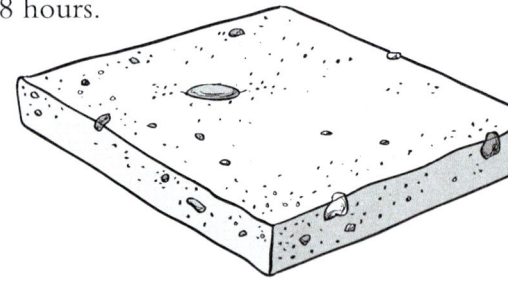

Part 2: Making Concrete

READING in Science Resources

9. Read "Making Concrete"
Read the article "Making Concrete." This article gives students a look at concrete production, from large-scale projects to their own schoolyard.

10. Use a questions/responses chart strategy
Before starting the reading, give students a few minutes to preview the text by looking at the photographs and captions. Tell students to choose one photograph and share with a partner what they think is happening and any personal connections they may be able to make. Ask students to think about what they will learn from reading this article. You might want to have students set up a two-column chart in their notebooks to jot down questions they have before reading the text.

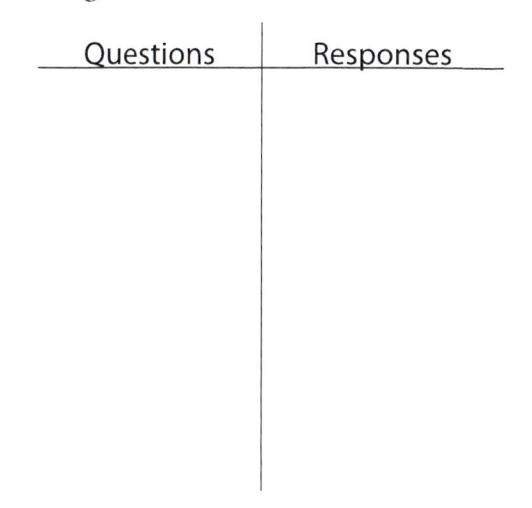

Questions	Responses

Read the article aloud or have students read independently or in guided-reading groups.

11. Discuss the reading
After the reading, give students a few minutes to jot down any responses they were able to find to their questions. Then have them share new information they learned from the reading with their table groups. Review any unfamiliar words, phrases, or ideas students might have. Discuss the reasons why local aggregates are used to make concrete [to make the concrete cost effective].

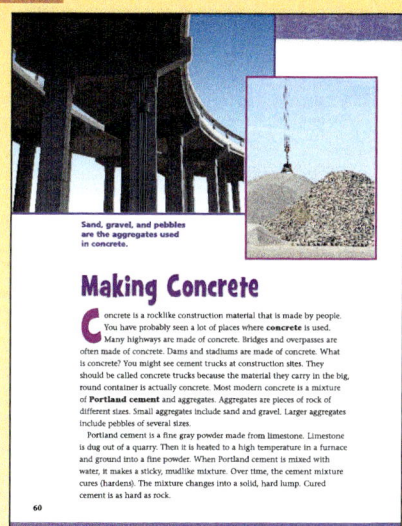

ELA CONNECTIONS

These suggested reading strategies address the Common Core State Standards for ELA.

RI 1: Refer to details and examples in a text when explaining what the text says explicitly and when drawing inferences from text.

RI 2: Determine the main idea of a text and explain how it is supported by key details; summarize the text.

W 7: Conduct short research projects that build knowledge through investigation of different aspects of a topic.

SL 1: Engage in collaborative discussions.

SL 4: Report on a text in an organized manner, using appropriate facts and relevant, descriptive details to support main ideas or themes; speak clearly at an understandable pace.

INVESTIGATION 4 – Natural Resources

Ask students to re-read the pages about the Roman concrete.

➤ *Why did scientists and engineers study the concrete used to make the Roman Colosseum? What did they discover?*

➤ *Why is this important to us today?*

Ask students to re-read this sentence from the reading.

➤ *Where did the aggregates come from to make the concrete foundation used to build your school?*

Ask students how they might go about finding out the answer to this question. This question could lead to a short research project students might conduct to find out more about how natural resources are used in their local community.

12. Observe the cured concrete

When the concrete has cured (24-48 hours), discuss the process with the class as they observe the stepping stone. Ask,

➤ *What earth materials are used to make concrete?* [Sand, gravel, cement (limestone), and water.]

➤ *What was the process we used?*

➤ *What effect does sand and gravel have in the concrete?* [The sand and gravel make the concrete stronger.]

➤ *Why do you think concrete is a good building material?* [The earth materials needed to make concrete are plentiful. Concrete is sturdy and lasts a long time.]

13. Review vocabulary

Review the vocabulary words that have been introduced in the investigation so far. Make sure that new words are on the word wall. Add any other words students might need.

This is a good opportunity to check student vocabulary using concept circles. Draw a circle on the board divided into four parts. Write the words cement, aggregate, limestone, and concrete in each section. Ask students to discuss (and/or write) about the connections these words have.

CROSSCUTTING CONCEPTS

Cause and effect

aggregate
cement
concrete

ELA CONNECTION

This suggested reading strategy addresses the Common Core State Standards for ELA.

SL 1: Engage in collaborative discussions.

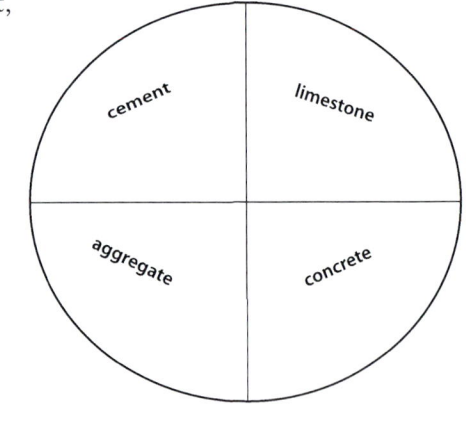

290

Full Option Science System

Part 2: Making Concrete

14. **Answer the focus question**
    Have students review the focus question.

    ➤ *How are natural resources used to make concrete?*

    Have students work independently to answer the question. Encourage them to use accurate vocabulary and include why local aggregates are used in the concrete-making process.

15. **Assess progress: notebook entry**
    Collect students' notebooks. After class, check to see that students understand how natural resources are used to make concrete and why local aggregates are used.

    ### What to Look For

    - Earth materials used to make concrete include Portland cement, water, sand, and gravel.
    - To make concrete cost effective, local aggregates are used.

WRAP-UP/WARM-UP

16. **Share notebook entries**
    Conclude Part 2 or start Part 3 by having students share notebook entries. Ask students to open their science notebooks to the most recent entry. Read the focus question together.

    ➤ *How are natural resources used to make concrete?*

    Ask students to pair up with a partner to discuss
    - natural resources and why are they important to us;
    - how concrete is made;
    - the benefits and costs of using natural resources such as rocks to make cement for such things as building roads.

EL NOTE

For students who need scaffolding, provide a sentence frame such as *Concrete is made with _____. We made concrete by _____. We discovered that _____.*

SCIENCE AND ENGINEERING PRACTICES

Constructing explanations and designing solutions

Obtaining, evaluating, and communicating information

Soils, Rocks, and Landforms Module—FOSS Next Generation

INVESTIGATION 4 – *Natural Resources*

MATERIALS *for*
Part 3: *Earth Materials in Use*

For each student

    1   Clipboard (optional) ★

    1   Pencil ★

    1   *FOSS Science Resources: Soils, Rocks, and Landforms*
- "Earth Materials in Art"
- "Where Do Rocks Come From?" (optional)

For the class

    1   Carrying bag ★

    1   Camera (optional) ★
- Extra pencils ★
- Self-stick notes ★

For embedded assessment
- *Performance Assessment Checklist*

For benchmark assessment

❏ • *Posttest*

❏ • *Assessment Record*

★ Supplied by the teacher.   ❏ Use the duplication master to make copies.

Part 3: Earth Materials in Use

GETTING READY for
Part 3: Earth Materials in Use

1. **Schedule the investigation**
   This part will take one session for the active investigation, one session for the reading, one session for the review, and one session for the posttest.

2. **Preview Part 3**
   Students go on a walk around the school and schoolyard, searching for earth materials in use. They search for various objects and structures and consider what natural resources were used to construct them. The focus question is **How do people use natural resources to make or build things?**

3. **Select the outdoor site**
   Plan the route that you will take students on and estimate how long it will take to walk. Remember to account for time to stop, observe closely, and record findings in notebooks.

   Scout the schoolyard ahead of time to locate interesting objects or structures. Here are some likely candidates.
   - Brick and mortar walls
   - Concrete blocks
   - Concrete sidewalks
   - Asphalt or blacktop sidewalks
   - Sculptures
   - Clay pots
   - Garden areas
   - Stones used in foundations or walls
   - Cobbles on roads
   - Stone benches or curbs

4. **Check the site**
   Check the outdoor site on the morning of the outdoor activity. Look for unexpected, unsightly, and distracting items, or anything else that could affect the activity in the area where students will be working.

Soils, Rocks, and Landforms Module—FOSS Next Generation

INVESTIGATION 4 – *Natural Resources*

TEACHING NOTE

This is a good opportunity for students to design a method for systematically recording observations.

5. **Prepare to take field notes**
   As students find earth materials in use, they record their findings in their science notebooks. They will set up the notebooks in class before heading outdoors. If necessary, place notebooks on a clipboard. Remember to take extra pencils in case of accidental breakage.

6. **Take a camera (optional)**
   If possible, take a camera to document what students see en route. Once you return to the classroom, a group of students could print out the pictures and make a poster to display what they found outdoors.

7. **Enlist additional adults**
   If possible, ask another adult to join you on the walk outdoors.

8. **Plan to read** *Science Resources*: **"Earth Materials in Art" and "Where Do Rocks Come From?" (optional)**
   Plan to read the article "Earth Materials in Art" and the optional reading "Where Do Rocks Come From?" during a reading period after completing the active investigation for this part and before the posttest.

9. **Plan assessment: performance assessment**
   Students observe how earth materials are used to build things. This is a good opportunity for you to observe students' scientific practices. Use the *Performance Assessment Checklist* to record how well students plan, organize, and record their observations.

10. **Plan benchmark assessment:** *Posttest*
    To help students prepare for the *Posttest*, plan a day for groups to review the important points from students' science notebooks. You might want to spend a little extra time on Investigation 4, given that students will not take an I-Check for this investigation. Later in the week, have students take the *Posttest*, working independently to respond to the items. Print or copy the assessment masters or schedule students on FOSSmap to take the *Posttest*. Review students' responses, using the coding guides found on FOSSweb.

Part 3: Earth Materials in Use

GUIDING the Investigation
Part 3: Earth Materials in Use

1. **Discuss the concrete stepping stone**
   Review how the class made concrete. Have students turn and talk to their neighbor to review how the concrete was made. Ask students to name other natural resources that people use to build things.

2. **Focus question: How do people use natural resources to make or build things?**
   Project or write the focus question on the board and have students copy it into their notebooks.

   ➤ *How do people use natural resources to make or build things?*

   Brainstorm a list of natural resources that might be used to make or build things. Write the earth materials students name in one column and other natural resources in a separate column, but don't name the columns at this time.

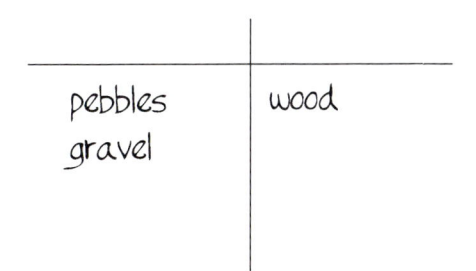

pebbles gravel	wood

3. **Introduce the investigation**
   Tell students that in order to answer the focus question, they will go on a walk to search for natural resources used as building materials. They will record in their notebooks what and where they find the natural resources in use. Encourage students to use their hand lenses to get a closer look at the earth materials they find.

4. **Set up notebooks for recording**
   Before going outdoors, ask groups to discuss effective ways to set up their notebooks in order to record what they might observe outdoors. You might want to provide an example.

   ➤ *If we go outside and see that sidewalks are made out of concrete, how should we record that in our notebooks?*

FOCUS QUESTION
How do people use natural resources to make or build things?

TEACHING NOTE
Students will consider all natural resources at the beginning of this part. Later, they will see that earth materials are stronger and more durable for building than other natural materials such as wood.

Soils, Rocks, and Landforms Module—FOSS Next Generation

INVESTIGATION 4 – *Natural Resources*

SCIENCE AND ENGINEERING PRACTICES

Planning and carrying out investigations

Constructing explanations and designing solutions

DISCIPLINARY CORE IDEAS

ESS3.A: Natural resources

CROSSCUTTING CONCEPTS

Structure and function

Materials for Step 7
- *Camera (optional)*

TEACHING NOTE

If you brought a camera, remember to take pictures of various objects along the way. Photos will enrich the discussion later on.

Have students share ideas for setting up their notebooks. They might make a table with room for text and a drawing. The column headings might be "object," "natural resource," "location," and "notes/drawing." Give students a few minutes to plan and organize their notes.

5. **Assess progress: performance assessment**
   Circulate from group to group. Listen to the group discussions and how students plan to organize their observations. Make notes on the *Performance Assessment Checklist* both while students are working now and later when they are outside making their observations.

   What to Look For
   - Students devise an effective means to record their observations. (Planning and carrying out investigations.)
   - Students offer explanations based on what they have learned throughout the module about the use of natural resources. (Constructing explanations and designing solutions; ESS3.A: Natural resources; structure and function.)

6. **Review outdoor safety rules**
   Before going outside, remind students of the rules and expectations for the outdoor walk. If you will be walking near roads, remind students that they must stay on the sidewalk.

7. **Go outdoors**
   Gather the necessary materials, get the appropriate clothing, line up, and head outdoors. Direct the class to head to the predetermined meeting spot in the usual orderly fashion. Once outdoors, gather the class to tell them exactly where you will go on your hunt. Take this opportunity to double-check that students know what they are looking for and what they need to record.

8. **Search the schoolyard**
   Walk the class around the schoolyard, examining various building materials. Include everything from the poles supporting the swing set, to the bricks on the building (and the mortar between the bricks), to the sidewalks, to the bench with wooden slats at the far end of the schoolyard. Take a break at each observation point and allow students to record the natural resources used to make the object, to speculate on how it was made, and to consider the durability of the item.

Part 3: Earth Materials in Use

9. **Discuss observations**

   Provide opportunities for students to make close observations of bricks, cement, concrete, and asphalt. It is important that they see and discuss the small earth materials used to make these products.

   On your walk, ask questions such as

   ➤ *Is anything in the concrete (asphalt, cement, bricks, etc.)?*

   ➤ *If the mortar had large pebbles in it, could the bricks lie flat?*

   ➤ *Why might roads have larger pebbles than sidewalks?* [It is less expensive to use coarse material for roads since they use a lot of paving materials.]

   ➤ *What is the condition of the wood on this bench (or table) compared to the concrete portions that support it? Which material will last longer?*

   ➤ *Why are the poles of the swing set made of metal, not wood or concrete?*

   ➤ *What natural resources are used to make the metal poles?*

   ➤ *Why are most schools made of steel and concrete rather than wood?*

   Visit as many observation points as possible before heading back indoors. Students may need some additional time to record their observations. Depending on the weather and students' attention spans, consider where to complete the notebooks—outdoors or back in the classroom.

10. **Have a sense-making discussion**

    Return to the classroom, and begin a discussion to answer the following questions (or similar questions that refer to objects from your schoolyard).

    ➤ *Why are certain natural resources used rather than others?* [If it needs to be very strong and long lasting or if it will be exposed to weathering, concrete or brick is better than wood.]

    ➤ *What was the condition of the wood on the bench? Why are most schools not made out of wood?*

    ➤ *If you were to design a sign with the school's name on it to be displayed in front of the school, what natural resources would you use to make it?*

    Ask students to turn and talk to a partner about some of the items they observed on the walk and how the material used to make the object was related to the function of the object.

SCIENCE AND ENGINEERING PRACTICES

Planning and carrying out investigations

Constructing explanations and designing solutions

> **TEACHING NOTE**
>
> Go to FOSSweb for Teacher Resources and look for the Science and Engineering Practices—Grade 4 chapter for details on how to engage students with the practice of constructing explanations and designing solutions.

CROSSCUTTING CONCEPTS

Structure and function

Soils, Rocks, and Landforms Module—FOSS Next Generation

INVESTIGATION 4 – Natural Resources

EL NOTE

For students who need scaffolding, provide a sentence frame such as: People use natural resources to build or make _____. In our schoolyard, we observed _____. I think _____ is better to _____, because _____.

CROSSCUTTING CONCEPTS

Structure and function

TEACHING NOTE

See the **Home/School Connection** for Investigation 4 at the end of the Interdisciplinary Extensions section. This is a good time to send it home with students.

11. Answer the focus question

Have students answer the focus question in their notebooks.

➤ *How do people use natural resources to make or build things?*

Encourage them to also discuss why earth materials are the best natural resource to use for things that need to be strong and last a long time.

Part 3: Earth Materials in Use

READING *in Science Resources*

12. Read "Earth Materials in Art"

This article with a variety of photographs explains how an artist uses clay to make tiles as pieces of art. Start by having students focus on the photograph of the woman cutting the clay. Write "observations," "inferences," and "personal response" on the board. Have students share what they can see in the photograph. Then have them share what they infer. Help generate responses by asking,

➤ *What can you assume based on what you see?*

➤ *What can you conclude about what is happening or will happen?*

Next, ask students what personal response they have from looking at the photograph. Ask,

➤ *What feelings does this photo make you have?*

➤ *What does it make you wonder about?*

Tell students that this article will give them more information about the artist in the photograph and how she makes clay tiles.

Read the story aloud or have students read independently or in guided reading groups. For reading strategies to support English learners and below-grade-level readers, see the Science-Centered Language Development chapter.

13. Use a 3-2-1 strategy

After the first read, have students read the article a second time and jot down notes in their notebooks. Use the 3-2-1 strategy to help students with comprehension. Have them write 3 important details; 2 interesting things; and 1 question.

14. Discuss the reading

Give students a few minutes to compare their notes with their group. Then, call for a few volunteers to share their questions. Note the questions on chart paper for possible future discussion or research topics. Have students take turns describing the process of making tiles out of clay. Encourage them to use the photographs as they retell the process. Ask students what other earth materials artists might use.

Focus students' attention on the text structures of the article. Ask,

➤ *What structures did the author use to organize the text?* [sequence, cause and effect]

➤ *What are some examples?*

Soils, Rocks, and Landforms Module—FOSS Next Generation

ELA CONNECTIONS

These suggested reading strategies address the Common Core State Standards for ELA.

RI 2: Determine the main idea of a text and explain how it is supported by key details; summarize the text.

RI 5: Describe the overall structure of information in a text.

RI 7: Interpret information presented visually, and explain how the information contributes to an understanding of the text.

W 8: Gather relevant information from experiences and print, and categorize the information.

SL 2: Paraphrase information presented orally.

SCIENCE AND ENGINEERING PRACTICES

Obtaining, evaluating, and communicating information

INVESTIGATION 4 – *Natural Resources*

15. Read "Where Do Rocks Come From?" (optional)

Read "Where Do Rocks Come From?" during a reading period after the active investigation. This final reading is optional but will provide review of concepts developed in the module dealing with volcanoes and fossils in sedimentary rocks. In this article, students learn about the rock cycle and igneous, sedimentary, and metamorphic rocks.

For reading strategies to support English learners and below-grade-level readers, see the Science-Centered Language Development chapter.

16. Discuss the reading

Ask,

➤ *How do **igneous** rocks form?* [They start out as melted rock in Earth's mantle. They can come to the surface in volcanoes and pour out as lava, or they can cool beneath the surface of Earth.]

➤ *How do **sedimentary** rocks form?* [They form from bits and pieces of recycled rocks and minerals. These sediments form layers and are pressed and cemented together over millions of years.]

➤ *How do **metamorphic** rocks form?* [These rocks form when one kind of rock changes into another kind of rock due to intense heat and pressure deep in Earth's crust or under lava.]

➤ *Explain how a metamorphic rock could change into a sedimentary rock.* [If a metamorphic rock weathered, chemically and physically, it could break apart and form sediments that could become a sedimentary rock.]

➤ *What kind of rocks often contain **fossils**?* [Sedimentary rocks.]

Part 3: Earth Materials in Use

WRAP-UP

17. Review the big ideas

Students created a list of the module's big ideas in Part 1 of this investigation and worked on stories and concept maps to show how those ideas are related. Have students review their lists of big ideas, stories, and concept maps. They should add the new ideas from this investigation to their lists.

Together review the word wall, saying the key vocabulary words out loud together.

18. Review Investigation 4

Distribute one or two self-stick notes to each student. Ask students to cut the note into three pieces, making sure that each piece has a sticky end. Ask students to take a few minutes to look back through their notebook entries for Parts 2 and 3 of this investigation to find the most important things they learned. Students should include at least one science and engineering practice, one disciplinary core idea, and one crosscutting concept. They should tag those pages with self-stick notes.

When students have completed their notebook review and placed their tags, lead a short class discussion to create a list of three-dimensional statements that summarize what students have learned in this investigation. Here are the big ideas that should come forward in review discussions.

- We conducted investigations dealing with properties of earth materials to make an important building product—concrete. (Planning and carrying out investigations; structure and function.)
- We gathered information from media about natural resources that are renewable and some that are nonrenewable. (Obtaining, evaluating, and communicating information.)
- Geoscientists study earth materials, in part to help humans use those resources wisely. (Obtaining, evaluating, and communicating information; scale, proportion, and quantity.)

You can have students record the class-generated list of key points or draw a line of learning under their last entry and compose a short paragraph responding to the investigative guiding question.

➤ *What makes rock a natural resource and how is this resource used by people?*

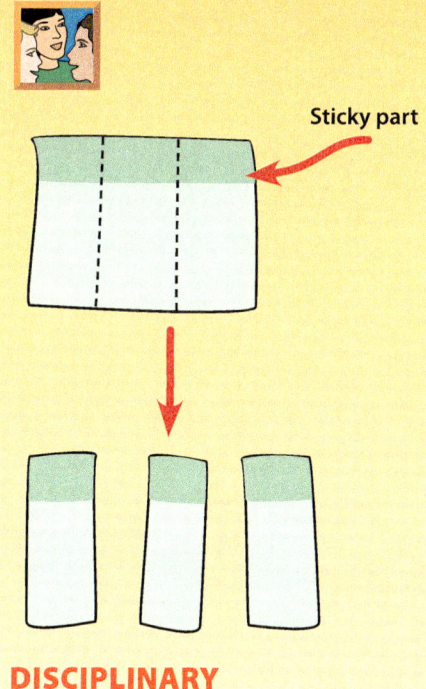

Sticky part

DISCIPLINARY CORE IDEAS

ESS3.A: Natural resources

ETS1.A: Defining and delimiting engineering problems

BREAKPOINT

Soils, Rocks, and Landforms Module—FOSS Next Generation

INVESTIGATION 4 – *Natural Resources*

19. Assess progress: *Posttest*

Give the *Postttest* assessment at least one day after the wrap-up review. Distribute a copy of the *Posttest* to each student. You can read the items aloud, but students should respond to the items independently in writing. Alternatively, you can schedule the *Posttest* on FOSSmap for students to take the test and then you can access data through helpful reports to look carefully at students' progress.

If students take the test on paper, collect the *Posttest*. Code the items using the coding guides you downloaded from FOSSweb.

Interdisciplinary Extensions

INTERDISCIPLINARY EXTENSIONS

Language Extension

- **Research natural resources in your region**
  Have students use the Internet to find what natural resources are in your area. They should research what the resources are, where they are located, how they are used and managed, what current issues are associated with their use, and how you can help make the resource sustainable. Have them prepare a poster to share.

Math Extension

- **Problem of the week**
  Two students were playing the game Rock Paper Scissors (rock crushes scissors, scissors cut paper, paper covers rock). They had agreed that at the end of each game, the loser would give the winner a rock from his or her collection.

  After playing many games, Student A had won three games and Student B had won the rest. When they stopped playing, Student B had three more rocks than she had when they began.

  What is the fewest number of games of Rock Paper Scissors they could have played?

  Notes on the problem. It is helpful to role-play a few rounds of the game. Ask two students to come to the front of the class, give each student four rocks, and have them play Rock Paper Scissors. After each game, the loser gives the winner a rock. Ask the class to come up with a way of keeping track of what's happening by making a table.

  This problem is not as easy as it may seem at first glance. Students often come up with a quick answer of "six games" because "Student A has won three," three more because Student B has three more than when she started. Students often don't consider that after each game a rock changes hands.

  One way to solve the problem is to make a table, like the one in the sidebar. First you have to pick the number of rocks they started with—any number greater than three will do. (An asterisk denotes the winner of each game.) At the end, Student A has won three games (first three), and Student B has three more rocks than when she started. They played at least nine games.

No. 25—Teacher Master

Game	STUDENT A	STUDENT B
Start	4	4
1	5*	3
2	6*	2
3	7*	1
4	6	2*
5	5	3*
6	4	4*
7	3	5*
8	2	6*
9	1	7*

Soils, Rocks, and Landforms Module—FOSS Next Generation

INVESTIGATION 4 – *Natural Resources*

Materials for plaster of paris "concrete"
- Metal spoon (from kit)
- Large paper cup
- Sand
- Gravel
- Water, 10–15 mL
- Stirring stick
- Plaster of paris

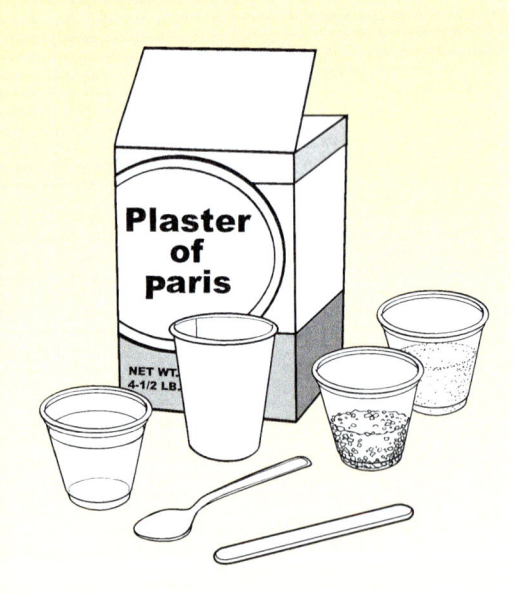

TEACHING NOTE
Encourage students to use the Science and Engineering Careers Database on FOSSweb.

Science Extensions

- ### Make "concrete" using plaster of paris
  As an alternative to making a concrete stepping stone with the class, you could have each student make a small round tile using sand, gravel, and plaster of paris (instead of cement). Plaster of paris is made from the mineral gypsum. Gypsum is heated to 150°C and then ground into a fine white powder. When the dry powder is mixed with water, it forms a paste that later hardens, but not as hard as cement.

  Plaster of paris is generally regarded as a safe material for routine use in very small quantities, but you should remind students to avoid breathing in the powder and to wash their hands after using it. It does generate heat when mixed with water, and in large quantities, this can be dangerous. For these reasons you should distribute the plaster powder rather than having students handle it independently.

  Here's a procedure for students to use.

  a. Get a large paper cup with a bottom diameter of about 6 centimeters (cm). Label the cup with your name.

  b. Put 1 spoon of gravel and 2 spoons of sand in the cup.

  c. Have the teacher add 1 heaping spoon of plaster of paris to the "aggregate." Stir the dry mixture with a stick.

  NOTE: When students add water to the mixture of sand, gravel, and plaster, it will become warm.

  d. Slowly and carefully add about 10 milliliters (mL) of water, constantly stirring with the stick until all the ingredients are mixed evenly.

  e. Place the paper cup in a safe place for the "concrete" tile to dry.

- ### Contact your local NRCS office
  Originally established by Congress in 1935 as the Soil Conservation Service (SCS), the Natural Resources Conservation Service (NRCS) has expanded to become a conservation leader for all natural resources, ensuring private lands are conserved, restored, and made more resilient to environmental challenges, like climate change. NRCS works with landowners through conservation planning and assistance designed to benefit the soil, water, air, plants, and animals, resulting in productive lands and healthy ecosystems. There are local NRCS offices in almost every county in the nation. To find the one closest to you, consult the website listed on FOSSweb.

Interdisciplinary Extensions

- **Invite a cement worker to class**
  Check with students to see who might have a parent or friend who works in construction and uses cement. Have students develop a list of questions to ask the guest speaker before he or she visits.

- **Take a field trip to an aggregate supplier**
  Natural aggregates include sand, gravel, and pebbles. Recycled aggregates consist mainly of crushed concrete and crushed asphalt pavement. Natural aggregate is very inexpensive, but it cannot be transported more than a few miles from its source without becoming prohibitively expensive because it is so heavy. Locate the nearest source of aggregate (sand and gravel) to your school. Plan a field trip.

- **Engage with the "Virtual Investigation: Natural Resources"**
  In this virtual investigation on FOSSweb, students review renewable and nonrenewable resources and how natural resources are used.

Art Extension

- **Research earth materials for decoration**
  Some cultures used pigments made by mixing crushed earth materials with animal fat to make paints and dyes. These paints could be used to decorate homes or skin. Rocks and minerals are also used to create jewelry. Have students use the Internet to research the use of earth materials for decoration, both in early times and the present day.

Home/School Connection

Students are encouraged to find what earth materials are used on the inside or outside of the place where they live. Print or make copies of teacher master 26, *Home/School Connection* for Investigation 4, and send it home before the reading in Part 3.

No. 26—Teacher Master

Soils, Rocks, and Landforms Module—FOSS Next Generation

INVESTIGATION 4 – Natural Resources

SOILS, ROCKS, AND LANDFORMS — *Assessment*

THE FOSS ASSESSMENT SYSTEM *for Grades 3–5*

"Assessment is like science. …To assess our students, we plan and conduct investigations about student learning and then analyze and interpret data to develop models of what students are thinking. These models allow us to predict the effect of additional teaching, addressing the patterns we notice in student understanding and misunderstanding. Assessment allows us to improve our teaching practice over time, spiraling upward" (*2016 Science Framework for California Public Schools, Kindergarten through Grade 12,* chapter 9, page 3).

An important rule of thumb in educational assessment is that assessments should be designed to meet specific purposes. One size does not fit all. The FOSS assessment system provides ample opportunities for both formative and summative assessment. Formative assessments provide short-term information about learning by making students' thinking visible in order to guide instructional decisions. Summative assessments provide valid, reliable, and fair measures of students' progress over a longer period of time, at the end of a module, or the end of the year. The purpose for the assessment determines the choice of instruments that you will use.

The FOSS assessment system is designed to assess students in cycles: short, medium, and long. The assessment tasks allow students to demonstrate their facility with three-dimensional understanding of science.

Short cycle. Embedded assessment opportunities are incorporated into each part of every investigation. These assessments use student-generated artifacts, including science notebook entries, answers to focus questions, response sheets, and **performance assessments**. Embedded assessments provide daily monitoring of students' learning and practices to help you make decisions about instructional next steps. Embedded assessments using science notebooks provide evidence of students' conceptual development. Performance assessments focus on science and engineering practices and crosscutting concepts, as well as disciplinary core ideas.

Contents

The FOSS Assessment System for Grades 3–5 .... **307**
Assessment for the NGSS **309**
Embedded Assessment **314**
Benchmark Assessment **318**
Next-Step Strategies **324**
FOSSmap and Online Assessment **328**
Sample Assessment Items **330**

▶ **NOTE**
For coding guides and student work samples, go to the Assessment section in *Teacher Resources* on FOSSweb.

Full Option Science System

SOILS, ROCKS, AND LANDFORMS — *Assessment*

I-Check opportunities occur every 1–2 weeks. These assessments are actually hybrid tools that provide summative information about students' achievement, and we have found they are even more powerful when used for formative assessment. Daily embedded assessments provide a quick snapshot of students' immediate learning, and I-Checks challenge students to put this learning into action in a broader context. Now students must think about the science and engineering practices, disciplinary core ideas, and crosscutting concepts they have been learning, and know when, where, and how to use them. I-Checks (short for "I check my own understanding") also provide opportunities for guided self-assessment, an important skill for future learning and development of a growth mindset. Properly executed feedback can help a student focus attention on areas that need strengthening. When a student responds to feedback, you can develop an even more precise understanding of the student's learning. A feedback/response dialogue can develop into a highly differentiated path of instruction tailored to the learning requirements of individual students.

Medium and long cycle. *Survey/Posttest*, **interim benchmark assessments, and portfolios** are tools provided for medium- and long-cycle assessment. Students take the *Survey* before instruction begins. This entry-level assessment provides you with information about students' prior knowledge and developing practices. It indicates modifications that might be needed for specific students or groups of students.

The *Posttest* is given at the end of a module. It provides summative information about students' three-dimensional learning. It also lets students compare their *Survey* responses to those on the *Posttest* and see how their understanding has grown. You can also use the *Posttest* for formative instructional evaluation by making notes about things you might want to focus on or do differently next time you teach the module.

Interim benchmark assessments can be given during a module when you want specific information about how students are progressing toward a specific performance expectation, at the end of the module, or at the end of the school year. These assessments specifically address NGSS performance expectations, and require students to apply what they have learned to new contexts. The interim assessment tasks can also inform instruction when used formatively to plan for next year's instruction.

Students can also collect work samples in a portfolio as they work through the module. At the end of each investigation, they can create derivative products to document their three-dimensional learning.

ASSESSMENT *for the* NGSS

A Framework for K–12 Science Education (National Research Council, 2012), *Next Generation Science Standards* (National Academies Press, 2013), *Developing Assessment for the Next Generation Science Standards* (National Research Council, 2014), and many state frameworks provide a new vision for science education. These documents emphasize the idea that science education should resemble the way that scientists work. Students plan and conduct investigations, gather data to construct explanations, and engage in argumentation to build their understanding of the natural world. They apply that knowledge to engineering problems and design solutions. Students are expected to construct and discuss explanations and model systems in more and more sophisticated ways as they move through the grades. Assessment plays an important role in this new vision of science education—assessment is the bridge between teaching and learning.

Several key points in these foundational documents provide guidance for a well-designed assessment system. FOSS has followed these guidelines to ensure a robust assessment system that provides valuable diagnostic information about students' learning.

Assessment tasks should consist of multiple components in order to measure all three dimensions of science and engineering learning. The FOSS assessment system provides multiple tools and strategies to assess the three dimensions: (1) science and engineering practices, (2) disciplinary core ideas, and (3) crosscutting concepts. These tools and strategies provide evidence about what students can do, their developing conceptual understandings, and the connections that they are making among disciplines. Entry-level assessments, given before instruction begins, show what students can do and what they know before they begin a new module.

Assessment systems should include formative and summative tasks. Formative assessment tasks are embedded in the curriculum at key stages in instruction. These tasks are designed to support teachers in collecting and analyzing data about students' conceptual understanding and growing practice. Notebook entries, answers to focus questions, response sheets, performance assessments, and oral presentations/interviews provide the information you need to decide what students need to do next to move toward a learning goal. FOSS suggestions for next-step strategies help you address students' developing conceptions and provide information for differentiated instruction as needed.

> *"Assessment plays an important role in this new vision of science education—assessment is the bridge between teaching and learning."*

SOILS, ROCKS, AND LANDFORMS — Assessment

Summative assessments are designed to provide valid, reliable, and fair measures of students' progress. The FOSS system includes three types of summative assessments: *Posttests*, interim assessments, and portfolios. These assessments include multicomponent tasks including open-ended constructed-response problems as well as some multiple-choice and short-answer items. *Posttests* are given at the end of a module and interim assessments can be given twice during a module or as an end-of-year assessment. Students can also collect work products at the end of each investigation for inclusion in a portfolio. Coding guides found on FOSSweb provide teachers with guidance when evaluating these assessments. All written assessments are consistent with grade-level writing and mathematics in the Common Core State Standards for ELA/Literacy and for Mathematics.

Assessment systems should support classroom instruction. The main purpose of the FOSS assessment system is to support classroom instruction—to provide the bridge between teaching and learning. Teachers need information daily about what students have learned or may be confused about. FOSS has developed a technique in which teachers spend only 10 minutes after a lesson, using a reflective-assessment practice (explained in detail later in this chapter), to gather data to determine instructional next steps. Are students ready to move on to the next lesson, or do they need some additional clarification? Our research has shown that the reflective-assessment practice provides evidence-based information that is crucial for differentiating instruction for all students—this practice can make a significant difference in students' overall achievement.

Assessment developers need to take a rigorous approach to the process of designing and validating assessments. The FOSS assessment system is based on a construct-modeling approach for assessment design. That means that we have done the research needed to describe a conceptual framework and learning performances that provide evidence of students' progressive learning (see the Framework and NGSS chapter), and we have done the technical work needed to ensure that assessment tasks provide valid, reliable, and fair evidence of students' learning. (See the Benchmark Assessment section for a more detailed description of the design behind the FOSS assessment system.)

> "Our research has shown that the reflective-assessment practice provides evidence-based information that is crucial for differentiating instruction for all students—this practice can make a significant difference in students' overall achievement."

Assessment for the NGSS

Assessment systems should include an interpretive system and locate students along a sequence of progressively more complex understanding. FOSS provides extensive support for interpreting assessment information. For each embedded assessment, specific information in the Getting Ready and Guiding the Investigation sections describes what students do and what to look for in student responses to assess progress. Coding guides, found on FOSSweb, are provided for each item on benchmark assessments (I-Checks and the *Survey/Posttest*, as well as intermin assessments). Samples of student work, especially for open-response questions, are also available on FOSSweb. Resources in this chapter and on FOSSweb provide you with information about how to use the coding guides, as well as what to do for next steps when students need to spend more time on a practice or core idea, or to look at it from a different perspective to see more connections.

You can use FOSSmap to have students take assessments online and generate a number of diagnostic and summary reports (delivered as PDFs). Some of these reports provide information about class progress; others provide individual students and parents with information about what students know and what they still need to work on. (See the FOSSmap and Online Assessment section in this chapter.)

The FOSS assessment system was developed over a period of 5 years with data from more than 500 teachers and their students. We know that teachers can employ this assessment system for the benefit of their students, and we know that students achieve more. Perhaps even more important is the change in classroom culture that occurs when assessment is thoughtfully employed as the bridge between teaching and learning. Assessment is no longer a stress factor for students or teachers. It encourages all to adopt a growth mindset—if I know where my strengths and weaknesses are and I continue to be thoughtful and work hard, I can make progress. It models what scientists do. Scientists use the information they have to argue for the best explanation, but they keep an open mind, so when new evidence emerges, they can incorporate that into their thinking, too. That's also what good curriculum and assessment are all about.

> *"Assessment is no longer a stress factor for students or teachers. It encourages all to adopt a growth mindset—if I know where my strengths and weaknesses are and I continue to be thoughtful and work hard, I can make progress."*

Soils, Rocks, and Landforms Module—FOSS Next Generation

SOILS, ROCKS, AND LANDFORMS — *Assessment*

NGSS Performance Expectations

"The NGSS are standards or goals, that reflect what a student should know and be able to do; they do not dictate the manner or methods by which the standards are taught.... Curriculum and assessment must be developed in a way that builds students' knowledge and ability toward the PEs [performance expectations]" (*Next Generation Science Standards*, 2013, page xiv). The FOSS assessment system includes embedded, performance, and benchmark assessments. The chart displayed on this and the next page provides an overview of these assessments across the three fourth-grade modules. These assessments help students build knowledge and ability in concert with active investigations and readings to meet the goals of the NGSS.

Grade 4 NGSS Performance Expectations	FOSS Module Embedded Assessment	Benchmark Assessment
4-PS3-1. Use evidence to construct an explanation relating the speed of an object to the energy of that object.	Energy • Inv 2, Part 2: notebook entry	Energy • *Investigation 4 I-Check* • *Survey/Posttest*
4-PS3-2. Make observations to provide evidence that energy can be transferred from place to place by sound, light, heat, and electric currents.	Energy • Inv 1, Part 1: notebook entry • Inv 4, Part 1: notebook entry	Energy • *Investigation 1 I-Check* • *Investigation 2 I-Check* • *Investigation 3 I-Check* • *Investigation 4 I-Check* • *Survey/Posttest*
4-PS3-3. Ask questions and predict outcomes about the changes in energy that occur when objects collide.	Energy • Inv 4, Part 3: response sheet	Energy • *Investigation 4 I-Check* • *Survey/Posttest*
4-PS3-4. Apply scientific ideas to design, test, and refine a device that converts energy from one form to another.	Energy • Inv 5, Part 3: performance assessment	Energy • *Investigation 1 I-Check* • *Investigation 3 I-Check* • *Survey/Posttest*
4-PS4-1. Develop a model of waves to describe patterns in terms of amplitude and wavelength and that waves can cause objects to move.	Energy • Inv 5, Part 1: response sheet	Energy • *Survey/Posttest*
4-PS4-2. Develop a model to describe that light reflecting from objects and entering the eye allows objects to be seen.	Energy • Inv 5, Part 2: notebook entry	Energy • *Survey/Posttest*
4-PS4-3. Generate and compare multiple solutions that use patterns to transfer information.	Energy • Inv 3, Part 3: notebook entry	Energy • *Investigation 3 I-Check*

Assessment for the NGSS

Grade 4 NGSS Performance Expectations	FOSS Module Embedded Assessment	FOSS Module Benchmark Assessment
4-LS1-1. Construct an argument that plants and animals have internal and external structures that function to support survival, growth, behavior, and reproduction.	**Environments** • Inv 1, Part 1: notebook entry • Inv 4, Part 1: performance assessment	**Environments** • *Investigation 1 I-Check* • *Survey/Posttest*
4-LS1-2. Use a model to describe that animals receive different types of information through their senses, process the information in their brain, and respond to the information in different ways.	**Environments** • Inv 2, Part 4: response sheet • Inv 3, Part 1: performance assessment	**Environments** • *Investigation 2 I-Check* • *Survey/Posttest*
4-ESS1-1. Identify evidence from patterns in rock formations and fossils in rock layers to support an explanation for changes in a landscape over time.	**Soils, Rocks, and Landforms** • Inv 2, Part 4: notebook entry	**Soils, Rocks, and Landforms** • *Investigation 2 I-Check* • *Survey/Posttest*
4-ESS2-1. Make observations and/or measurements to provide evidence of the effects of weathering or the rate of erosion by water, ice, wind, or vegetation.	**Soils, Rocks, and Landforms** • Inv 1, Part 2: response sheet • Inv 1, Part 3: performance assessment • Inv 2, Part 1: notebook entry • Inv 2, Part 2: performance assessment • Inv 2, Part 3: response sheet	**Soils, Rocks, and Landforms** • *Investigation 1 I-Check* • *Investigation 2 I-Check* • *Survey/Posttest*
4-ESS2-2. Analyze and interpret data from maps to describe patterns of Earth's features.	**Soils, Rocks, and Landforms** • Inv 3, Part 1: notebook entry • Inv 3, Part 2: response sheet	**Soils, Rocks, and Landforms** • *Investigation 3 I-Check* • *Survey/Posttest*
4-ESS3-1. Obtain and combine information to describe that energy and fuels are derived from natural resources and their uses affect the environment.	**Energy** • Inv 5, Part 3: notebook entry	**Energy** • *Survey/Posttest* **Soils, Rocks, and Landforms** • *Survey/Posttest*
4-ESS3-2. Generate and compare multiple solutions to reduce the impacts of natural Earth processes on humans.	**Soils, Rocks, and Landforms** • Inv 3, Part 4: notebook entry	**Soils, Rocks, and Landforms** • *Investigation 3 I-Check* • *Survey/Posttest*
3-5-ETS1-1. Define a simple design problem reflecting a need or a want that includes specified criteria for success and constraints on materials, time, or cost.	**Energy** • Inv 1, Part 4: performance assessment	**Energy** • *Investigation 1 I-Check*
3-5-ETS1-2. Generate and compare multiple possible solutions to a problem based on how well each is likely to meet the criteria and constraints of the problem.	**Energy** • Investigation 3 **Soils, Rocks, and Landforms** • Inv 3, Part 3: performance assessment	**Energy** • *Investigation 2 I-Check* • *Investigation 3 I-Check*
3-5-ETS1-3. Plan and carry out fair tests in which variables are controlled and failure points are considered to identify aspects of a model or prototype that can be improved.	**Environments** • Inv 1, Part 2: response sheet • Inv 3, Part 3: response sheet **Soils, Rocks, and Landforms** • Inv 1, Part 3: performance assessment • Inv 2, Part 2: performance assessment	**Energy** • *Investigation 3 I-Check* • *Investigation 4 I-Check*

Soils, Rocks, and Landforms Module—FOSS Next Generation

SOILS, ROCKS, AND LANDFORMS — *Assessment*

EMBEDDED *Assessment*

Assessment is the bridge that connects teaching and learning. Assessing students on a regular basis gives you valuable information that guides instruction and keeps families and other interested members of the educational community informed about students' progress.

Embedded assessments are suggested for most investigation parts. You will find a description of what and when to assess in the **Getting Ready** section of each part of each investigation. For example, here is the Getting Ready assessment step for Part 1 of the first investigation.

> **20. Plan assessment: notebook entry**
> In Step 17 of Guiding the Investigation, students answer the focus question in their notebooks, and in Step 18 they begin thinking about why soils may differ, based on location. Look for evidence of students' understanding of soil composition and prior knowledge about why different soils have different kinds and sizes of earth materials. Record your observations on a copy of *Embedded Assessment Notes*.

As you progress through the lesson and it is time for students to create a work product to be assessed, you will find a step in **Guiding the Investigation**. It provides a bulleted list of what to look for when you review the students' work for disciplinary core ideas, science and engineering practices, and crosscutting concepts.

> **19. Assess progress: notebook entry**
> Have students hand in their notebooks open to the page on which they answered the focus question. Review students' notebooks after class and check to see that they know what makes up soil. You might also want to take a quick look at students' soil location guesses to help you prepare for the next set of lessons.
>
> **What to Look For**
> - *Based on direct observation, students write that soil is composed of different sizes of rock (e.g., sand, gravel, pebbles) and humus (decaying plants and animals).*

Embedded assessment is enacted through two strategies: science notebook entries and teacher observation of science practices.

Embedded Assessment

Performance Assessments

Assessing the three dimensions envisioned in the NRC *Framework* and the NGSS performance expectations challenges students to be engaged in science and engineering practices in order to build disciplinary core ideas bridged by crosscutting concepts. This is an everyday occurence in the FOSS curriculum. One part in each investigation has been designated as a performance assessment for you to formatively check students' progress for all three dimensions at the same time. You peek over students' shoulders while they are in the act of doing science or engineering. You take note of what they are doing and discussing and sometimes conduct short interviews. Observing the rich conversation among students and the actions they are taking to investigate phenomena or design solutions to problems provides important information about student progress. At times, you might step in with a 30-second interview to ask a few carefully crafted questions to learn more about students' deeper conceptual understanding and practices. The What to Look For bullets in each performance assessment step will help you focus on pertinent science and engineering practices, disciplinary core ideas, and crosscutting concepts. You can record student progress on the *Performance Assessment Checklist* for each part.

Science Notebook Entries

Making good observations and using them to develop explanations about the natural world is the essence of science. This process calls for critical thinking and honed communication skills. Science notebook entries are designed specifically to help you understand the practices, crosscutting concepts, and scientific explanations and knowledge that students are developing.

Four kinds of notebook entries serve as assessments for learning. Each part of each investigation is driven by a **focus question**. Each part usually concludes with students writing an answer to the focus question in their notebooks. Their answers reveal how well they have made sense of the investigation and whether they have focused on the relevant actions and discussions. Prepared **notebook sheets** or **free-form notebook entries** provide information about how students make and organize observations and how they think about analyzing and interpreting data. Finally, **response sheets** provide more formal embedded-assessment data once in each investigation. These assessments focus on specific scientific knowledge, practices, or crosscutting concepts that students often struggle with, giving you an additional opportunity to help students untangle concepts that they might be overgeneralizing or have difficulty differentiating.

> **NOTE**
> You only need 10 minutes after a lesson to review student work and gather evidence of learning. See the reflective-assessment practice on the next page.

SOILS, ROCKS, AND LANDFORMS — Assessment

Using the Reflective-Assessment Practice

Successful teachers incorporate a system of continuous formative assessment into their standard teaching practices. One of the keys to formative assessment is frequency. The more often you can gather evidence about students' progress, the more able you will be to guide each student's path to understanding. The **reflective-assessment practice** provides a proven method for gathering that information, in a way that takes little time, but has a big impact on students' learning.

Reflective-Assessment Practice

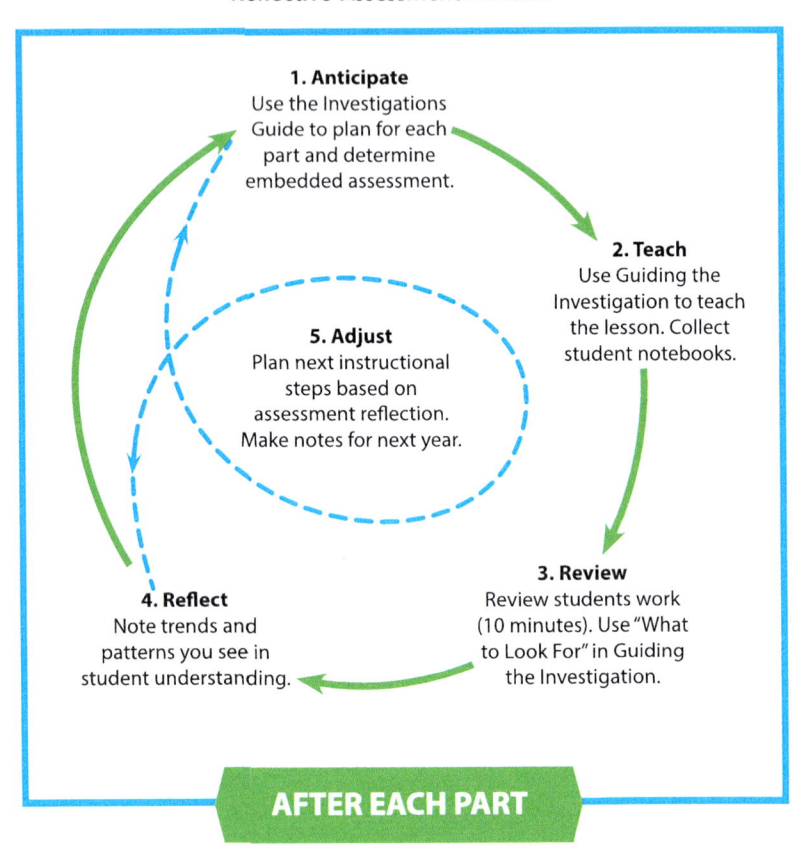

1. **Anticipate** — Use the Investigations Guide to plan for each part and determine embedded assessment.

2. **Teach** — Use Guiding the Investigation to teach the lesson. Collect student notebooks.

3. **Review** — Review students work (10 minutes). Use "What to Look For" in Guiding the Investigation.

4. **Reflect** — Note trends and patterns you see in student understanding.

5. **Adjust** — Plan next instructional steps based on assessment reflection. Make notes for next year.

AFTER EACH PART

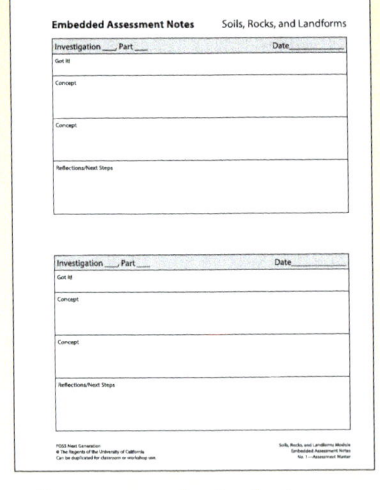

You can record notes for two lessons on each page.

Print or make copies of **Embedded Assessment Notes** to record students' progress with the embedded assessments in each part of an investigation (see the What to Look For bullets). If you don't think you can review every student's work in 10 minutes, choose a random sample. Many teachers are pleasantly surprised to discover how many students they can review in that short amount of time. The important thing is that you are looking at student work on a daily basis as often as possible and looking for patterns that reveal strengths and areas that need additional support.

Full Option Science System

Embedded Assessment

1. **Anticipate.** Check the Getting Ready section for the suggested embedded assessment for the part you are planning. Before class, fill in the investigation and part number along with the date on *Embedded Assessment Notes*. Check the assessment step in Guiding the Investigation, and fill in the things you are looking for in student thinking. Limit your assessment to one or two important ideas.

2. **Teach.** Follow the steps in Guiding the Investigation.

3. **Review.** Collect the notebooks at the end of class. Have students turn in their notebooks *open to the page you will be reviewing.* (This may sound trivial, but it will save you a lot of time.) If you didn't write in your "What to look fors" before teaching the lesson, do that now by checking the assessment step in Guiding the Investigation. Use *Embedded Assessment Notes* to record what you observe. Make a tally mark for each student who "got it"; write in names and notes for students who need help. Spend no more than 10 minutes on this review.

> **TEACHING NOTE**
>
> We encourage you to stop after 10 minutes even though you may be tempted to spend more time. If you keep this process quick and easy you are more likely to use it frequently. Frequency is key to the success of this practice.

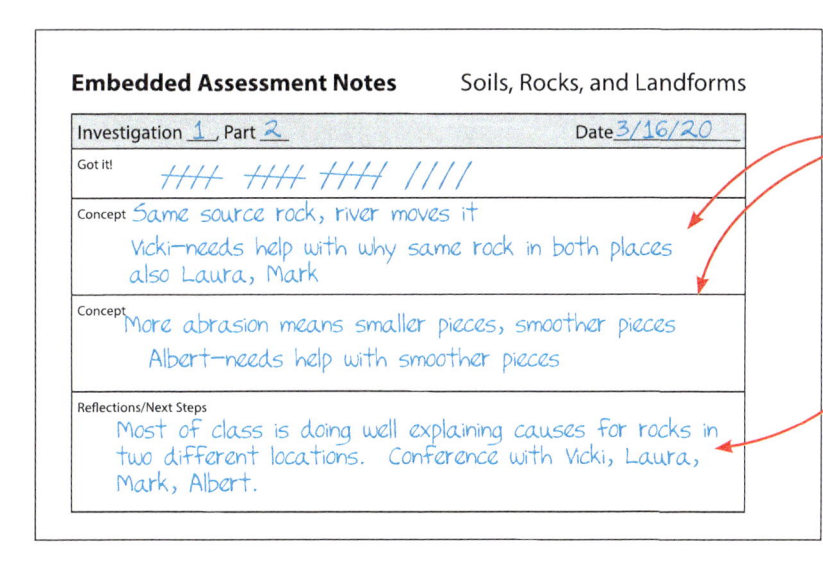

Embedded Assessment Notes — Soils, Rocks, and Landforms

Investigation _1_, Part _2_   Date _3/16/20_

Got it! ┼┼┼┼ ┼┼┼┼ ┼┼┼┼ ////

Concept: Same source rock, river moves it
 Vicki—needs help with why same rock in both places
 also Laura, Mark

Concept: More abrasion means smaller pieces, smoother pieces
 Albert—needs help with smoother pieces

Reflections/Next Steps:
 Most of class is doing well explaining causes for rocks in two different locations. Conference with Vicki, Laura, Mark, Albert.

Add what to look for and make notes as you review student work.

Write reflections and next steps here after 10 minutes of review.

4. **Reflect.** Take 5 minutes to summarize the trends and patterns (highlights and challenges) you saw, and record notes in the "Reflections/Next Steps" section.

5. **Adjust.** In the same section of the sheet, describe the next steps you will take to clarify any problems, or note highlights you saw in students' progress. Some suggestions appear later in this chapter and accompany the coding guides. This is the defining factor in formative assessment. You must take some action to help students improve. If you do this process frequently, the next steps required should take only a few minutes of class time when the next part begins.

SOILS, ROCKS, AND LANDFORMS — Assessment

BENCHMARK Assessment

The FOSS benchmark assessments are carefully designed to help you look at students' progress across the module. Each of these assessments includes items that incorporate the three dimensions described in the *NRC Framework*: science and engineering practices, disciplinary core ideas, and crosscutting concepts. Benchmark assessments provide a broader focus on students' learning than do embedded assessments. Benchmark assessments require students to determine when and how pieces of knowledge, and larger conceptual relationships need to be recalled or applied.

Survey

The *Survey* is administered a few days before instruction begins. Students are often uneasy about having to take a "test" when they haven't yet had the instruction they need in order to do well. It is important for them to know that the *Survey* will help you determine what they already know and what they need to learn, so that you can plan extra support when needed—students will not be graded on this assessment. Help students view the *Survey*, as well as all of the assessments, as learning tools. At the end of the module, have students compare their answers from the *Survey* to their answers on the *Posttest*. Students have few opportunities to see how their knowledge changes.

I-Checks

At the end of each investigation, students take an I-Check benchmark assessment. I-Checks (short for "I check my own understanding") provide students with an opportunity to demonstrate their learning. When you return the I-Checks, students have another opportunity to reflect on their understanding through self-assessment activities.

If students complete an I-Check using paper and pencil, determine their progress by reviewing each item using the coding guides provided on FOSSweb. When you code the open-response items, we recommend that you code them one item at a time; for example, code item 8 for all students, then move on to item 9, then 10, and so on. Even though you have to shuffle papers more, you will find that it takes less time overall to code the assessments. Proceeding one item at a time, rather than one student at a time, allows you to establish a mind-set for each item, think about the whole class's performance on that item, and plan next steps you might take. Another benefit to following this procedure is that you will code more consistently.

▶ **NOTE**
Alternatively, students can take the *Survey*, I-Checks, and *Posttest* online. Most of the items will be automatically coded for you. You can print students' answers from the *Survey* for comparison with the *Posttest*. See the FOSSmap and Online Assessment section in this chapter.

Benchmark Assessment

I-Checks are most valuable when you use them as formative assessments. Studies have shown that students learn more when they take part in evaluating their own responses. To do this, code students' tests, but do not put any marks on them. Thinking about the content of the questions comes to a halt when students see marks on their papers (especially grades) made by their teachers. Using the self-assessment strategies described later in this chapter is an important step in developing a class culture focused on growth mindset that values assessment as a learning process.

Posttest

Students take the *Posttest* after all the investigations are completed. It can be administered in any of the ways described for the other benchmark assessments. After coding the *Posttests*, return them and the *Surveys* to students. Have students compare their *Survey* and *Posttest* responses. Discuss the changes that have occurred.

Use the *Posttests* for formative evaluation of the module. Make notes about things you might want to focus on or do differently the next time you teach the module.

When Grades Are Required

"If we are always focusing on formative assessment, how do we give grades?" is a frequent question. Our recommendation is that you use derivative products for giving grades. That way students always know when they will be graded, and they always have a chance to improve their work before turning something in for a grade. For example, if you need to grade a notebook entry, have students rewrite that entry on a separate sheet of paper to be turned in for the grade. In this process, they can work with other students or use a class discussion to improve their entry before turning it in.

When grading I-Checks, rather than figuring a percentage, choose three or four items that students reflect on through self-assessment strategies, then have the opportunity to rewrite before turning in for a grade. When you follow this process, students are always clear about when they will be graded, and know that they always have the opportunity to make their work better, which helps develop a growth mindset.

If you must base grades on a percentage, first subtract 1 from each code (except zero, no negative numbers) to transform codes to points. Then you can add points and determine a percentage. We suggest 80% should be an A, rather than the traditional 90%, because these assessments are designed for diagnostic purposes, not minimum mastery.

Soils, Rocks, and Landforms Module—FOSS Next Generation

SOILS, ROCKS, AND LANDFORMS — *Assessment*

Portfolios

Students can choose work at the end of each investigation to include in a summative portfolio for end-of-course presentations or grades. Give students a *Portfolio Checklist* and a folder to hold the checklist and their work samples. At the end of each investigation, suggest an item or two that they might add to the portfolio to demonstrate their three-dimensional learning. Students can photograph and print notebook pages, or create a derivative product from the notebook page to be included in the portfolio. As always, notebooks should not be graded—they should be a risk-free environment for students to record their thinking. If students are going to include something from their notebook, they should have a chance to improve it before adding it to the portfolio. A sample *Portfolio Checklist* can be found with the other FOSS assessment charts, or you could make your own with students.

Interim Assessments

These assessments directly target specific NGSS performance expectations. They are not meant to be diagnostic for daily instruction using FOSS curriculum, rather they are generic tasks that students should be able to answer given any curriculum used to teach the NGSS. They are a good tool for students to use to practice for state or district tests. Interim assessments can be given after students complete an investigation focusing on a particular performance expectation, or you can give them at the end of the year for program evaluation, especially when a state or district test is not given at that grade level. (These tasks will evolve and be updated as we learn more about large-scale assessments.)

These tasks begin with a scenario that students read to set the context for the practices and items that will be involved in the task. Teachers can opt in many cases to include a hands-on experience, and in a few cases a computer-simulation. Each task then consists of a number of related items that students answer. Most are constructed-response items, but a few may be multiple-choice, multiple-answer, or short answer.

For third graders, we suggest that you use these items as a teaching and learning experience. That means that you support students by reading the text aloud, and have the class work through the task together. You provide any support the students need, short of giving them the answers. After students complete the tests, use them formatively, like the I-Checks, to evaluate students' progress. Keep in mind that the purpose of these assessments in third grade is to introduce students to these more complex tasks in a way that develops a growth mindset.

Benchmark Assessment

Benchmark Assessment Design

The foundation of the FOSS assessment system rests on three pillars: a conceptual framework, a student-progress model, and the NGSS performance expectations. It is important to remember that the performance expectations are a sampling of expected student proficiencies and that a full curriculum will always necessarily include more. The **conceptual framework** for this module can be found in the Framework and NGSS chapter. It is a segment from a larger learning progression that FOSS has carefully engineered for grades K–8. The concepts you see in the framework span grades K–8, and the bullets designate learning that is important for third graders. The **student-progress model** is displayed in the table below. It describes in general terms how FOSS uses four levels to categorize student learning. The **NGSS performance expectations** also guide assessment design, providing a sampling of end goals and always reminding us that learning science is a three-dimensional enterprise that includes science and engineering practices, disciplinary core ideas, and crosscutting concepts.

Level	Description
Strategic (4)	Performance on an item or assessment shows exceptional understanding of three-dimensional learning. Students are able to apply their knowledge of practices, disciplinary core ideas, and crosscutting concepts to explain novel phenomena or solve new problems. Students at this level continue to build a network of knowledge, practice, and crosscutting concepts to bridge disciplines. They can apply all of those to real-world phenomena and design problems.
Conceptual (3) (minimum goal for all students)	Performance on an item or assessment shows well-developed understanding of three-dimensional learning. Students are making connections among practices, core ideas, and crosscutting concepts in order to answer more complex questions and solve more complex problems. To get to the next level (strategic), students need to continue to build connections and be able to transfer this knowledge from the classroom to real-world phenomena.
Recognition (2)	Performance on an item or assessment shows developing understanding of three-dimensional learning. Students have built a foundational repertoire of pieces of knowledge and practices, and use academic language with greater facility. To get to the next level (conceptual), students need to continue to add knowledge about practices, core ideas, and crosscutting concepts, and then build connections among those pieces to form more complex understandings about phenomena.
Notions (1)	Performance on an item or assessment shows relatively little understanding of three-dimensional learning. Students may include some scientific vocabulary or recall of simple facts or procedures, but there is little evidence of the impact of instruction. To get to the next level (recognition), students need to develop practices, begin to incorporate academic language in their communication, and construct more pieces of knowledge that help explain phenomena.

SOILS, ROCKS, AND LANDFORMS — Assessment

Through our own research (ASK Project, 2003–2009) and that of others (most notably, Dylan Wiliam and Carol Dweck), FOSS developers have concluded that the most important assessment goal is to support learning in a way that helps students develop a growth mindset. That means that the focus must be on formative assessment.

It also means that we don't use traditional scoring. Instead, we code student responses to categorize the level at which students have demonstrated competence. For instance, at first glance it looks as though students are being awarded a point for an incorrect answer. In fact, their work is in the level 1 category. They have attempted to answer the question, but the answer shows little influence of instruction. You will also notice that some items can be coded no higher than a level 2, even though the student-progress model has four levels. That particular item asks students only to demonstrate that they can perform a simple practice or have a piece of knowledge (a level 2 proficiency) essential to a broader understanding. When it is important to know what pieces students have or don't have in order to determine how to help them improve, a simpler question can help identify differentiated instruction that is necessary for some students. These less-complex questions may not include all three dimensions, but are important in terms of providing diagnostic formative information.

Coding Benchmark Assessments

Answer sheets and coding guides for all benchmark assessments are available on FOSSweb. A two-page spread has been dedicated to each page of each assessment. Here we show an example for coding a multiple-answer and an open-response question.

Recording Benchmark Data

Use *Assessment Record* to record students' responses on the benchmark assessments (*Survey*, I-Checks, and *Posttest*). Follow this procedure if you plan to use the I-Checks for formative assessment.

1. Record students' responses on *Assessment Record* or on a spreadsheet. For multiple-choice and multiple-answer items, record the *letter* of the response rather than the code. (Later, if you need a total score, you can replace the letter responses with numbers.) For multiple-answer items, check the coding guide; for short-answer questions, record the word or the code, depending on the answer patterns you want to see. Do not make any marks on students' papers. (Students will do that later.)

> **NOTE**
> Coding guides for interim assessments are found on FOSSweb.

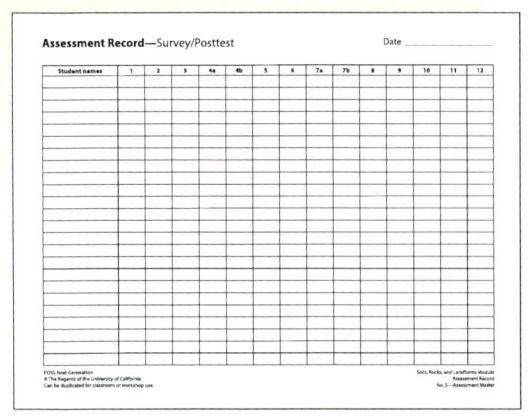

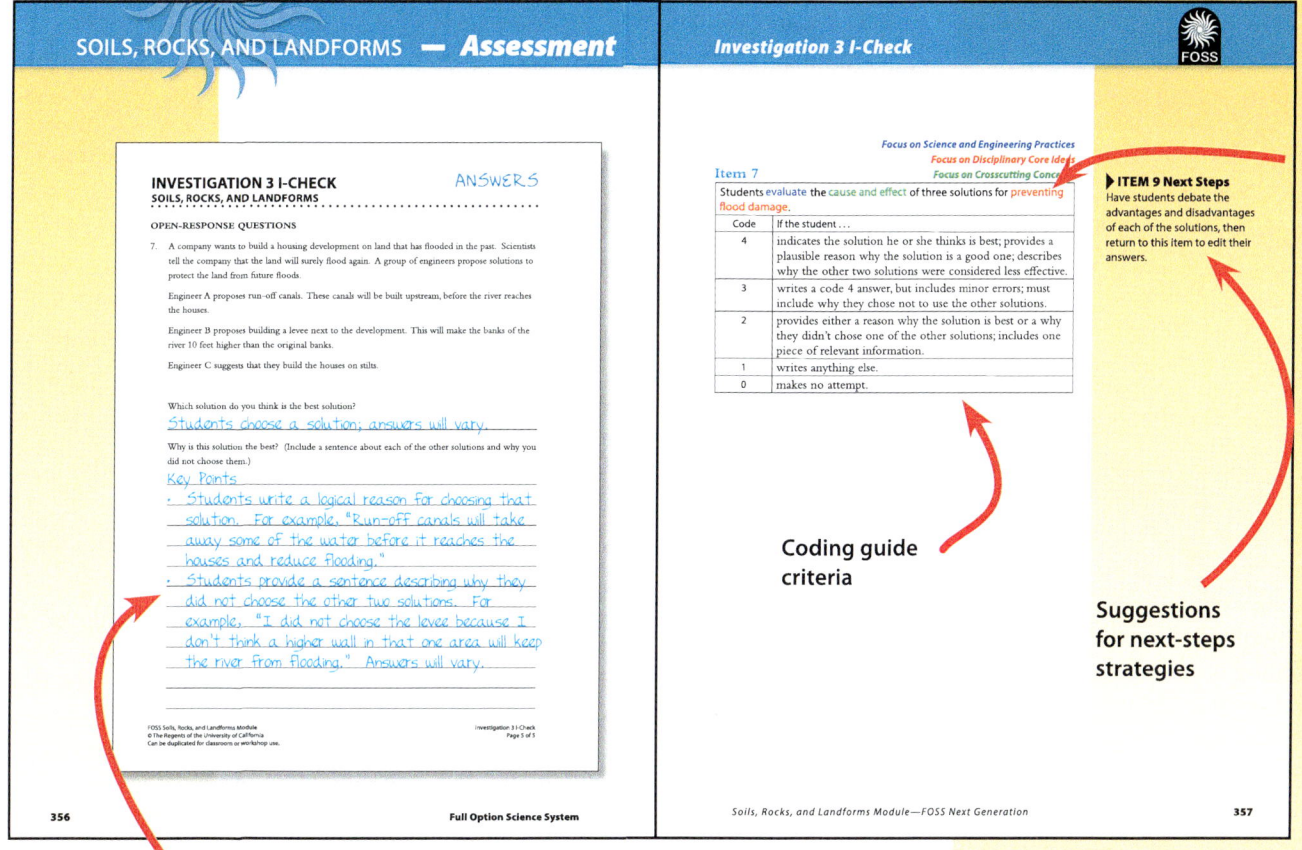

Benchmark Assessment

Answer sheet with key points that students should include in their responses

Coding guide criteria

Suggestions for next-steps strategies

Item description identifying dimensions of learning required to respond

2. Look for patterns of right and wrong answers, so that you can determine which questions will need follow-up reflection and discussion by students.

3. Review students' open-response answers. Record codes on *Assessment Record*. Make notes about concepts or ideas you want to bring back for students' reflection in self-assessment activities.

4. Choose three or four items from the assessment that you want to discuss with students during class time. Determine the next steps you want to introduce. (See the following section on next-step strategies.)

▶ **NOTE**
If students take the benchmark assessments online, FOSSmap does most of the coding for you. You need to code open-response items and check codes for short-answer items. Then you can run a variety of class and individual diagnostic reports.

Soils, Rocks, and Landforms Module—FOSS Next Generation 323

SOILS, ROCKS, AND LANDFORMS — Assessment

NEXT-STEP Strategies

The ASK Project (Assessing Science Knowledge) was funded by the National Science Foundation in 2003. For 6 years, the FOSS development team worked with nine centers around the United States, including more than 500 teachers and their students as research partners. Based on the evidence provided by the assessments, we learned very quickly that assessment is worth doing only if follow-up action is taken to enhance understanding. Self-assessment provides students the opportunity to be responsible for their own learning and is a very effective tool for building students' scientific knowledge and practice.

Self-assessment is more than reading correct answers to the class and having students mark whether they got the right answer. Self-assessment must provide an opportunity for students to reflect on their current thinking and judge whether that thinking needs to change. This kind of reflective process also helps students develop a better understanding of what is expected in terms of well-constructed responses.

Self-assessment requires deep, thoughtful engagement with complex ideas. It involves students in whole-class or small-group discussion, followed by critical analysis of their own work. For this reason, we suggest that you focus your probing discussions on three or four questions from an I-Check, rather than on the entire assessment. The techniques described here are meant to give you a few strategies for entering the process of self-assessment. There is no single right way to engage students in this process, but it works best when you change the process from time to time to keep it fresh. The strategies listed here are sorted into two groups: (1) strategies for whole-class feedback, and (2) strategies for individual-student and small-group feedback.

Strategies for the Whole Class

Multiple-choice discussions. Students sit in groups of three or four, depending on how many possible answers there were for a given question. You assign an answer to each student (not necessarily the answer he or she chose). Each student is responsible for explaining to the group whether the assigned answer is correct and why or why not.

Multiple-choice corners. When the class is equally split on what students have chosen as the correct response, or only a few students got the correct answer, have them meet in different corners of the room. Those who chose A go to one corner, those who chose B go to another corner, and so on. Each corner group needs to come up with

> **TEACHING NOTE**
>
> If students have taken an assessment online, print out their responses (the Student Responses Report) and use a projection system and the assessment masters or the PDF to show students the items under discussion.

Next-Step Strategies

an argument to convince the other corners that their answer is correct. As in a class debate, students are allowed to disagree with themselves if they become convinced their position is flawed or the reasoning of another group is more convincing. They then move to that corner and continue by helping their new group shape its argument. (Don't be surprised if you find all students migrating to one corner before the presentation of arguments even begins!)

Key points. Begin this strategy by discussing the item in question. After it is clear that students understand what is intended by the item prompt, call on individuals or groups to suggest key points that should be included in a complete answer. Write the key points on the board as phrases or individual words that will scaffold students' revision, rather than complete sentences they might mindlessly copy. When students return to their responses, they can number each of the key points they originally included in their answers, then add anything they missed.

Revision with color. Another way that students can revise their answers after a key-points discussion is to use colored pens or pencils and the three C's. As they read over their responses, they *confirm* correct information by underlining with a green pen; they *complete* their responses by adding information that was missing, using a blue pen; and they *correct* wrong information, using a red pen.

Review and critique anonymous student work. Use examples of student work from another class, or fabricate student work samples that emulate the problems students in your class are having. Project the work, using an overhead projector, a document camera, or an interactive whiteboard. Have students discuss the strengths and weaknesses of the responses. This is a good strategy to use when first getting students to write in their notebooks. It helps them understand expectations about what and how much to write.

Line of learning. Many teachers have students use a line of learning to show how their thinking has changed. When students return to original work (embedded or benchmark) to revise their understanding of a concept, they start by drawing and dating a line of learning under the original writing. The line of learning delineates students' original, individual thinking from their thinking after a class or group discussion has helped them reconsider and revise their thoughts.

Find the problems. Review the codes you have recorded and let students know how many items they got right and how many need to be corrected. Give them time in class to work in groups or individually to find their mistakes and correct them. They can turn in the assessment again, or write a few sentences in their notebooks explaining how their thinking has changed.

Soils, Rocks, and Landforms Module—FOSS Next Generation

SOILS, ROCKS, AND LANDFORMS — *Assessment*

> **NOTE**
> You can make inexpensive whiteboards by using card stock and plastic sheet protectors. Students use whiteboard marking pens to write answers. Old socks make great erasers.

Group consensus/whiteboards. Have students in each group (or pairs in each group) work together to compare their answers on selected I-Check questions or key points of a notebook entry. They first create a response that the group agrees is the best answer. Groups write their responses on a whiteboard. When you give a signal, one student in each group holds up his or her whiteboard and compares answers. The class discusses any discrepant answers.

Class debate. A student volunteers an answer to an item on an assessment (usually one that many students are having trouble with or one that elicits a persistent misconception). That student is in charge of the debate. He or she puts forth an answer or explanation. Other students agree or disagree, and must provide evidence to back up their thinking. Students are allowed to disagree with themselves if they hear an argument during the discussion that leads them to change their thinking. You can ask questions to keep the discussion on track, but otherwise you should stay on the sidelines.

Critical competitor. Use the critical-competitor strategy when you want students to attend to a specific detail. You need to present students with two things that are similar in all but one or two aspects. You can use any medium: two drawings, two pieces of writing, or a combination (such as a diagram compared to a description). The point is to compare two pieces of communication or representations in some way that will help students focus on an important detail they might be missing.

Sentence frames. After completing other self-assessment activities, have students consider all the items on the assessment and write a short reflection, using sentence frames. This strategy directs students to choose one or two items that they would like to tell you more about.

I used to think ___, but now I think ___.

I should have gotten this one right, but I just ___.

I know ___, but I'm still not sure about ___.

The most important thing to remember about ___ is ___.

Can you help me with ___?

I shouldn't have gotten this one wrong, because I know ___.

I'm still confused about ___.

Next time, I will remember to ___.

Now I know ___.

Full Option Science System

Next-Step Strategies

Strategies for Individual Students and Small Groups

Feedback notes. As you read through students' notebooks, add self-stick notes with comments or questions that help guide students to further reflect on and improve their understanding.

Response log. Set up a response log at the back of students' science notebooks (before or after the index or glossary if those are used). Fold a notebook page in half, or draw a line down the center of the page. Have students write "Teacher Feedback" at the top of the left side of the page and "My Responses" at the top of the right side of the page. When you want a student to think about something in his or her notebook, write your note in the "Teacher Feedback" column (or students can move a self-stick note from another page to the response log). Students then respond in the right column, either addressing your comment there or telling you which page to turn to in order to see how they have responded.

Conferences. Use silent-reading time or other times when students work independently to confer with small groups or individual students.

Centers. Set up a center at which students can continue to explore their ideas and refine their thinking. You might pair students so that a student who understands the concept well works with another student who needs some help.

Reteach or clarify a concept. Set up a modified investigation in which a small group of students works with the concept again.

> **NOTE**
> You can also create feedback notes for embedded assessments using FOSSmap.

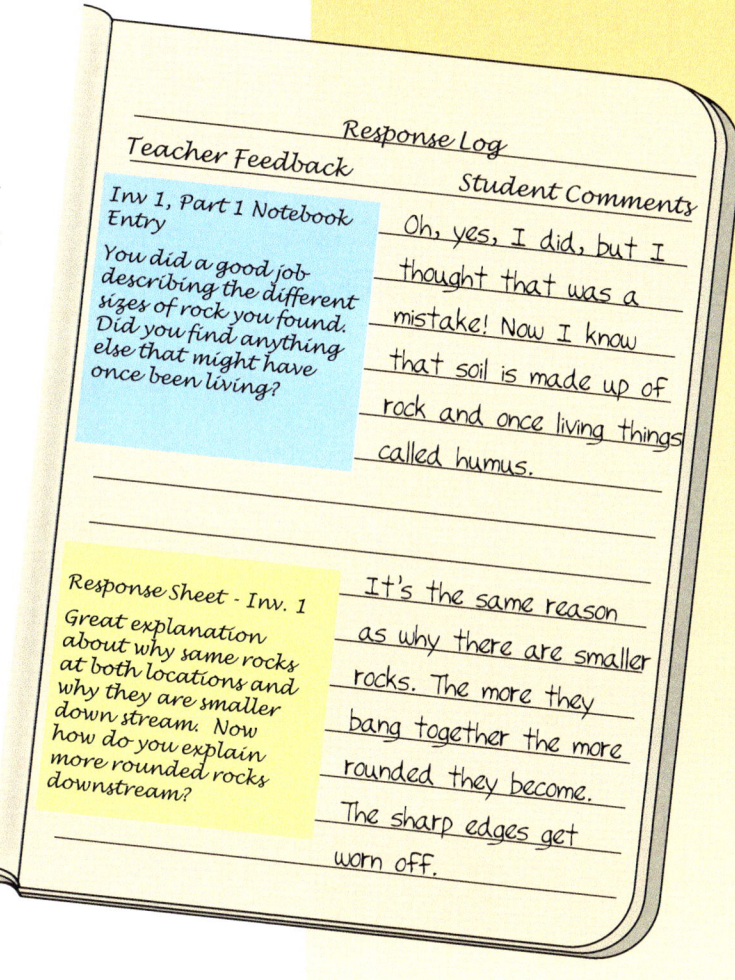

Soils, Rocks, and Landforms Module—FOSS Next Generation

SOILS, ROCKS, AND LANDFORMS — *Assessment*

FOSSMAP *and Online Assessment*

FOSSmap (fossmap.com) is the assessment management program designed specifically for teachers using the FOSS Program in grades 3–5. This user-friendly system allows you to open online assessments for students, to review codes for student responses, and to run reports to help you assess student learning. FOSSmap was developed at the Lawrence Hall of Science in conjunction with the Berkeley Evaluation and Assessment Research (BEAR) center at the University of California, as part of a 5-year research and development project funded by the National Science Foundation. It is based on the tools developed in the Assessing Science Knowledge (ASK) project.

Navigation page and a sample Embedded Assessment Report

Embedded-assessment data can be entered into FOSSmap to provide evidence of differentiated instruction, to run reports for formative analysis, and to print notes to provide feedback in student notebooks. It is also a tool for teacher reflection and instructional improvement from year to year.

FOSSmap allows you to give students access to the **Online Assessment** system (fossmap.com/icheck). Students log into this system to take the benchmark assessments (*Survey*, I-Checks, and *Posttest*). Responses are automatically sent to the FOSSmap teacher program where most are automatically coded. You will need to check short answers (mainly for correct answers that include inventive spelling), and to code open-response items. Students can answer open-response items on the computer or using paper and pencil, depending on how extensive their typing skills are.

Full Option Science System

FOSSmap and Online Assessment

FOSSmap Reports

The **Embedded Assessment Report** is a record you can run for each embedded assessment your students complete. It lists the names of students who "got it" as well as those that need some individual help. It is also a good place to record things that you might want to put more emphasis on or do slightly differently the next time you teach the module.

The **Code Frequency Chart** tells you at a glance which items were problems for the class. The red-bar items are the ones you want to take back to students for self-assessment activities.

The **Class by Item Report** shows the detail of each item and students' responses. You can go directly to the problem items (identified using the Code Frequency Chart) and see what the problem is. This report displays students' names for each response, with a brief description of what each code means in terms of full or partial credit. The report helps you decide what steps need to be taken next.

The **Student Responses Report** provides a printout of individual students' responses to all items answered online (including open-response items if they were typed into the system). This report is used for student self-assessment activities.

The **Student by Item Report** (a good report to send home to parents) lists all the items on a test and shows how individual students responded to each item. It also provides the correct answer and a description of what the student knows or needs to work on, based on the evidence to be inferred from the item.

The **Class Item Codes Report** provides a spreadsheet that can be opened in any spreadsheet program. It gives you a list of the students, the maximum code for each item, and the code each student received on each item. You can use this sheet if you want to convert codes into scores in order to determine percentage correct if that is needed for giving grades. To do that, you need to subtract 1 from each code, so that you are not actually awarding a point for wrong answers. Remember though, that FOSS assessments are designed to be diagnostic and not minimum mastery, so you may need to adjust your cut points for giving ABC grades. For example, instead of 90% being an A, you may decide that 80% is a better cut point for an A.

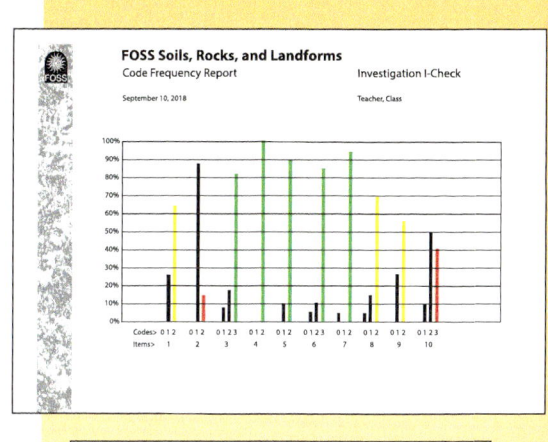

Code frequency, Class by Item, and Student by Item reports

Soils, Rocks, and Landforms Module—FOSS Next Generation

SOILS, ROCKS, AND LANDFORMS — Assessment

SAMPLE ASSESSMENT ITEMS

INVERSTIGATION 2 I-CHECK
SOILS, ROCKS, AND LANDFORMS

ANSWERS

5. This picture shows a cross section of rock layers. Which layer is the oldest?

   (Mark the one best answer.)
   - ○ **A** Layer 1
   - ○ **B** Layer 2
   - ○ **C** Layer 3
   - ● **D** Layer 4

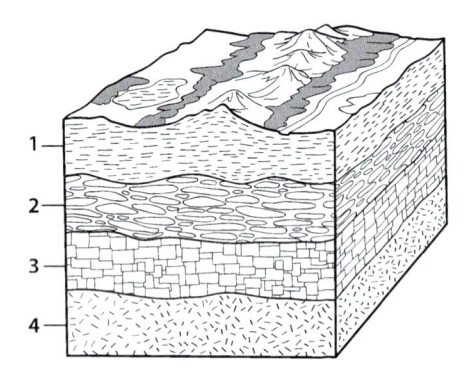

6. A scientist found fossils in a mountainous desert area that led her to believe that the area had once been covered by an ocean. Mark **X** next to fossils below that provide evidence for the argument that water covered the area at some time long ago.

   _____ A    _X_ B 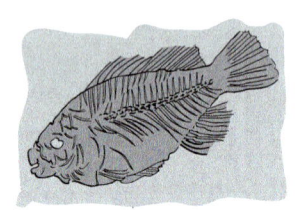   _X_ C

   _X_ D    _____ E    _____ F

FOSS Soils, Rocks, and Landforms Module
© The Regents of the University of California
Can be duplicated for classroom or workshop use.

Sample Assessment Items

Item 5

Focus on Science and Engineering Practices
Focus on Disciplinary Core Ideas
Focus on Crosscutting Concepts

Students analyze rock layers; they understand the pattern that older rock is layed down first and therefore would be on the bottom.

Code	If the student...
2	marks D.
1	marks A, B, C, or more than one answer.
0	makes no attempt.

ITEM 5 Next Steps

Have students reread "Fossils Tell a Story" in *FOSS Science Resources*, then revisit this item.

Item 6

Focus on Science and Engineering Practices
Focus on Disciplinary Core Ideas
Focus on Crosscutting Concepts

Students identify evidence that supports the claim that an area was once covered by the ocean.

Code	If the student...
2	marks fossils B, C, and D.
1	marks any other way.
0	makes no attempt.

ITEM 6 Next Steps

Have students review the *Fossils* video, then discuss each of the fossils in this item to determine which provide evidence that an ocean once covered the area.

NOTE

Look for these resources in the Assessment section in *Teacher Resources* on FOSSweb.

- Assessment chapter
- Assessment masters
- Coding guides
- Assessment charts

Soils, Rocks, and Landforms Module—FOSS Next Generation

SOILS, ROCKS, AND LANDFORMS — Assessment

SAMPLE ASSESSMENT ITEMS

INVESTIGATION 3 I-CHECK *ANSWERS*
SOILS, ROCKS, AND LANDFORMS

OPEN-RESPONSE QUESTIONS

7. A company wants to build a housing development on land that has flooded in the past. Scientists tell the company that the land will surely flood again. A group of engineers propose solutions to protect the land from future floods.

   Engineer A proposes run-off canals. These canals will be built upstream, before the river reaches the houses.

   Engineer B proposes building a levee next to the development. This will make the banks of the river 10 feet higher than the original banks.

   Engineer C suggests that they build the houses on stilts.

   Which solution do you think is the best solution?
   Students choose a solution; answers will vary.

   Why is this solution the best? (Include a sentence about each of the other solutions and why you did not choose them.)
   Key Points
   - *Students write a logical reason for choosing that solution. For example, "Run-off canals will take away some of the water before it reaches the houses and reduce flooding."*
   - *Students provide a sentence describing why they did not choose the other two solutions. For example, "I did not choose the levee because I don't think a higher wall in that one area will keep the river from flooding." Answers will vary.*

Sample Assessment Items

Focus on Science and Engineering Practices
Focus on Disciplinary Core Ideas
Focus on Crosscutting Concepts

Item 7

Students evaluate the cause and effect of three solutions for preventing flood damage.

Code	If the student . . .
4	indicates the solution he or she thinks is best; provides a plausible reason why the solution is a good one; describes why the other two solutions were considered less effective.
3	writes a code 4 answer, but includes minor errors; must include why they chose not to use the other solutions.
2	provides either a reason why the solution is best or a why they didn't chose one of the other solutions; includes one piece of relevant information.
1	writes anything else.
0	makes no attempt.

▶ **ITEM 9 Next Steps**
Have students debate the advantages and disadvantages of each of the solutions, then return to this item to edit their answers.

▶ **NOTE**
Look for these resources in the Assessment section in *Teacher Resources* on FOSSweb.
- Assessment chapter
- Assessment masters
- Coding guides
- Assessment charts

SOILS, ROCKS, AND LANDFORMS — *Assessment*

TEACHER NOTES

TEACHER NOTES